Radius and Ulna

Other books in the Musculoskeletal Trauma Series:

Humerus
Edited by E. Flatow and C. Ulrich

Tibia and Fibula
Edited by C. Court-Brown and D. Pennig

Femur
Edited by C. Court-Brown

Series Editors

C. Court-Brown MD, FRCSEd (Orth)
Consultant Orthopaedic Surgeon
The Royal Infirmary of Edinburgh
Edinburgh, UK

D. Pennig MD
Professor and Director
Department of Trauma, Hand and Reconstructive Surgery
St Vinzenz Hospital
Cologne, Germany

Musculoskeletal Trauma Series

Radius and Ulna

Edited by

M. M. McQueen MD, FRCSEd (Orth)
Consultant Orthopaedic Surgeon
The Royal Infirmary of Edinburgh
Edinburgh, UK

J. B. Jupiter MD
Professor of Orthopaedics
Harvard Medical School
Director, Orthopaedic Hand Service
Massachusetts General Hospital, USA

OXFORD   AUCKLAND   BOSTON   JOHANNESBURG   MELBOURNE   NEW DELHI

Butterworth-Heinemann
Linacre House, Jordan Hill, Oxford OX2 8DP
225 Wildwood Avenue, Woburn, MA 01801-2041
A division of Reed Educational and Professional Publishing Ltd

 A member of the Reed Elsevier plc group

First published 1999

British Library Cataloguing in Publication Data
Radius and ulna. – (Musculoskeletal trauma series)
  1. Radius – Fractures   2. Ulna – Fractures
  I. McQueen, Margaret M.   II. Jupiter, Jesse B.
  617.1'5'7

ISBN 0 7506 0835 8

Library of Congress Cataloguing in Publication Data
Radius and ulna/edited by M. M. McQueen, J. B. Jupiter.
  p.   cm. – (Muskuloskeletal trauma series)
  Includes bibliographical references and index.
  ISBN 0 7506 0835 8
  1. Radius – Fractures – Treatment.   2. Ulna – Fractures – Treatment.
  3. Forearm – Fractures – Treatment.   I. McQueen, Margaret M.
  II. Jupiter, Jesse B.   III. Series.
  [DNLM: 1. Radius Fractures.   2. Ulna Fractures.   WE 820 R129]
  RD557.R33
  617.1'57–dc21   98–52736
  DNLM/DLC   CIP

ISBN 0 7506 0835 8

Printed and bound in Great Britain by The Bath Press, Bath

Contents

List of Contributors   vi

Series Foreword   vii

Series Preface   viii

Preface   ix

1   **Epidemiology of fractures of the radius and ulna**   1
    M. M. McQueen

2   **Classification of fractures of the radius and ulna**   12
    M. M. McQueen

3   **Forearm anatomy**   37
    D. Ring and J. B. Jupiter

4   **Fractures of the distal radius**   63
    M. M. McQueen

5   **Forearm fractures**   84
    T. Rüedi

6   **Isolated fractures of the bones of the forearm**   97
    C. S. Mudgal and J. B. Jupiter

7   **Complications of forearm fractures in adults**   119
    D. Ring and J. B. Jupiter

8   **Fractures of the forearm in children**   138
    K. Graham

9   **Fractures of the radial head**   159
    C. Geel

10   **Fractures of the olecranon**   169
    B. J. Holdsworth

Index   183

List of Contributors

Christopher W. Geel MD FACS
Associate Professor, Director of Orthopaedic
Trauma
Department of Orthopaedics
550 Harrison Street
Syracuse
New York 13202
USA

H. Kerr Graham MD FRCS(Ed) FRACS
Professor of Orthopaedic Surgery
Royal Children's Hospital and University of
Melbourne
Flemington Road
Parkville 3052
Victoria
AUSTRALIA

Brian J. Holdsworth BSc MB FRCS
Consultant in Orthopaedics and Skeletal Trauma
University Hospital
Queen's Medical Centre
Nottingham
NG7 2UH
UK

Jesse B. Jupiter MD MS(Orth) MCh(Orth)
Professor of Orthopaedics
Harvard Medical School
Director, Orthopaedic Hand Service
Massachusetts General Hospital
15 Parkman Street, WAC 527
Boston, MA 02114
USA

Margaret M. McQueen MD FRCS Ed(Orth)
Consultant Orthopaedic Surgeon
The Royal Infirmary of Edinburgh
Lauriston Place
Edinburgh
EH3 9YW
UK

Chaitanya S. Mudgal MD MS(Orth) MCh(Orth)
Harvard Combined Orthopaedics Program
Massachusetts General Hospital
15 Parkman Street, WAC 527
Boston, MA 02114
USA

David Ring MD
Resident
Harvard Combined Orthopaedic Residency
Massachusetts General Hospital
ACC 527
15 Parkman Street
Boston, MA 02114
USA

Thomas P. Rüedi MD FACS
Professor of Surgery, Director
Department of Surgery
Kantonsspital
7000 Chur
SWITZERLAND

Series Foreword

Surgeons who deal with musculoskeletal injuries tend, on the whole, to use the methods of treatment practised at the units where they trained and served their apprenticeships. The more adventurous and intellectual of them also study other methods, but they are inevitably influenced by their earlier experiences. Those who trained where operative fixation was the norm are inclined to look down on those whose first choice of treatment is conservative; and the latter may in turn feel (if only unconsciously) that the 'always operate' merchants do so because 'fixation is fun'. Indeed, who can deny the joy of using skilled hands effectively.

What trauma surgeons need is a balanced view and here, in this series, we have precisely that. I was tempted to apply the initials NATO meaning, in the present context, the North Atlantic Trauma Organisation. Certainly the contents are a happy blend of North American and European views on fractures. Charles Court-Brown and Dietmar Pennig, the series editors, have earned great renown on both sides of the Atlantic for their work on the understanding and management of musculoskeletal trauma. They have collected a most distinguished set of authors and have been careful to ensure that the text does not read as if it was written by a committee; though it is clear that they have provided guidelines and minimized overlap,

they have wisely allowed each contributor to retain something of his national linguistic flavour.

The series consists of four volumes: Court-Brown and Pennig have themselves edited the volume on the tibia and fibula; the other three volumes are on the humerus (edited by E. Flatow and C. Ulrich), the radius and ulna (edited by M. McQueen and J. Jupiter), and the femur (edited by C. Court-Brown and J. Chapman). Each volume provides a comprehensive description of all the injuries discussed. Epidemiology, classification and fractures of the shafts as well as the upper and lower ends of each bone are described in detail; management includes authoritative accounts and clear illustrations of both operative and conservative treatment. Summaries of the literature are a notable feature of the text, and the lists of references will satisfy even the greediest reader.

This is the work which every trauma surgeon has been waiting for, using it to revise their knowledge of a particular field or to give practical guidance in approaching the individual patient. I wish it had been available when I was in active practice. I would have treated my patients better and slept more soundly at night.

The late Alan Apley

The death of Alan Apley after the publication of the first book in this series has deprived the orthopaedic community of one of its great teachers. He had a profound influence on all surgeons of our generation and all orthopaedic surgeons are in

his debt. Urbane, witty, charming and incisive he correctly questioned our more aggressive approach but always provided encouragement and support.

C. Court-Brown and D. Pennig

Series Preface

Our aims in editing a series of four books dealing with limb fractures were twofold. There has been an explosion of interest in the management of fractures in the last decade with the result that the large textbooks dealing with the management of fractures have tended to become reference books and we believe that there is a need for a series of short, concise texts describing the management of fractures in different bones. We have therefore been pleased to help with the production of four books detailing the epidemiology, classification, treatment and complications of fractures of the femur, tibia, humerus and forearm.

Our second goal was to try to combine the talents of North American and European editors and authors. Textbooks have tended to be based in one continent, often to their detriment. This collaboration has not always been easy or even possible but we are happy with the results of the enterprise and we are grateful to the volume editors and the individual authors for their contributions. We hope that the series will be of interest to all orthopaedic surgeons interested in trauma.

C. Court-Brown and D. Pennig

Preface

The forearm represents a critical anatomic unit of the upper limb, permitting the effector organ of the upper limb, the hand, to be placed in any position to either grasp or support an object. As Kapandji has pointed out, of the 7 degrees of freedom found in the articulations of the upper limb extending from the shoulder to the hand, forearm rotation (pronation-supination) represents one of the most vital motions, as it is essential for control of hand orientation. It permits the hand to grasp an object found anywhere within the ability of the shoulder to position the hand, to carry the object to the mouth, or to allow the hand to assist in the care of the individual. The freedom of forearm rotation has proven to be among the most important in man's evolutionary adaptation to his environment, allowing the use of tools, and ultimately civilization, to develop as we know it today.

The forearm maintains a stable link between elbow and wrist serving as an origin for many of the muscles that insert on the hand. The articulation is one of two bones, one rotating around the other, joined proximally and distally by stable radioulnar joints, and bound together in its mid substance by an interosseous membrane. This is far from a simple articulation, as witnessed by the complexities when all three of these units are disrupted, as seen in the problematic Essex-Lopresti injury. In fact, the longitudinal axis of rotation of the forearm passes uniquely through these structures; yet, during forearm rotation, the radius, being the curved bone, rotates around the axis which is parallel to neither the radius nor ulna.

Acute traumatic injuries pose a threat to this unique and sophisticated articulation. The fact that the forearm extends and includes components of the elbow as well as the wrist makes injuries, skeletal, soft tissue or both, a potential threat to the integrated function of the entire upper limb. Furthermore, chronic disorders can jeopardize not only the rotation of the forearm but also the stability and strength required for normal hand and upper extremity function.

This text represents a comprehensive perspective on the assessment and management of both acute and chronic conditions of the forearm. We have assembled an international group of experts who have contributed to our understanding of the intricacies of traumatic injuries to the forearm. The reader will find comprehensive discussions regarding problems involving the proximal, mid and distal aspects of the forearm articulation. Furthermore, the reader will find a thorough presentation of the assessment and management of chronic conditions of the forearm, including infection, failure to unite, malunion, and other conditions affecting forearm rotation. Throughout there is an emphasis on practical and proven techniques to aid the reader in the management of these conditions.

This text should provide the surgeon involved in the management of musculoskeletal trauma, as well as those specializing in the upper limb, a useful reference for virtually any condition that may arise in the course of a clinical practice.

M. M. McQueen and J. B. Jupiter

1

Epidemiology of fractures of the radius and ulna

M. M. McQueen

Introduction

Little has been written about the epidemiology of fractures of the radius and ulna except for fractures of the distal end, which have received more attention because of their relative frequency.

There are some incontrovertible facts. Firstly, fractures are common and upper limb fractures are more common than those in the lower limb with a ratio around 6:4 (Garraway *et al.*, 1979; Court-Brown and Brewster, 1996). Fractures of the distal radius are the commonest single type of fracture treated in the Edinburgh Orthopaedic Trauma Unit, with hip fractures running a close second.

Secondly, the number of fractures is increasing faster than the predicted rate allowing for the increase in the age of the population. This has been demonstrated for fractures of the proximal femur (Jarnlo *et al.*, 1989; Rockwood *et al.*, 1990; Nilsson *et al.*, 1991) and the distal radius (Bengner and Johnell, 1985). Equally, children's fractures are increasing in number, probably due to lifestyle changes (Landin, 1983).

However, it is often difficult to compare epidemiological data as reports are frequently based on selected groups of patients from an undefined population. In many countries, people with specific injuries go to predetermined centres such as the level I trauma centres in the USA. Surgeons working in these centres will not see the broad spectrum of injuries occurring in the whole population, while surgeons working outwith these centres will not deal with more complex injuries.

Knowledge of epidemiology is important in planning training and services. Training should concentrate to a large extent on conditions that the surgeon is likely to encounter: there is little point in spending a large amount of time being trained to manage a condition which may rarely or never be seen again in practice. Equally, service provision should be allocated in proportion to the frequency of injuries.

This chapter contains an epidemiological analysis of all fractures of the radius and ulna that presented to the Orthopaedic Trauma Unit of the Royal Infirmary of Edinburgh over a 3-year period from January 1990 to December 1992. The Unit manages all adult fractures occurring in a defined population of approximately 700 000 and therefore reflects the true epidemiology of fractures of the radius and ulna in a westernized country. All patients over the age of 14 years were included. The fractures were classified using the AO system (Müller *et al.*, 1990) and open fractures were graded by Gustilo's classification (Gustilo and Anderson, 1976; Gustilo *et al.*, 1984). The latter system was designed for diaphyseal fractures but was also used for this study for open metaphyseal fractures despite the possible limitations. The causes of injury were separated into several basic types: falls from standing height; falls from a height; road traffic accidents either as pedestrian or car occupant; sporting injuries; and direct blows or assaults. The only significant difference from North American centres is the absence of gunshot injuries.

During the 3-year period 2812 fractures of the radius and ulna were treated. Of these, 526 (19%) were proximal fractures (olecranon, radial head), 145 (5%) were diaphyseal fractures and 2141 (76%) were distal radial fractures. Of the 2812 fractures, 872 occurred in males and 1940 in females, giving an average annual incidence of 91 cases per 100 000 males and 196 cases per 100 000 females.

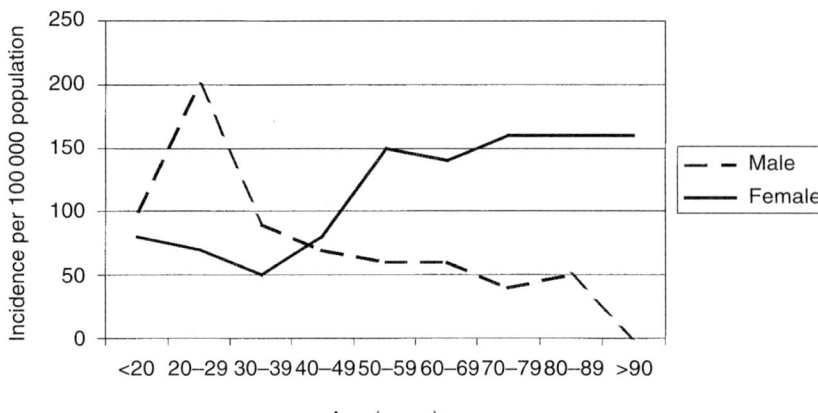

Figure 1.1 Average annual age- and sex-specific incidence of proximal radial and ulna fractures 1990–1992

Proximal radius and ulna fractures

There were 526 fractures of the proximal end of the radius and ulna treated during the 3-year period: 273 patients were female and 253 were male with an average age of 44 years (range 14–98 years). Women were generally older with an average age of 53 years compared to 34 years for men. The age- and sex-related incidence is shown in Figure 1.1 and this reflects a bimodal distribution. Males show an L-shaped distribution and females a J-shaped distribution. This corresponds to the patterns found by Buhr and Cooke (1959) for proximal radius fractures in Oxford which they describe as the pre-wage earning pattern for males and the post-wage earning pattern for females.

Sixty-nine per cent of these fractures were caused by a fall from standing height. The next most common reasons were a fall from a height (9%) and a road traffic accident (9%), most occurring as vehicle occupants. Five per cent occurred playing sports, 4% were a result of a direct blow and 3% followed an assault, with a few other miscellaneous causes making up the total.

There are some interesting age differences related to the different injury types. It is no surprise that most sporting injuries occur in a younger age group with a mean age of 26 years (range 14–50 years). Falls from a height generally occur in a younger group than falls from standing height. The oldest mean age is in the road traffic accident group involving pedestrians, at 57 years (range 21–83 years). This is a similar finding to that for tibial fractures (Court-Brown and McBirnie, 1995) and illustrates the vulnerability of older people whose reflexes are slower, making avoidance of cars less likely.

Open fractures in the proximal forearm are rare. Only 8 (1.5%) patients during this 3-year period had an open fracture, 6 being Gustilo grade 1 and 2 grade 2. There were no Gustilo grade 3 open fractures. All but one were open fractures of the olecranon.

Eighty-two patients (16%) had other injuries. Fifty-four had other upper limb injuries, 31 had lower limb injuries and 6 had associated injuries such as a head or chest injury. Six patients had multiple injuries with an ISS >15. In addition, 17 patients had an associated dislocation of the elbow with 1 patient having an Essex-Lopresti type injury. There were no predominant AO types in the dislocations.

Of the 526 fractures, the majority (392) were AO type B fractures, i.e articular fractures of one bone with or without an extra-articular fracture of the other. Most of the rest were type A, i.e. extra-articular of one or both bones. A small number (15) were the more complex type C fractures and these were distinguished by their occurrence in patients with a mean age of 59 years compared to mean ages in the early forties for types A and B. This suggests that complex fractures occur more in older patients because the osteoporotic nature of their bone is more vulnerable to severe injury rather than because they are victims of high-energy injury.

The distribution of AO types and subtypes is shown in Table 1.1. The three commonest types are the A2 or extra-articular proximal radius (radial neck) fracture, the B1 or intra-articular proximal ulna (olecranon) fracture and the B2 or intra-articular (radial head) fracture.

The largest group is the B2 fractures or radial head intra-articular fractures. These occur in a rela-

Table 1.1   Proximal radial and ulnar fractures

	No.	*%*
A1.1	2	0.4
A1.2	3	0.6
A1.3	3	0.6
A2.1	1	0.2
A2.2	104	19.8
A2.3	0	0.0
A3.1	4	0.8
A3.2	3	0.6
A3.3	0	0.0
Total	120	
B1.1	141	27.6
B1.2	5	0.9
B1.3	4	0.8
B2.1	173	33.5
B2.2	30	5.7
B2.3	34	6.5
B3.1	0	0.0
B3.2	2	0.4
B3.3	2	0.4
Total	391	
C1.1	2	0.4
C1.2	1	0.2
C1.3	0	0.0
C2.1	3	0.6
C2.2	2	0.4
C2.3	2	0.4
C3.1	2	0.4
C3.2	1	0.4
C3.3	2	0.2
Total	15	0.1

from standing height. Falls from a height (8%), sporting injuries (5%) and direct blows (5%) occur much less commonly than falls from standing height. Eight per cent of radial head fractures occur in road traffic accidents, mostly as pedestrians.

The next most common group is the B1 fractures: the isolated proximal ulnar fractures. The majority (141) were B1.1 or unifocal proximal ulnar fractures of which, excluding 14 isolated coronoid fractures, 127 were olecranon fractures. The annual age- and sex-related incidence (Figure 1.3) show that this is predominantly a fracture of elderly women with a large peak starting in the eighth decade. The average age of the whole group was 48 years (range 14–98 years) with almost equal numbers of men and women. Slightly over half sustained their injury in a fall from standing height and 13% in a fall from a height. Fourteen per cent sustained their olecranon fractures due to a direct blow and 13% in a road traffic accident.

The third-largest group is the AO type A2 extra-articular proximal radial or radial neck fracture, of which there were 104. The average age was 43 years (range 14–86 years) with twice as many women as men. The risk of middle-aged and elderly women sustaining a radial neck fracture is up to 6 times that of men of similar age, but is identical between the sexes up to the age of 40 years (Figure 1.4).

Diaphyseal fractures of the radius and ulna

In the 3-year period there were 149 fractures of the radial and ulnar shafts. The majority were men

tively young age group with an average age of 39 years (range 14–89 years) and are distributed between the sexes. Figure 1.2 shows equal peaks of young men and middle-aged women but the latter peak is at a relatively young age, explaining the younger mean age of this group. The majority of these patients (74%) sustained their injuries in falls

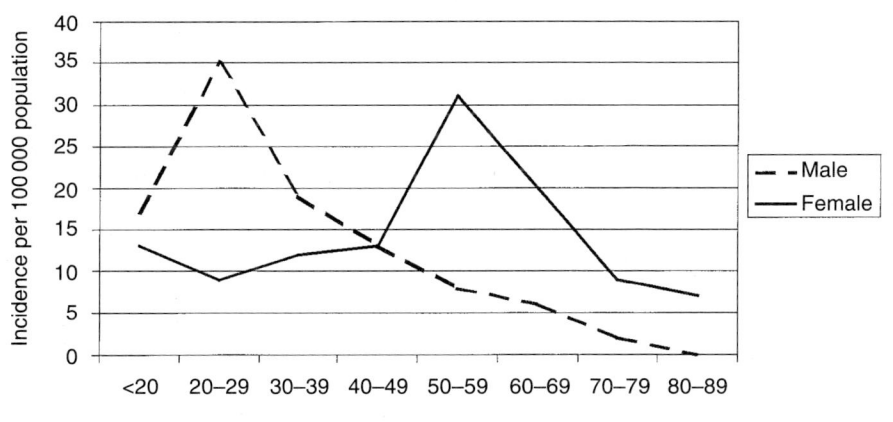

Figure 1.2 Average annual age- and sex-specific incidence of radial head fractures 1990–1992

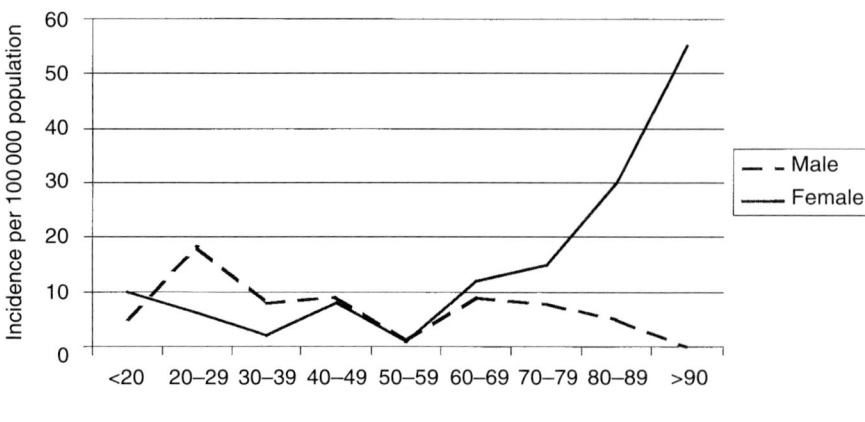

Figure 1.3   Average annual age- and sex-specific incidence of olecranon fractures 1990–1992

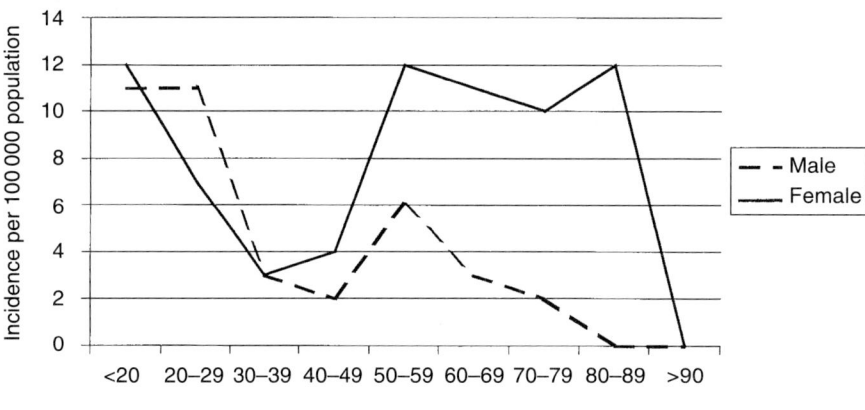

Figure 1.4   Annual age- and sex-specific incidence of radial neck fractures 1990–1992

(108) with 41 women. There was a marked difference in the mean age of men at 29 years (range 14–69 years) compared to women at 60 years (range 14–94 years). The age- and sex-specific incidences (Figure 1.5) show the largest peak in adolescent boys and an L-shaped or pre-wage earning pattern in men (Buhr and Cooke, 1959). In women the opposite is the case with a J-shaped post-wage earning pattern (Buhr and Cooke, 1959) implying an osteoporotic fracture pattern. The peak in women starts at an older age than for the proximal end of the radius and ulna. This is a less common injury with a total incidence of approximately one-quarter of that at the proximal end.

The commonest cause of a diaphyseal fracture of the radius and ulna is a fall from standing height (35%) followed by a direct blow (30%). Nineteen per cent of fractures occurred in road traffic accidents involving pedestrians compared to only 4% in road traffic accidents involving vehicle occupants. Eight per cent were caused by sport and the small remainder were either a result of a fall from a height or due to miscellaneous reasons.

The majority of fractures (59%) were AO class A, i.e. simple fractures of the ulna, radius or both bones. Thirty-nine per cent were of type B with wedge fractures of one or both bones, and a very small proportion (2%) were the more complex type C injuries. Seventy-two fractures involved the ulna only, 37 the radius only and 40 were fractures of both bones. There were 15 Monteggia fractures and 8 Galeazzi fractures.

Of the 72 isolated ulnar fractures, the commonest reason for injury was a direct blow (38%), the so-called 'nightstick' injury. Thirty-one per cent were caused by a simple fall, 14% were caused by pedestrian road traffic accidents and none occurred playing sports. The average age of the patients was 43 years (range 18–94 years) with a prepon-

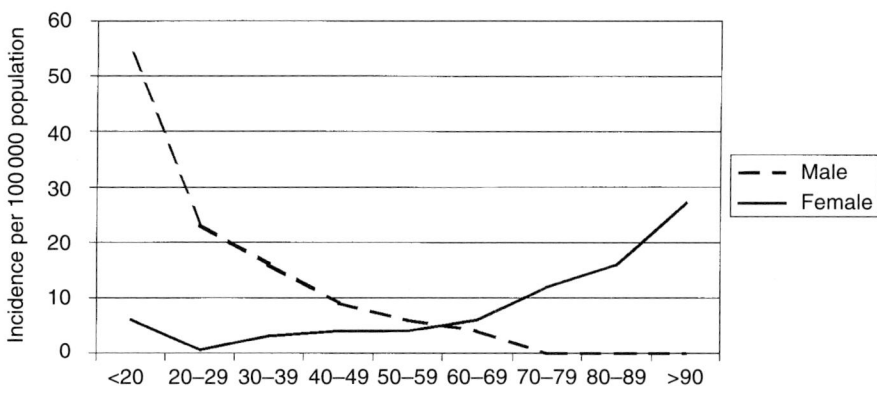

Figure 1.5 Average annual age- and sex-specific incidence of forearm diaphyseal fractures 1990–1992

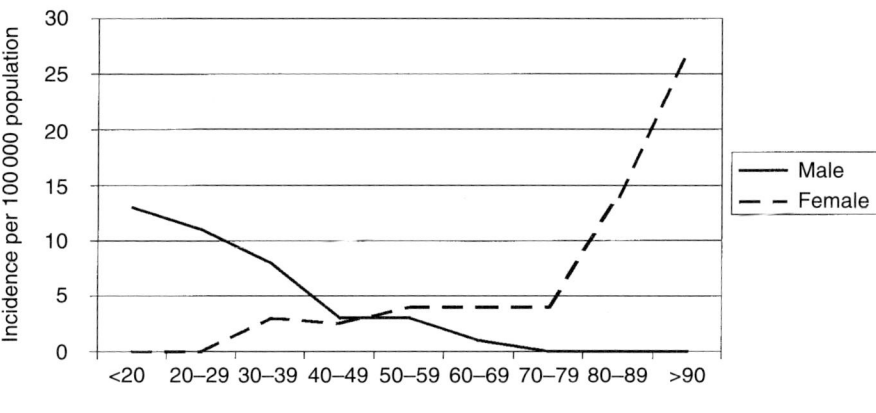

Figure 1.6 Average annual age- and sex-specific incidence of isolated ulnar shaft fractures 1990–1992

derance of males (63%). The annual age- and sex-specific incidence (Figure 1.6) shows a typical bimodal distribution with the risk of sustaining an isolated ulnar fracture in elderly women being almost twice that of young men.

Of the 37 isolated radial fractures, 32% were caused by a direct blow and 30% by a simple fall. Fourteen per cent were caused by a pedestrian road traffic accident and 11% by sport. The average age of patients was 34 years (range 14–87 years), and 84% were male. In contrast to isolated ulnar diaphyseal fractures, the risk of isolated radial diaphyseal fractures in young men is almost 6 times that of elderly women (Figure 1.7), implying a much smaller influence of osteoporosis.

Forty-three per cent of fractures of both bones of the forearm were caused by a fall from standing height, 22% by pedestrian and road traffic accidents and 19% by sports. Only 5% occurred as a result of a direct blow. The average age of patients

was 31 years (range 14–73 years) and 78% were male. The risk of a young male sustaining this fracture was five times that of an elderly female (Figure 1.8). It would seem therefore that isolated ulnar shaft fractures are the 'osteoporotic' fracture of the forearm diaphysis while young men can sustain any of the three types.

Fractures of the distal end of the radius and ulna

There were 2141 fractures of the distal end of the radius and ulna treated during the 3-year period: 515 male patients with an average age of 43 years (range 14–92 years) and 1626 female patients with an average age of 65 years (range 14–100 years), and an overall incidence of 68 per 100 000 males and 206 per 100 000 females. The annual age- and

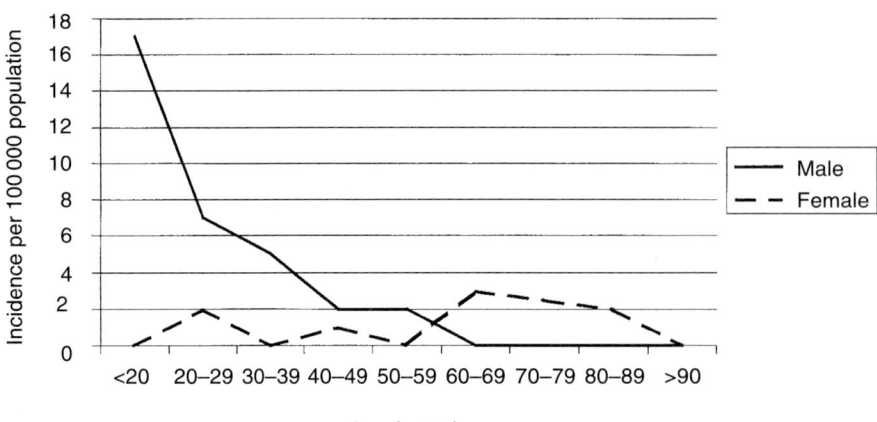

Figure 1.7 Average annual age- and sex-specific incidence of isolated radial shaft fractures 1990–1992

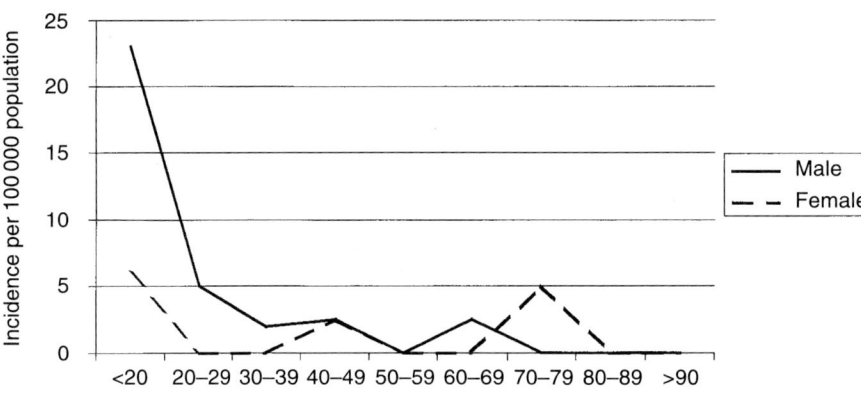

Figure 1.8 Average annual age- and sex-specific incidence of fractures of both bones of the forearm 1990–1992

sex-specific incidences are shown in Figure 1.9. This shows a large peak in females from the age of 40 years onwards which is only reflected 40 years later in the 10th decade in males. Thus the shape of the distribution is J-shaped or a post-wage earning pattern for both sexes. This contrasts with the distribution published almost 40 years ago in Oxford (Buhr and Cooke, 1959) where the curve was L-shaped for men and J-shaped for women.

Over the 40-year time period, the overall incidence has risen by over 100-fold among elderly women. It is unlikely that the increased numbers in Edinburgh are due to differences in demographics between two similar cities in the UK. Equally it would appear that capture of data is reliable in both centres. These differences are more likely to be due to a true increase in the incidence of distal radial fractures among elderly women. Over a 40-year period a marked increase in incidence is also seen in men in their 90s in whom the frequency of

distal radial fractures is approaching that of women of the same age in Oxford 40 years ago. This may be due to the longer survival of men who then become vulnerable to falling and sustaining osteoporotic fractures. This trend has also been observed in Scandinavia in both fractures of the distal radius (Bengner and Johnell, 1985; Robertsson *et al.*, 1990) and of the proximal femur (Johnell *et al.*, 1984).

Most fractures occurred after a fall from standing height (Table 1.2). The average age of this group was 64 years with a large majority of women. The rest of the types of injury make up less than one-fifth of the total. Table 1.2 also shows that younger men tend to break their wrists falling from a height or playing sport. As in other parts of the forearm and in the tibia (Court-Brown and McBirnie, 1995), wrist fractures of pedestrians in road traffic accidents are more likely to be in older people than those sustained as vehicle

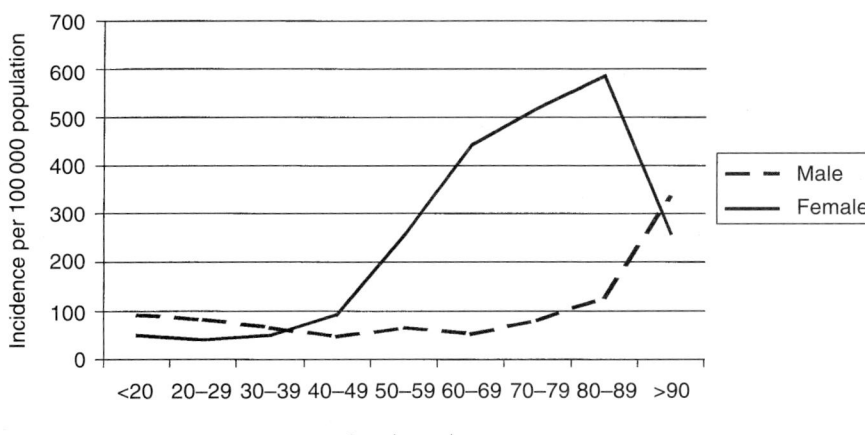

Figure 1.9 Average annual age- and sex-specific incidence of fractures of the distal radius and ulna 1990–1992

Table 1.2 Mode of injury of fractures of the distal radius and ulna

	No.	%	Age	% female
Fall	1736	81	64	85
Fall from height	142	7	49	41
Sport	157	7	32	32
RTA (pedestrian)	16	1	55	56
RTA (vehicle occupant)	47	2	35	25
Assault/direct blow	43	2	43	53

occupants. There is a marked seasonal variation with peak incidences occurring in the winter months (Figure 1.10).

The general fitness of patients with distal radial fractures clearly has an influence on the treatment methods employed. This was assessed in this population by the ability of patients to do their own shopping independently. Of 1105 women over the age of 60 years, 883 (80%) had this ability. Of 113 men, 84 (74%) were capable of doing so. Thus

although the bulk of distal radial fractures occur in the middle-aged to elderly population, this is generally a fit and active group.

Ninety per cent of distal radial fractures were split between AO type A or extra-articular fractures and AO type C or complete intra-articular fractures, with a slight preponderance of type A fractures (Table 1.3). The commonest single type of fracture was the A3.2 (extra-articular with metaphyseal comminution) at 26% of the whole group of distal radial fractures, closely followed by the C2.1 (simple intra-articular with metaphyseal comminution) at 24% of the total. Thus these two fracture types, which are either extra-articular or minimally intra-articular but have metaphyseal comminution and are likely to be unstable, made up half of the total number of distal radial fractures.

The next commonest fractures were the A2.1 (15%), the B1.1 (7%) and the C1.2 (7%), all of which are simple, uncomminuted fractures. The C3.2 fracture was also 7% of the total and, of

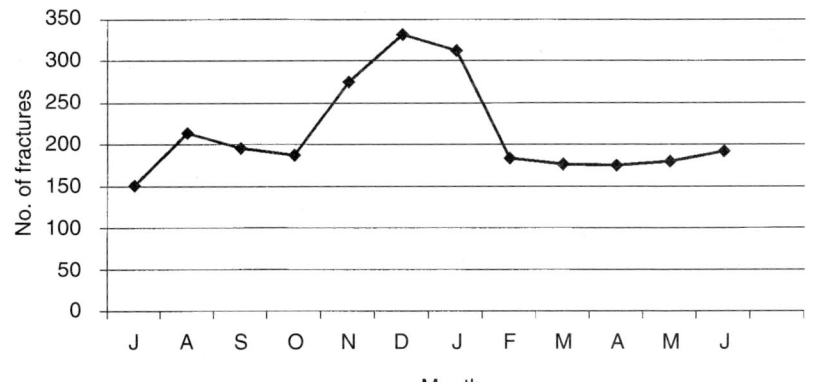

Figure 1.10 Seasonal variation in incidence of distal radial fractures

Table 1.3 AO distribution of fractures of the distal radius and ulna

	No.	%
A1.1	1	<1
A1.2	1	<1
A1.3	0	0
A2.1	346	15
A2.2	25	1
A2.3	15	<1
A3.1	43	2
A3.2	578	26
A3.3	20	1
Total	1029	48
B1.1	151	7
B1.2	11	<1
B1.3	3	<1
B2.1	9	<1
B2.2	1	<1
B2.3	1	<1
B3.1	11	<1
B3.2	16	<1
B3.3	16	<1
Total	219	10
C1.1	11	<1
C1.2	143	7
C1.3	5	<1
C2.1	508	24
C2.2	31	1
C2.3	12	<1
C3.1	25	1
C3.2	151	7
C3.3	7	<1
Total	893	42

these, 25 (1%) had articular displacement requiring open or closed reduction and internal fixation.

The age- and sex-specific incidences for the more common fractures (Figures 1.11–1.16) show the large peaks in incidences for females from the age of 50 years onwards for the more complex A3.2, C2.1 and C3.2 fractures. There is a small peak of young males for the C3.2 fractures which reflects the small incidence of high-energy severe articular injuries.

The pattern is different for the less complex and non-comminuted fractures (Figures 1.14–1.16). There remains a peak of women from 40 years of age onwards in the A2.1 fractures but there is also a large increase in risk in the very elderly males for the A2.1, C1.2 and B1.1 fractures, none of which have metaphyseal comminution and are therefore less likely to be osteoporotic fractures. This suggests that the time lag in the increase in incidence of males sustaining this injury is not related to deterioration of their bone quality. It may be because men in that age group fall over more. Equally, women have much less risk of sustaining a non-comminuted fracture because of their osteoporotic tendencies.

Twenty-three (1%) of the 2141 fractures were open, 14 being Gustilo grade 1, 8 grade 2 and 1 grade 3. Twelve of the 23 (52%) had diaphyseal extension of their fractures compared to 42 (2%) of the remaining 2118 closed fractures, implying that diaphyseal extension is a significant risk factor for open fractures of the distal radius.

Epidemiology of forearm fractures in children

The Edinburgh Orthopaedic Trauma Unit treats adults and adolescents over 14 years of age, so no data are available for the younger age group. However, several studies have been published on

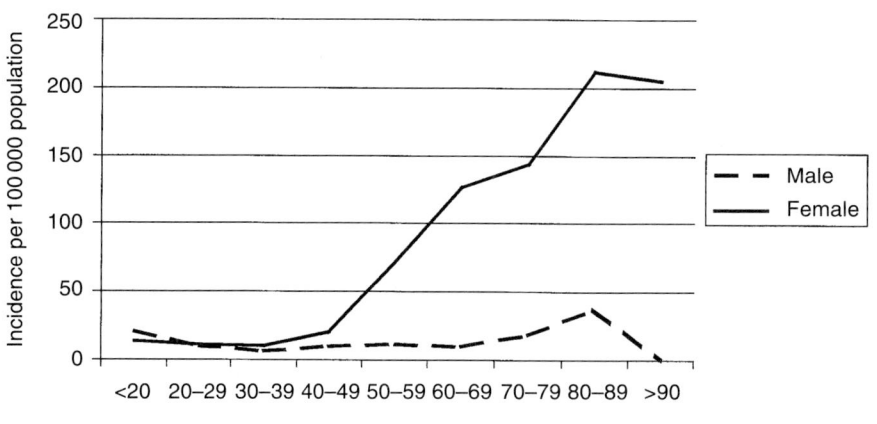

Figure 1.11 Average annual age- and sex-specific incidence of AO type A3.2 fractures of the distal radius and ulna 1990–1992

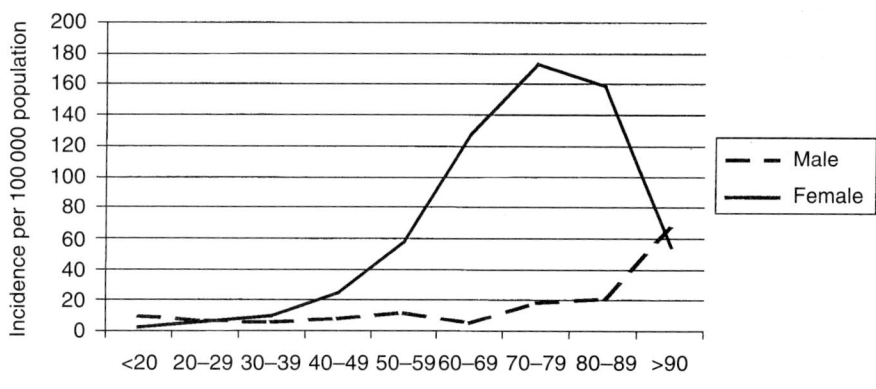

Figure 1.12 Average annual age- and sex-specific incidence of AO type C2.1 fractures of the distal radius and ulna 1990–1992

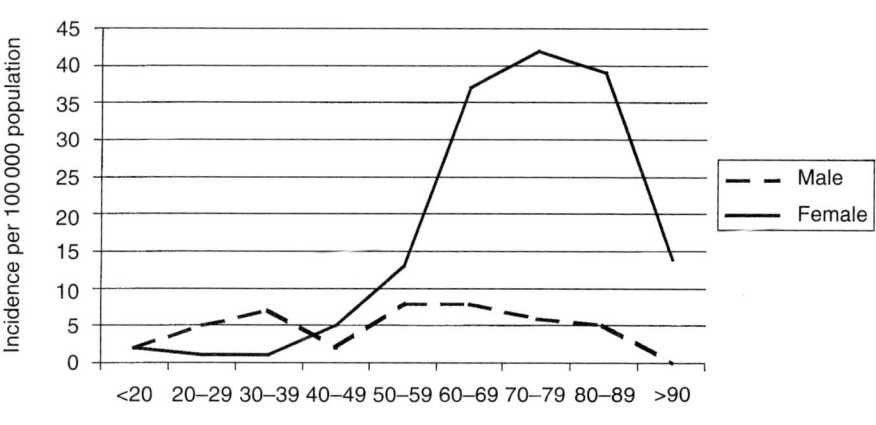

Figure 1.13 Average annual age- and sex-specific incidence of AO type C3.2 fractures of the distal radius and ulna 1990–1992

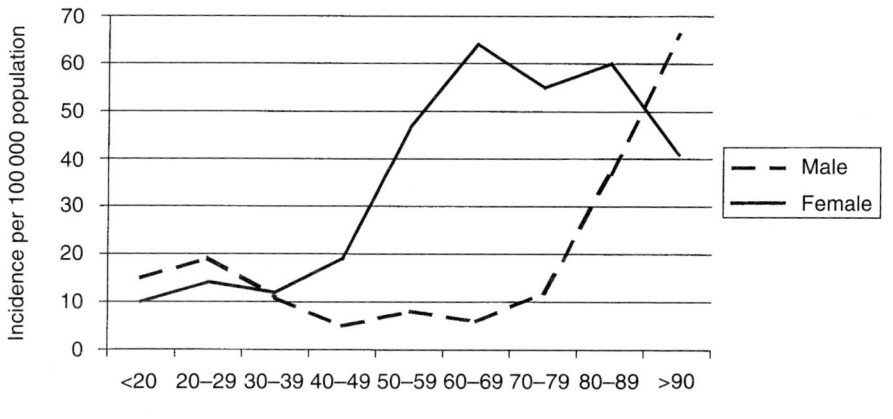

Figure 1.14 Average annual age- and sex-specific incidence of AO type A2.1 fractures of the distal radius and ulna 1990–1992

large numbers of children with fractures of all types which have some relevance to forearm fractures (Landin, 1983; Cheng and Shen, 1993). Landin found that the risk of fracture of any type increased with age up to the age of 11–12 years in girls and 13–14 years in boys and that fractures were more common in boys of all ages. This study accumulated data from three decades and during

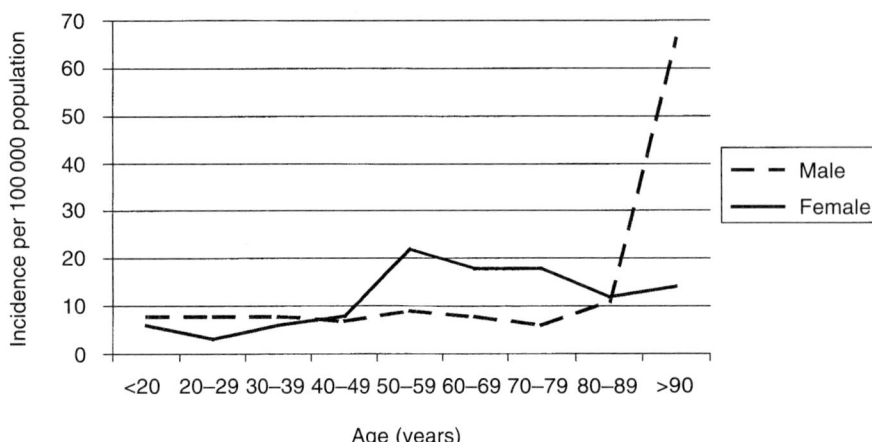

Figure 1.15 Average annual age- and sex-specific incidence of AO type C1.2 fractures of the distal radius and ulna 1990–1992

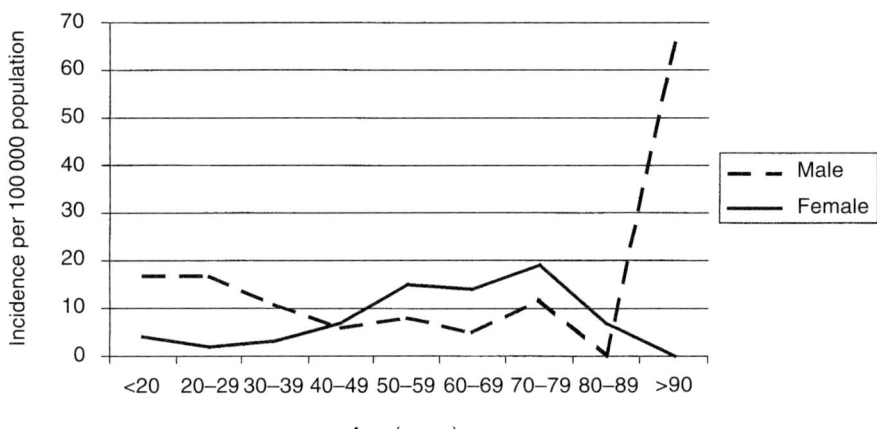

Figure 1.16 Average annual age- and sex-specific incidence of AO type B1.1 fractures of the distal radius and ulna 1990–1992

that period the risk of fracture in childhood has doubled. The commonest cause of injury in childhood was playing (24%), sport (21%) and road traffic accidents (12%), and the first two had increased by a factor of five over the time period studied.

As in adults, the commonest fracture in children is of the distal end of the radius and ulna (Landin, 1983; Cheng and Shen, 1993), being approximately 20% of the total. When the fractures are split into age groups it is evident that distal radius fractures occur more commonly between 8 and 16 years of age, peaking at the age of 12 years for girls and 14 years for boys (Bailey *et al.*, 1989). The latter authors have related these ages to the peak velocity of growth for girls and boys and postulate that increases occur at these stages

because of relatively reduced bone density at times of peak growth.

Children's fractures have a seasonal distribution opposite to that of adults with a peak in the summer months (Landin, 1983; Cheng and Shen, 1993; Masterson *et al.*, 1993). This has previously been related to the increased activity of children during school holidays but there is also a strong independent correlation with average monthly sunshine hours (Masterson *et al.*, 1993).

Landin and Danielsson (1986) studied 589 fractures of the elbow in children under the age of 16 years. Although the commonest were fractures of the distal humerus, 14% were radial neck fractures. Twenty per cent of olecranon fractures were associated with another elbow fracture and occurred most often in boys.

References

Bailey, D.A., Wedge, J.H., McCulloch, R.G. *et al.* (1989) Epidemiology of fractures of the distal end of the radius in children as associated with growth. *J. Bone Joint Surg.*, **71A**, 1225–31

Bengner, U. and Johnell, O. (1985) Increasing incidence of forearm fractures. A comparison of epidemiological patterns 25 years apart. *Acta Orthop. Scand.*, **56**, 158–60

Buhr, A.J. and Cooke, A.M. (1959) Fracture patterns. *Lancet*, **1**, 531–6

Cheng, J.C. and Shen, N.Y. (1993) Limb fracture pattern in different paediatric age groups: a study of 3350 children. *J. Orthop. Trauma*, **7**, 15–22

Court-Brown, C.M. and Brewster, N. (1996) Epidemiology of open fractures. In *Management of Open Fractures* (C.M. Court-Brown, M.M. McQueen and A. Quaba, eds) London: Martin Dunitz, pp. 25–36

Court-Brown, C.M. and McBirnie, J. (1995) The epidemiology of tibial fractures. *J. Bone Joint Surg.*, **77B**, 417–21

Garraway, W.M., Stauffer, R.N., Kurland, L.T. and O'Fallon, W.M. (1979) Limb fractures in a defined population. 1. Frequency and distribution. *Mayo Clin. Proc.*, **54**, 701–7

Gustilo, R.B. and Anderson, J.T. (1976) Prevention of infection in the treatment of one thousand and twenty-five open fractures of long bones. *J. Bone Joint Surg.*, **58A**, 453–8

Gustilo, R.B., Mendoza, R.M. and Williams, D.N. (1984) Problems in the management of type III (severe) open frac-tures: a new classification of type III open fractures. *J. Trauma*, **24**, 742–6

Jarnlo, G.B., Jakobsson, B., Ceder, L. and Thorngren, K.G. (1989) Hip fracture incidence in Lund, Sweden 1966–1986. *Acta Orthop. Scand.*, **60**, 278–82

Johnell, O., Nilsson, B., Obraut, K. and Sernbo, L. (1984) Age and sex patterns of hip fractures – changes in 30 years. *Acta Orthop. Scand.*, **55**, 290–2

Landin, L.A. (1983) Fracture patterns in children. Analysis of 8,682 fractures with special reference to incidence, aetiology and secular changes in a Swedish urban population 1950–1979. *Acta Orthop. Scand. Suppl.*, **202**, 1–109

Landin, L.A. and Danielsson, L.G. (1986) Elbow fractures in children. An epidemiological analysis of 589 cases. *Acta Orthop. Scand.*, **57**, 309–13

Masterson, E., Borton, D. and O'Brien, T. (1993) Victims of our climate. *Injury*, **24**, 247–8

Müller, M.E., Nazarian, S., Koch, P. and Schatzker, J. (1990) *The Comprehensive Classification of Fractures of Long Bones.* Berlin: Springer-Verlag

Nilsson, R., Hofman, O., Berglund, K. *et al.* (1991) Increased hip fracture incidence in the county of Ostergotland, Sweden 1940–1986, with forecasts up to the year 2000: an epidemiological study. *Int. J. Epidemiol.*, **20**, 1018–24

Robertsson, G.O., Jonsson, G.T. and Sigurjonsson, K. (1990) Epidemiology of distal radius fractures in Iceland in 1985. *Acta Orthop. Scand.*, **61**, 457–9

Rockwood, P.R., Horne, J.G. and Cryer, C. (1990) Hip fractures: a future epidemic? *J. Orthop. Trauma*, **4**, 388–93

2

Classification of fractures of the radius and ulna

M. M. McQueen

Introduction

Classification systems have been employed for many years from the first use of eponyms for fractures. Thus in the radius and ulna we have Monteggia's and Galeazzi's fractures, Colles', Smith's and Barton's fractures and many others. The use of eponyms can be confusing and misleading, and more sophisticated classification systems have superseded them as knowledge of various fractures grows and is refined.

The ideal classification system is unattainable. It is said that a classification system should define the morphology of a fracture, aid in its surgical treatment, encompass its aetiology and predict its prognosis. It should also be sufficiently simple to use in daily clinical practice and should have intra- and inter-observer reliability! This is clearly impossible since many of these outcomes are multi-factorial. For instance, the prognosis of a fracture depends on the patient, the surgeon and the surgical technique involved more than it does on its morphology. In practice most systems are either morphological or aetiological. Nevertheless, classification systems still have an important role to play in highlighting the importance of different fracture types and characteristics and in aiding communication between surgeons, either verbal or written.

The simplest classification systems are those which differentiate between simple types of injuries such as undisplaced and displaced fractures or open and closed fractures. However, as understanding of the pathophysiology of fractures has increased, there has been a demand for more sophisticated classification systems. The most comprehensive published so far is the AO classification of long bone fractures (Müller et al.,

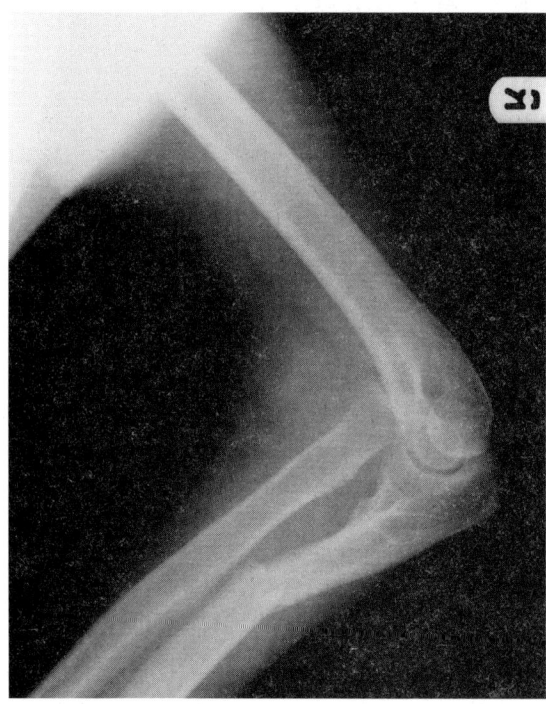

Figure 2.1 An AO type A1.3 fracture of the diaphysis of the radius and ulna. The ulnar fracture is simple and there is an associated dislocation of the radial head. This is also a Bado's type 1 Monteggia fracture

1990) which was designed to be a systematic classification of fractures of the whole skeleton.

In the AO classification, the location of the fracture is coded numerically. A number between 1 and 9 specifies each long bone or group of bones. The radius and ulna are specified as no. 2. The location within the long bone is then specified by a number between 1 and 3 where 1 is the proximal

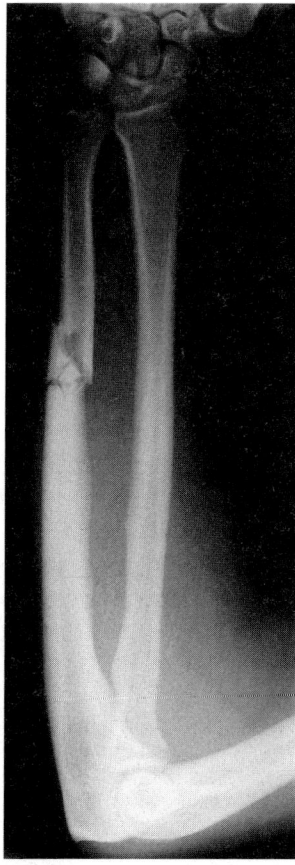

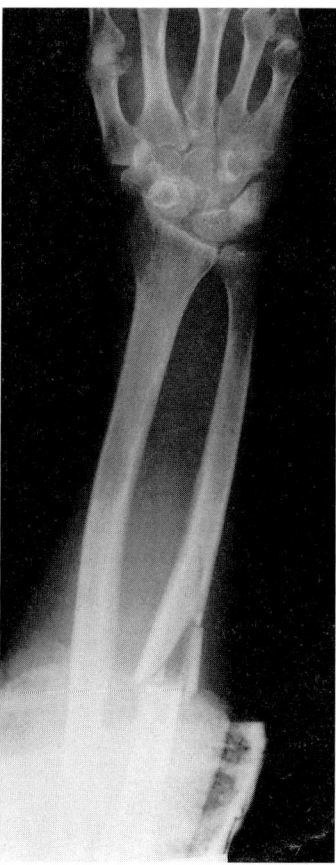

Figure 2.2 An AO type B1.2 fracture of the radial and ulnar diaphysis. This is a wedge fracture of the ulna with the radius intact

Figure 2.3 An AO type C1.3 fracture of the diaphysis of the ulna. The fracture is irregular with no contact between the intact cortices

metaphyseal segment, 2 is the diaphysis and 3 the distal metaphyseal segment. Thus the distal radius is 23. The fractures in each segment are then classified as A, B or C. In the diaphysis, type A is a simple fracture with two fragments (Figure 2.1), type B is a fracture with a wedge fragment (Figure 2.2) and type C is a complex fracture where after reduction there is no contact between the two main fragments (Figure 2.3). In the metaphyseal fractures, type A are extra-articular (Figure 2.4). Type B are partial articular where a section of the articular surface is fractured but the remainder is intact and attached to the metaphysis and diaphysis, or where two bones are involved and there is an intra-articular fracture of one with or without an extra-articular fracture of the other (Figure 2.5). Type C are complete articular fractures with a disrupted articular surface detached from the metaphysis (Figure 2.6).

Each fracture type is then divided into groups and subgroups, with each subgroup a variant of the type. The variation depends on a variety of factors including increasing comminution and associated dislocation of the direction of the fracture plane. Some subgroups also have a number of minor variants.

It has been claimed that the AO system is complex, but the underlying principles are systematic and logical, and it can be used with reliability provided that attention is paid to detail both in interpreting X-rays and to the classification system.

The AO system has been criticized for poor inter- and intra-observer reliability, but often with inexperienced observers (Johnstone *et al.*, 1993; Andersen *et al.*, 1996), and others have acknowledged the importance of experience in achieving consistency (Kreder *et al.*, 1996). The system is also criticized for not considering displacement, especially in joint injuries which will clearly have

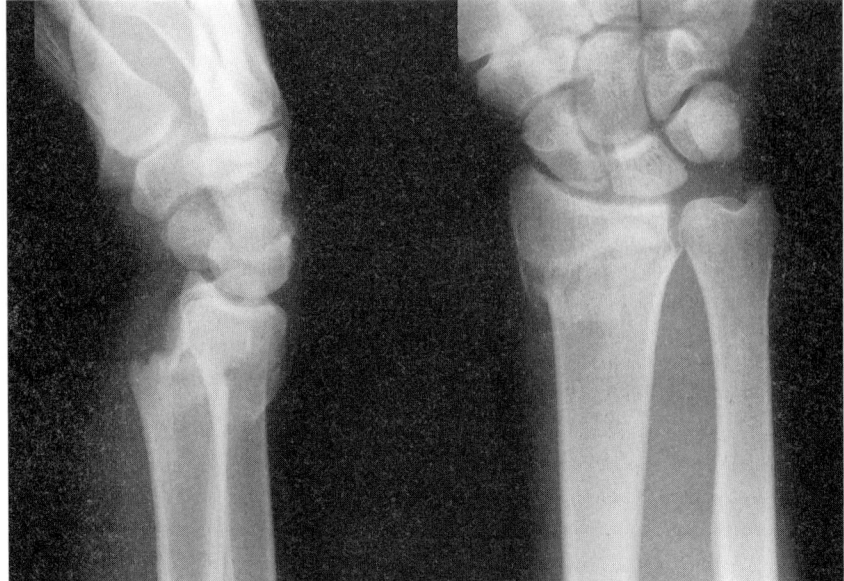

Figure 2.4   An extra-articular fracture of the distal radius with dorsal metaphyseal comminution. This is an AO type A3.2. It is also a Frykman 6 (involvement of distal radioulnar joint plus ulnar styloid fracture) and a Fernandez type I (extra-articular bending fracture)

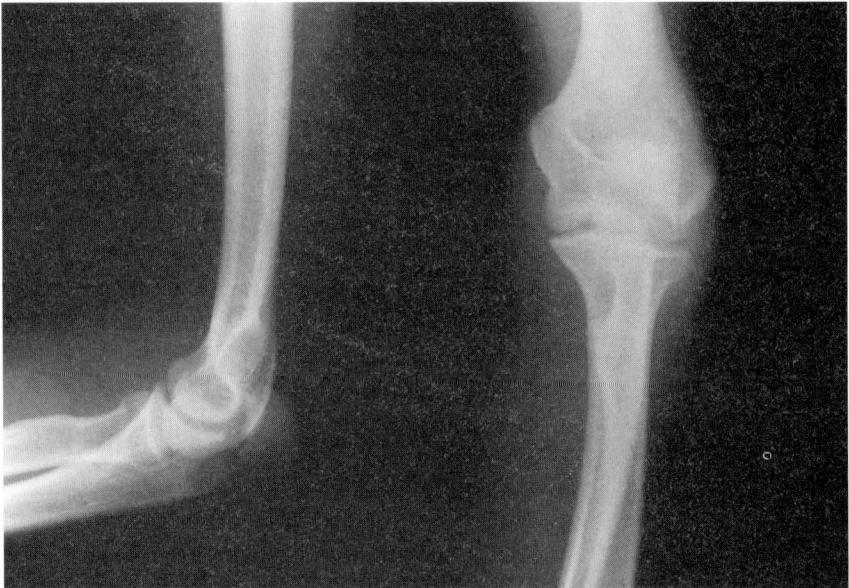

Figure 2.5   An AO type B1.1 fracture of the proximal ulna. There is an intra-articular fracture of the ulna with an intact radius. This is also a Colton group 1 olecranon fracture

an influence on the management and prognosis of a fracture.

Classification of children's fractures

Epiphyseal injuries

Although previous authors have classified injuries to the epiphyseal plate, the system most widely used today is that of Salter and Harris (1963) (Figure 2.7) which is a more sophisticated version of the first classification of epiphyseal injuries which was produced by Poland in 1898.

The Salter–Harris type I injury occurs when there is a complete separation of the physis from the metaphysis or epiphysis. It is usually the result of a shearing or avulsion force and tends to occur in early childhood.

The Salter–Harris type II injury is the most common epiphyseal plate injury. The line of sepa-

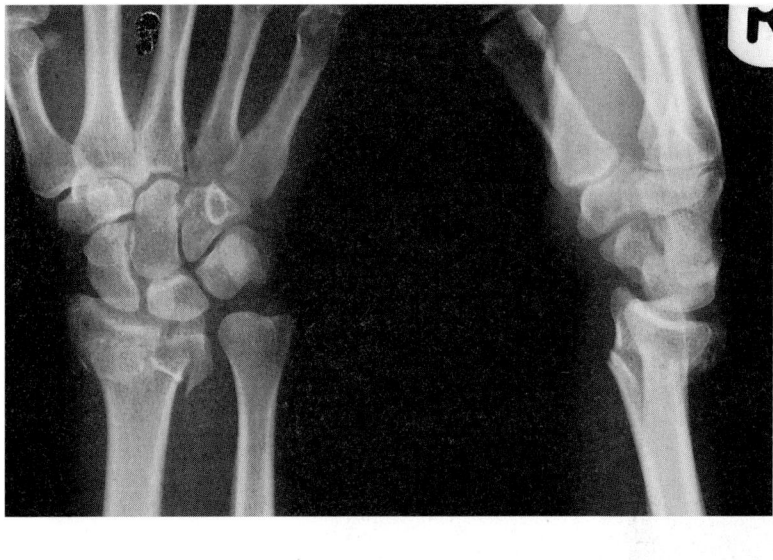

Figure 2.6 A complete articular fracture of the distal radius with a sagittal articular fracture line and metaphyseal comminution: the AO type C2.1

ration extends along the epiphyseal plate and out through a portion of the metaphysis, leaving a triangular shaped metaphyseal fragment still attached to the epiphyseal plate. This occurs most often in children over 10 years of age and is the result of a shearing or avulsion force.

The type III injury is less common and consists of an intra-articular fracture extending from the joint surface to the epiphyseal plate and out along the plate to its periphery. The type IV injury extends from the joint surface through the epiphysis, the epiphyseal plate and a portion of the metaphysis, producing a complete split.

The fifth Salter–Harris type of fracture results from a severe crushing force applied through the epiphysis to one area of the epiphyseal plate. Displacement is unusual and X-rays may appear normal. Growth arrest is almost inevitable after this injury.

A further type of epiphyseal injury with a high likelihood of growth arrest was described by Rang in 1969. Injury to the periphery of the plate or its overlying periosteum may cause bony bridging between the metaphysis and the diaphysis. This causes premature closure on one side of the growth plate, leading to angular deformities.

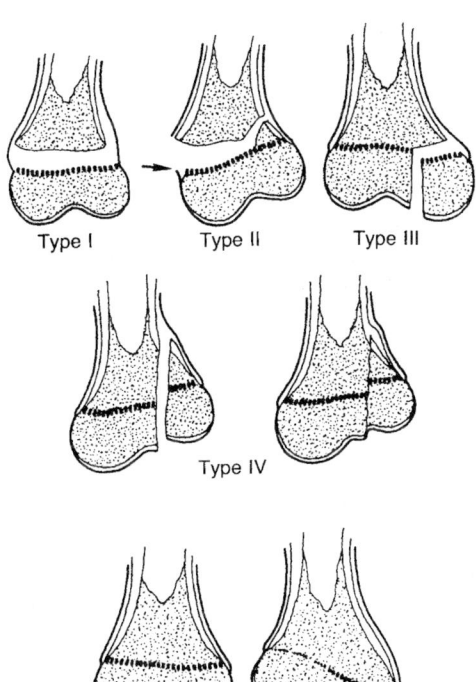

Figure 2.7 The Salter–Harris classification of epiphyseal fractures

Type I	Fracture through the physis. This tends to occur in younger children and is associated with birth injuries. The periosteal attachments may or may not remain intact
Type II	This is the most common type. The fracture line is mainly through the physis but a metaphyseal fragment of variable size remains attached to the epiphysis. This greatly aids reduction
Type III	The fracture line passes through the epiphysis to exit through the physis. A potentially unstable fracture
Type IV	The fracture line passes vertically through both metaphysis and epiphysis. The fracture tends to displace proximally
Type V	Axial compressive force, frequently causing partial or complete growth arrest

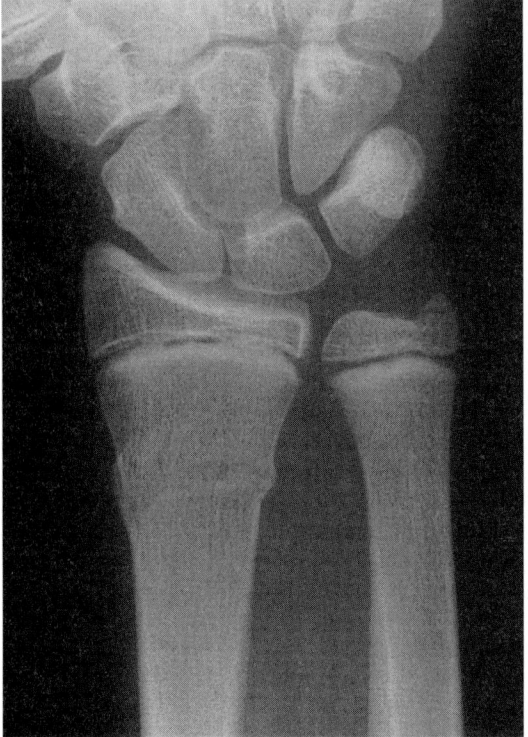 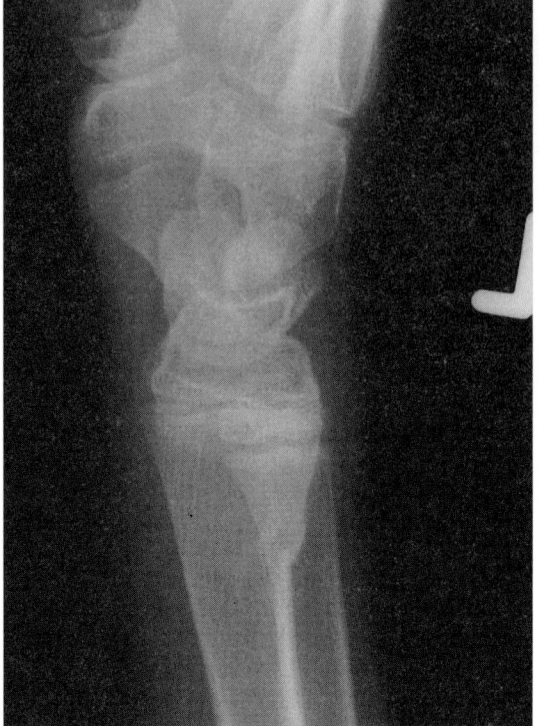

Figure 2.8   A buckle or torus fracture of the distal radius

Non-physeal injury

Although many children's fractures which do not involve the epiphyseal area would fit into an adult classification system, some types are specific to young bone. These are generally incomplete fractures due to the ability of developing bones to undergo plastic deformation. Thus we see greenstick fractures, torus or buckle fractures and bowing. A bending force usually causes a greenstick fracture where the tension side fractures and the compression side merely bends but the continuity of the bone is intact. A buckle fracture usually occurs at the metaphysis when compression causes the bone to buckle rather than completely fracture, thus causing a stable injury (Figure 2.8). Traumatic bowing occurs in younger bone when plastic deformation results in response to injury.

Classification of soft tissue injury

Classification of the extent of soft tissue injury is more difficult than for bony injury because it is more difficult to visualize the extent of damage to the soft tissues by imaging techniques. It may be that increasing expertise in MRI scanning and ultrasound techniques will allow better classification systems in the future, but currently no classification method is based on these techniques.

Existing soft tissue classification systems are subjective, often based on an educated guess as to the state of the soft tissues underlying the skin, and can therefore be unreliable. Nevertheless it is important to attempt to classify the extent of soft tissue damage in both closed and open fractures since it is one of the major determinants of fracture outcome.

Classification of soft tissue injury in open fractures

Classification of soft tissue injury in open fractures is of particular importance because of the significant influence that the extent of the soft tissue injury has on outcome. Clearly a puncture wound associated with a fracture is a much more benign injury than a fracture associated with severe contamination and massive soft tissue loss.

Table 2.1   Gustilo classification of open fractures

Type I		Clean wound of less than 1 cm in length
Type II		Wound larger than 1 cm in length without extensive soft tissue damage
Type III		Wound associated with extensive soft tissue damage. Usually longer than 5 cm
		Open segmental fracture
		Traumatic amputation
		Gunshot injury
		Farmyard injury
		Fracture associated with vascular repair
		Fracture more than 8 hours old
Subtype	IIIA	Type III wound with adequate periosteal cover
	IIIB	Presence of significant periosteal stripping. Wound usually contaminated
	IIIC	Vascular repair required to revascularize leg

The Gustilo classification system (Gustilo and Anderson, 1976; Gustilo *et al*., 1984) is most commonly used in clinical practice. It divides open fractures into three types depending on the size of the skin wound, the amount of underlying soft tissue damage and the extent of contamination. In 1984 Gustilo and his co-authors divided the type III open fracture into three subgroups based on the presence of periosteal stripping and major vascular injury. Table 2.1 defines the complete Gustilo classification, which has gained widespread support because of its prognostic value (Court-Brown *et al*., 1991; Sanders *et al*., 1993). However, some difficulties remain. Many surgeons classify an injury prior to debridement and can underestimate its severity because the skin wound is small although underlying soft tissue damage may be extensive. This is well illustrated in a study where inter-observer reliability of the system was criticized (Brumback and Jones, 1994) on the basis of a videotape of 12 open fractures and questionnaires despite the fact that the underlying tissues were not demonstrated in a number of fractures. A further difficulty arises with classifying open metaphyseal fractures which may have a different prognosis from similar diaphyseal injuries. Nevertheless, the Gustilo classification system remains the most useful in modern practice.

Other classification systems for open fractures are similar to the Gustilo system. Oestern and Tscherne's (1984) types O1 to O3 are the equivalent of Gustilo's three types, but their type O4 is subtotal or total amputation.

The AO classification for open fractures (Müller *et al*., 1991) is also somewhat similar to Gustilo's but places more emphasis on the importance of

Table 2.2   The AO open integument classification designed for use with the long bone classification

IO1	Skin wound from inside out
IO2	Skin wound from outside in. Less than 5 cm in length. Contused edges
IO3	Skin wound greater than 5 cm in length. Increased contusion. Devitalized wound edges
IO4	Larger wound. Full-thickness contusions. Extensive open degloving or skin loss

degloving. The four types are shown in Table 2.2. The first three are essentially similar to Gustilo's three types although the type IO4 is probably similar to the Gustilo IIIB.

Classification of soft tissue injury in closed fractures

Unfortunately, if the skin is closed, little attention is paid to the soft tissue injury associated with fractures despite the fact that the amount of soft tissue injury has a significant effect on the outcome of a fracture.

The best-known classification of closed fractures based on associated soft tissue injury is probably that of Tscherne (Oestern and Tscherne, 1984). This system is illustrated in Figure 2.9. It is based on the amount of abrasion or contusion in skin and muscle, the severity of the fracture, the presence of degloving, damage to major vessels and the presence of an acute compartment syndrome, and has some influence on the prognosis in terms of fracture union (Court-Brown *et al*., 1990). However, unless the fracture is treated by open methods such as plating, quantification of muscle damage can only be an educated guess.

The AO group (Müller *et al*., 1991) has devised a classification system for soft tissue injury in closed fractures with three sections to assess skin, muscle and neurovascular injury (Table 2.3). It defines soft tissue injury more precisely than Tscherne's system, although the same reservations about assessment of muscle injury apply. The prognostic value of this system has not as yet been fully tested.

Table 2.3   The AO closed integument classification

IC1	No skin lesion
IC2	No skin laceration but contusions present
IC3	Circumscribed degloving injury
IC4	Extensive closed degloving injury
IC5	Skin necrosis present

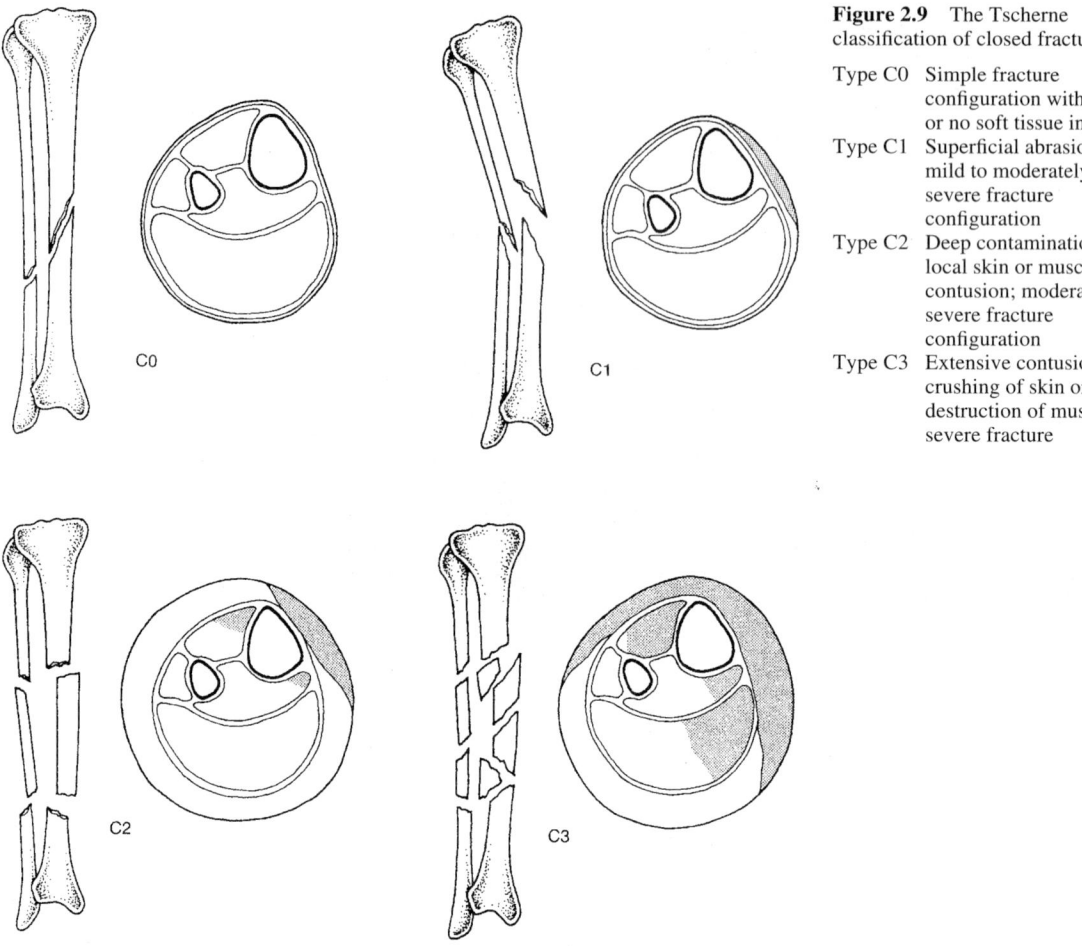

Figure 2.9 The Tscherne classification of closed fractures

Type C0 Simple fracture configuration with little or no soft tissue injury

Type C1 Superficial abrasion; mild to moderately severe fracture configuration

Type C2 Deep contamination with local skin or muscle contusion; moderately severe fracture configuration

Type C3 Extensive contusion or crushing of skin or destruction of muscle; severe fracture

Classification of fractures of the proximal radius and ulna

Radial head and neck fractures

Mason classification

Mason proposed a classification of radial head fractures in 1954 which is still commonly used today. He concentrated on intra-articular fractures and did not include extra-articular radial neck fractures. He classified the fractures into three types based on radiological findings. Type 1 consist of undisplaced fractures of the radial head, which Mason described as fissure fractures, or marginal sector fractures without displacement (Figure 2.10). Type 2, or marginal sector fractures with displacement, include those fractures in which a segment of the border of the radial head is separated from the rest of the radial head by impaction, depression or tilting. Type 3 fractures

are those in which there is complete disorganization of the head of the radius with comminution.

Mason related his classification to outcome and showed that the type 1 fracture had minimal long-term problems. He treated all his type 3 fractures by excision with some long-term loss of motion, but it was in the type 2 cases that long-term problems ensued with non-operative management.

The Mason classification may be criticized for being too simplistic. Some fractures, such as the radial neck fracture with an undisplaced articular segment, are difficult to classify with this system. Associated ligamentous injury including elbow dislocation is ignored. Johnston added a type 4 to Mason's classification in 1962 to correct this omission. This is defined as any radial head fracture with elbow dislocation. This concept was further refined by Davidson and his co-authors (1993), who noted significant correlation between fracture type and medial ligament injury with

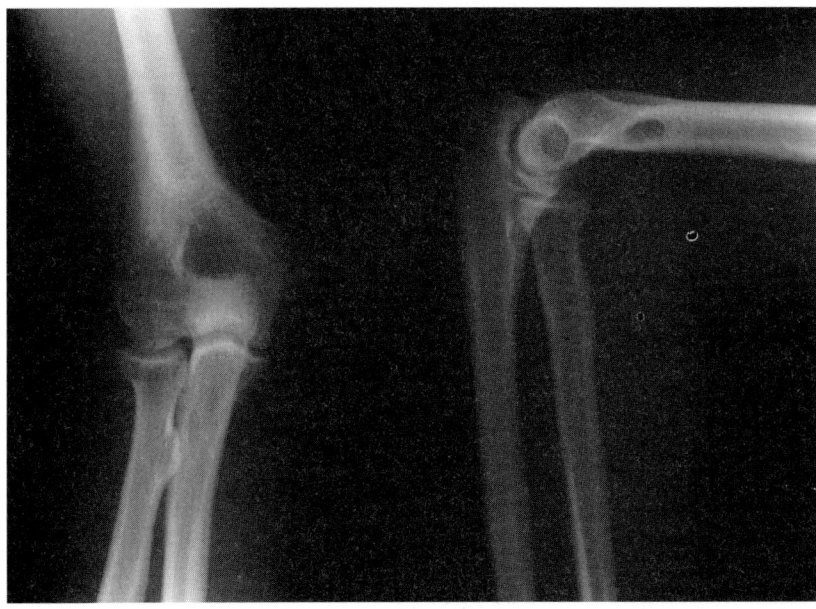

Figure 2.10 An undisplaced fissure fracture of the radial head: a Mason type 1

Mason type 2 fractures which they extended to include tilted radial neck fractures commonly showing instability to valgus stress. The majority of the Mason type 3 fractures dislocated to valgus stress.

Until the advent of the AO classification, there was no universally accepted classification of radial neck fractures, especially for adults. A system of classification of children's radial neck fractures was outlined by Newman (1977) with five types based on the displacement of the radial head relative to the radial shaft, two of which were associated with elbow dislocation.

Olecranon fractures

In 1974, Colton proposed a simple classification system for olecranon fractures to serve as a basis for the selection of the correct treatment for each injury type. This classification is based on fracture morphology and is divided into four main groups (Figure 2.11):

1. Avulsion group
2. Oblique group
3. Fracture dislocation group
4. Unclassified group.

The avulsion group (group 1) contains those fractures in which a transverse fracture line separates a small proximal fragment of the olecranon process (Figure 2.5). This is likely to be a 'pull-off' fracture due to the action of the triceps with elbow extended, and is often associated with a fall onto the outstretched hand. The oblique pattern (group 2) occurs when the fracture line starts near the deepest part of the trochlear notch, runs dorsally and distally and exits on the subcutaneous border of the proximal ulna. Group 2 is the only group which is subdivided depending on the amount of comminution, ranging from type 2a which has no comminution to type 2d where there is extensive comminution of a central depressed fragment. It is probably due to bending force with synergistic flexor and extensor activity, often while falling onto the elbow. The fracture dislocation group (group 3) occurs when the olecranon fracture is at or just proximal to the tip of the coronoid process so that there is a plane of instability through the fracture and the radiohumeral joint. Anterior or posterior elbow dislocation or valgus subluxation with radial head fracture may result. The mechanism of injury is identified as a heavy force, possibly with valgus force also. The unclassified group (group 4) contains fractures which are so severely comminuted that no pattern is discernible. Colton goes on to recommend treatment based on the type of fracture but makes no claims to forecast prognosis.

A further system was recommended by Horne and Tanzer (1981), and is also based on treatment methods. Their type 1 is very similar to Colton's with the addition of extra-articular fractures involving the tip of the olecranon only. Their type 2 and 3 fractures are very similar to Colton's, but the classification is less detailed.

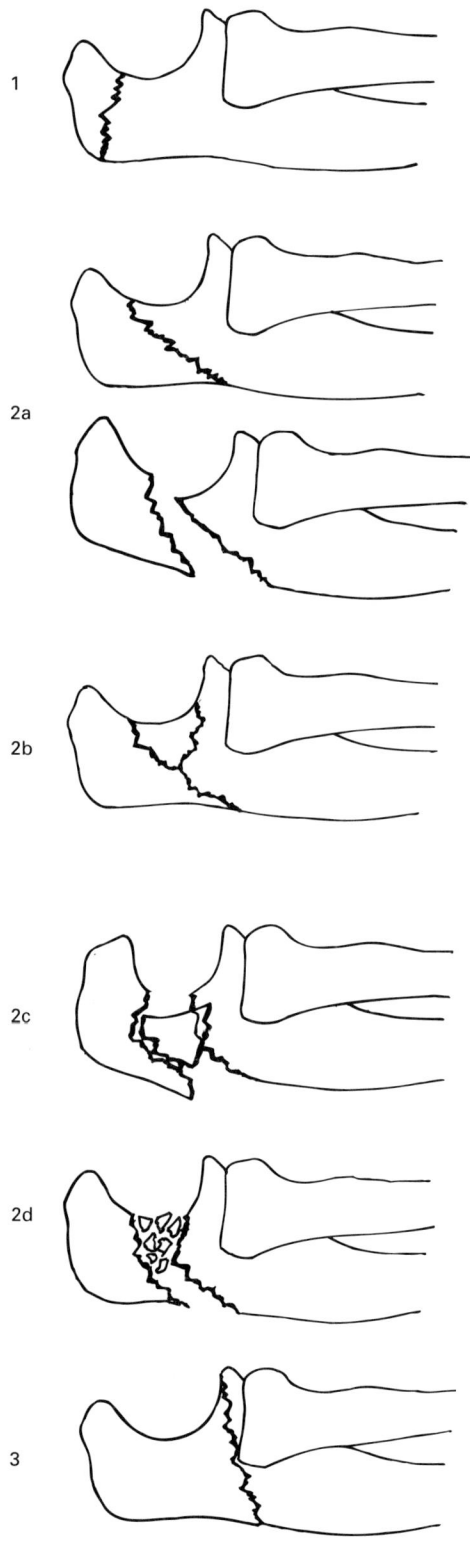

1

2a

2b

2c

2d

3

Figure 2.11   Colton's classification of olecranon fractures comminution

Group 1   Avulsion fracture
Group 2   Oblique fracture with subgroup depending on comminution and displacement. A sagittal split (B) may occur in any of these fractures
Group 3   The olecranon fracture is at the tip of the coronoid process and may result in a fracture dislocation of the elbow

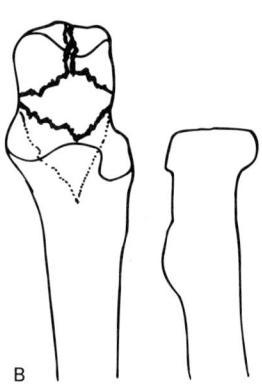

B

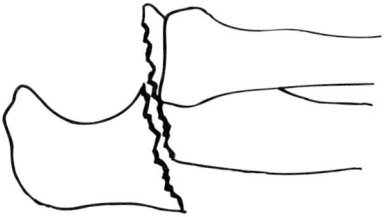

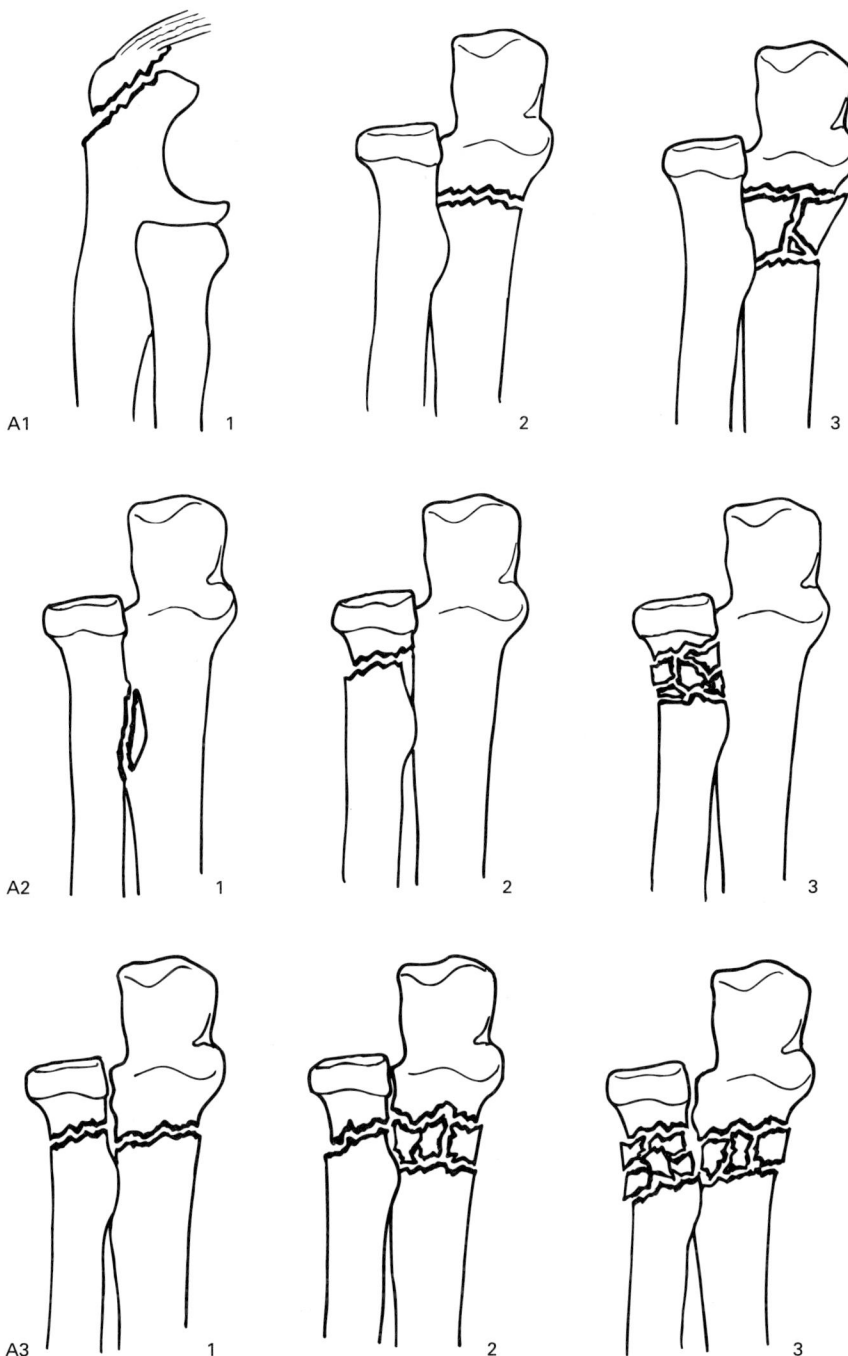

Figure 2.12 The AO classification of fractures of the proximal radius and ulna

Group A1	Extra-articular fracture of the ulna, radius intact	
Subgroup	A1.1	triceps avulsion
	A1.2	metaphyseal simple
	A1.3	metaphyseal multifragmentary
Group A2	Extra-articular fracture of the radius, ulna intact	
Subgroup	A2.1	avulsion biceps tuberosity
	A2.2	neck simple
	A2.3	neck multifragmentary
Group A3	Extra-articular fracture of both bones	
Subgroup	A3.1	simple of both bones
	A3.2	multifragmentary of one bone, simple of the other
	A3.3	multifragmentary of both bones

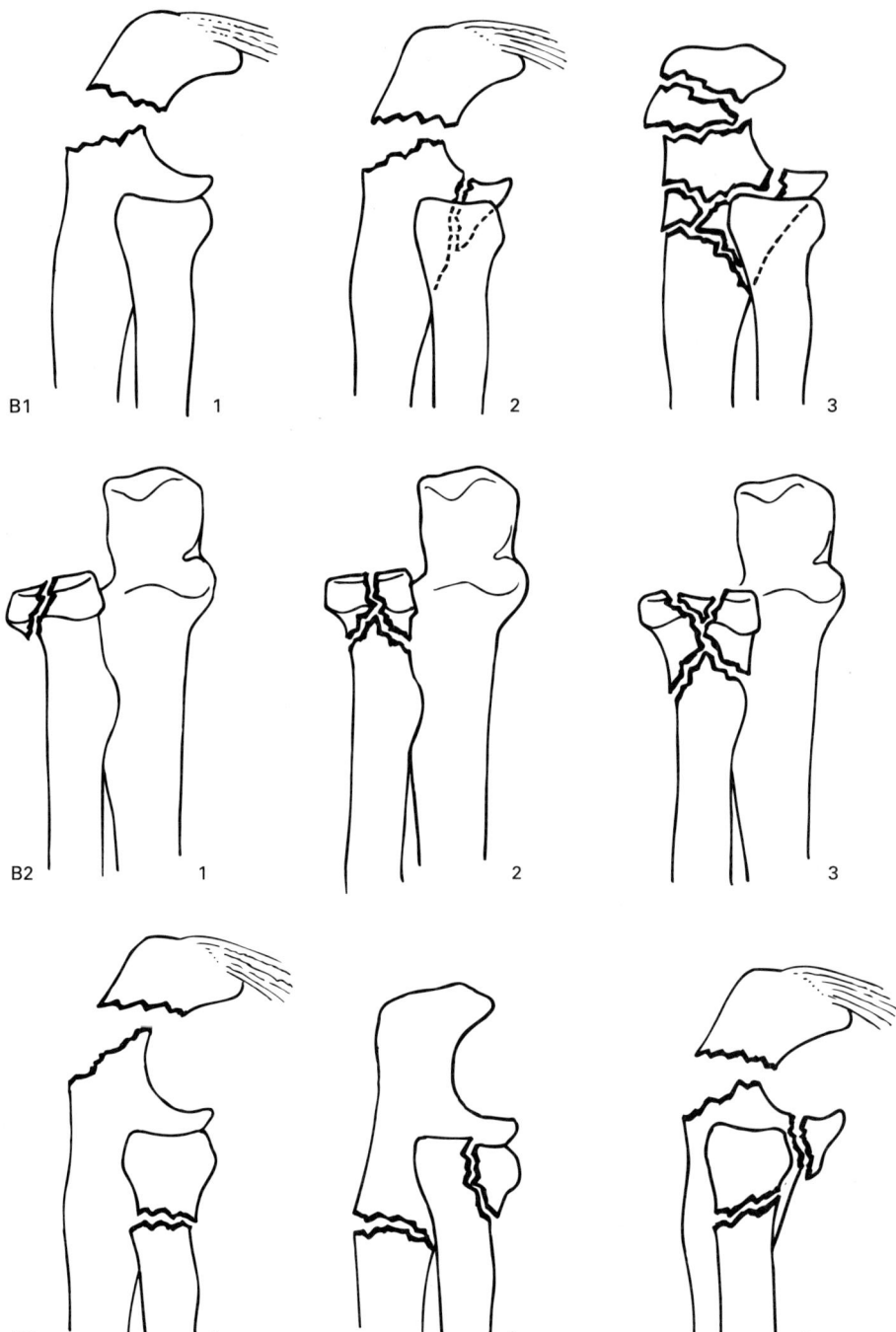

Figure 2.12   *cont.*

Group B1   Articular fracture of the ulna, radius intact
Subgroup   B1.1   unifocal
           B1.2   bifocal simple
           B1.3   bifocal multifragmentary
Group B2   Articular fracture of the radius, ulna intact
Subgroup   B2.1   simple

B2.2   multifragmentary without depression
B2.3   multifragmentary with depression
Group B3   Articular fracture of one bone with extra-articular
           fracture of the other
Subgroup   B3.1   ulna, articular simple
           B3.2   radius, articular simple
           B3.3   articular multifragmentary

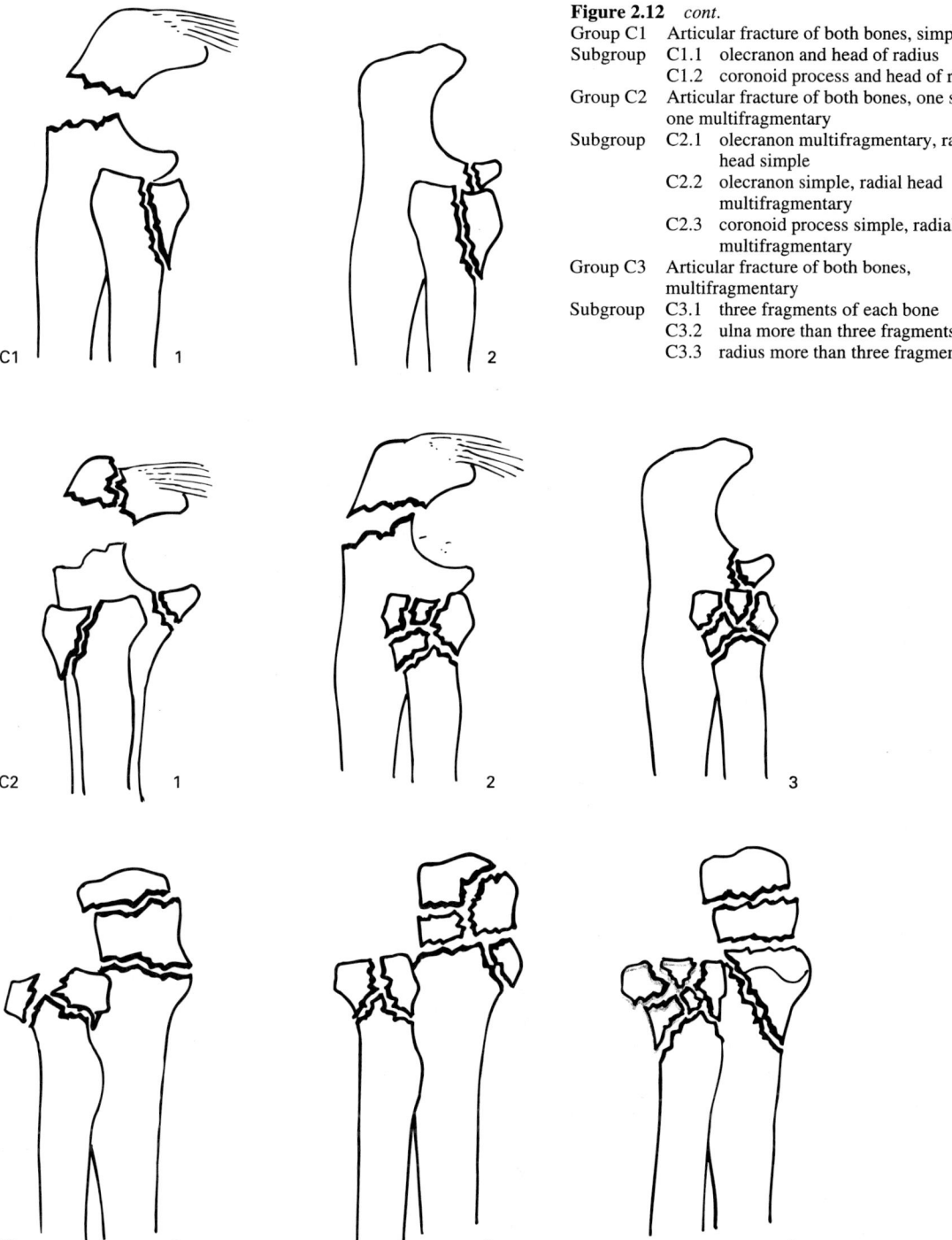

Figure 2.12 *cont.*

Group C1   Articular fracture of both bones, simple
Subgroup C1.1 olecranon and head of radius
            C1.2 coronoid process and head of radius
Group C2   Articular fracture of both bones, one simple,
            one multifragmentary
Subgroup C2.1 olecranon multifragmentary, radial
                 head simple
            C2.2 olecranon simple, radial head
                 multifragmentary
            C2.3 coronoid process simple, radial head
                 multifragmentary
Group C3   Articular fracture of both bones,
            multifragmentary
Subgroup C3.1 three fragments of each bone
            C3.2 ulna more than three fragments
            C3.3 radius more than three fragments

Coronoid fractures

There is only one classification system that has concentrated exclusively on this uncommon frac- ture (Morrey, 1993), which groups the fracture depending on the extent of involvement: type 1 is a fracture of the tip of the coronoid; type II involves a single or comminuted fracture of about half the coronoid process; and a type III fracture involves more than half of the coronoid. All three may be associated with elbow dislocation, although the likelihood of this increases as the fracture type increases. This classification system has also been shown to have some prognostic value in terms of function.

The AO classification for proximal radius and ulna fractures

The AO classification system broke the mould of previous classification systems by considering both the radius and ulna together. Because of this, the classification system was modified with the two bones being considered as a single articulation, with the radius being the lateral and the ulna the medial component (Figure 2.12).

The type A fractures are extra-articular (Figure 2.13). The A1 fractures involve only the ulna, the A2 fractures only the radius and the A3 fractures both bones, but still extra-articular. This therefore encompasses the radial neck fracture and fractures of the ulna distal to the trochlear notch with and without comminution.

The type B fractures are partial articular fractures, defined in the proximal forearm as an articular fracture of one bone with or without an extra-articular fracture of the other. Group B1 are articular fractures of the ulna with the radius intact (Figure 2.5), group B2 are articular fractures of the radius with the ulna intact and group B3 are articular fractures of one bone with an extra-articular fracture of the other bone. In general, as one progresses within each group there is increasing comminution. Coronoid fractures appear in sub-

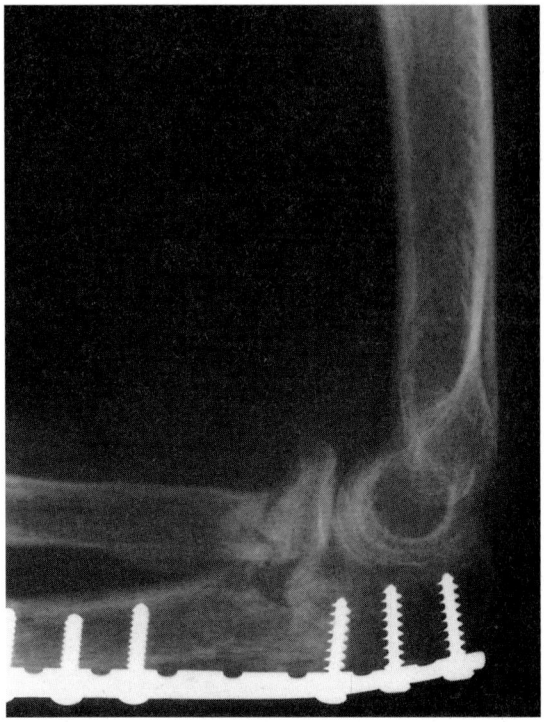

Figure 2.13 An extra-articular fracture of the proximal radius and ulna. This is an AO type A3.2. The ulna has been reduced and internally fixed

divisions of the subgroup B1.1. Displacement is only taken into account in group B2, the radial head fractures.

The type C fractures are articular fractures of both bones and as such are considered complete articular fractures (Figure 2.14) with increasing

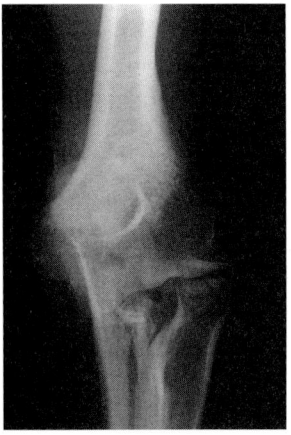
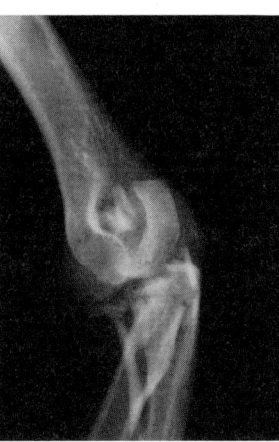

Figure 2.14 An articular fracture of both the proximal radius and ulna. This is an AO type C3.2

complexity as one advances down the system. At no point is an association with elbow dislocation considered.

This classification is by far the most detailed morphological system available at the present time. Although there is increasing severity of fracture between and within the groups and there is likely to be at least some loose correlation to functional outcome, this has not as yet been fully investigated.

Diaphyseal fractures of the radius and ulna

Until the advent of the AO classification, there was no classification system published specifically for diaphyseal fractures of the forearm although eponyms in common usage were applied to the unstable fractures with radioulnar joint dislocations, namely the Monteggia and Galeazzi fractures.

The AO classification for the forearm diaphyseal fractures is based on the same types as the other diaphyseal fractures (Figure 2.15). Type A are the simple fractures, type B are wedge fractures and type C are complex fractures (Figures 2.1–2.3). The groups are then split into three with ulna fractures being group 1, radius fractures group 2 and fractures of both bones being group 3. The exceptions are the type C fractures where the group is determined by whichever is the complex fracture, e.g. a complex ulnar and simple radial fracture is group 1. The subgroups are determined on the basis of a combination of two or more of the following: direction and complexity of the fracture line; involvement of the radioulnar articulations; and the level of the fracture in the bone. Most of the groups contain a version of the Monteggia or Galeazzi fracture (Figure 2.1).

Monteggia fractures were classified by Bado in 1967, whose system is still widely used 30 years later. He described four types (Figure 2.16). The first has an anterior dislocation of the radial head with a fracture of the ulnar diaphysis at any level (Figure 2.1). Bado's type 2 has a posterior or posterolateral dislocation of the radial head with posterior angulation of the ulnar diaphyseal fracture. Type 3 has a lateral or anterolateral dislocation of the radial head with a fracture of the ulnar metaphysis. Bado's type 4 consists of an anterior dislocation of the radial head with fractures of the proximal thirds of both the radius and ulna. He also describes several equivalents possessing characteristics mostly very similar to his type 1 (Table 2.4).

Table 2.4 Bado's Monteggia equivalents

Type 1
Anterior dislocation of the radial head (in children, the 'pulled elbow syndrome')
Fracture of the ulnar diaphysis and fracture of the proximal third or neck of the radius
Fracture of the neck of the radius
Fracture of the ulnar diaphysis and olecranon with anterior dislocation of the radial head
Posterior dislocation of the elbow and fracture of the ulnar shaft ± proximal radius fracture
Proximal radius fracture

Type 2
Epiphyseal fracture of the dislocated radial head
Radial neck fracture

Type 3 and 4
No equivalents

An alternative classification for children's Monteggia fractures has been suggested (Letts *et al.*, 1985). These authors' types A, B and C are variants of Bado's type 1, allowing for bending and greenstick fractures of the ulna. Types D and E are equivalent to Bado's types 2 and 3.

Bado's original system was intended to be descriptive and he made no attempt to correlate it to outcome, although it has been suggested that type 1 lesions have better results than the other three groups (Reckling, 1982).

Fractures of the distal end of the radius and ulna

Fractures of the distal end of the radius have probably the greatest number of eponyms attached to them. However, these eponyms, although commonly used, may cause confusion and do not describe the injury accurately. As a result of this, many classification systems have been devised for this fracture, some simple and some complex, although none have achieved the almost impossible goal of detailing fracture patterns and accurately predicting outcome. However, many classification systems include a number of the following criteria: the morphology of the fracture; the direction and amount of displacement; the presence and extent of intra-articular involvement whether it be radiocarpal or radioulnar; the presence and extent of comminution; the relative stability of the fracture; associated injuries; and the likely outcome.

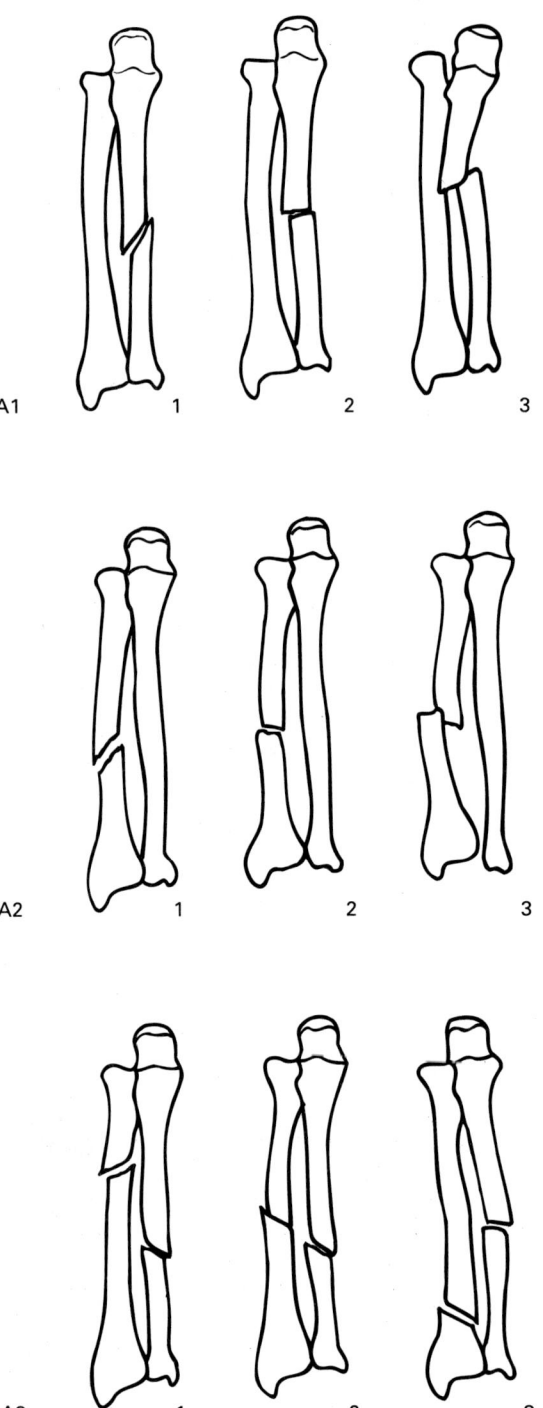

A1   1   2   3

A2   1   2   3

A3   1   2   3

Figure 2.15   The AO classification of fractures of the diaphysis of the radius and ulna

Group A1   Simple fracture of the ulna, radius intact
Subgroup   A1.1   oblique
            A1.2   transverse
            A1.3   with dislocation of the radial head
                    (Monteggia)
Group A2   Simple fracture of the radius, ulna intact
Subgroup   A2.1   oblique
            A2.2   transverse
            A2.3   with dislocation of the distal radioulnar
                    joint (Galeazzi)
Group A3   Simple fracture of both bones
Subgroup   A3.1   radius proximal zone
            A3.2   radius middle zone
            A3.3   radius distal zone
Group B1   Wedge fracture of the ulna, radius intact
Subgroup   B1.1   intact wedge
            B1.2   fragmented wedge
            B1.3   with dislocation of the distal radioulnar
                    joint (Galeazzi)
Group B2   Wedge fracture of the radius, ulnar intact
Subgroup   B2.1   intact wedge
            B2.2   fragmented wedge
            B2.3   with dislocation of the distal radioulnar
                    joint (Galeazzi)
Group B3   Wedge fracture of one bone, simple or wedge
            fracture of the other
Subgroup   B3.1   ulnar wedge, simple fracture of the radius
            B3.2   radial wedge, simple fracture of the ulna
            B3.3   radial and ulnar wedges
Group C1   Complex fracture of the ulna
Subgroup   C1.1   bifocal, radius intact
            C1.2   bifocal, radius fractured
            C1.3   irregular
Group C2   Complex fracture of the radius
Subgroup   C2.1   bifocal, ulna intact
            C2.2   bifocal, ulna fractured
            C2.3   irregular
Group C3   Complex fracture of both bones
Subgroup   C3.1   bifocal
            C3.2   bifocal of one, irregular of other
            C3.3   irregular

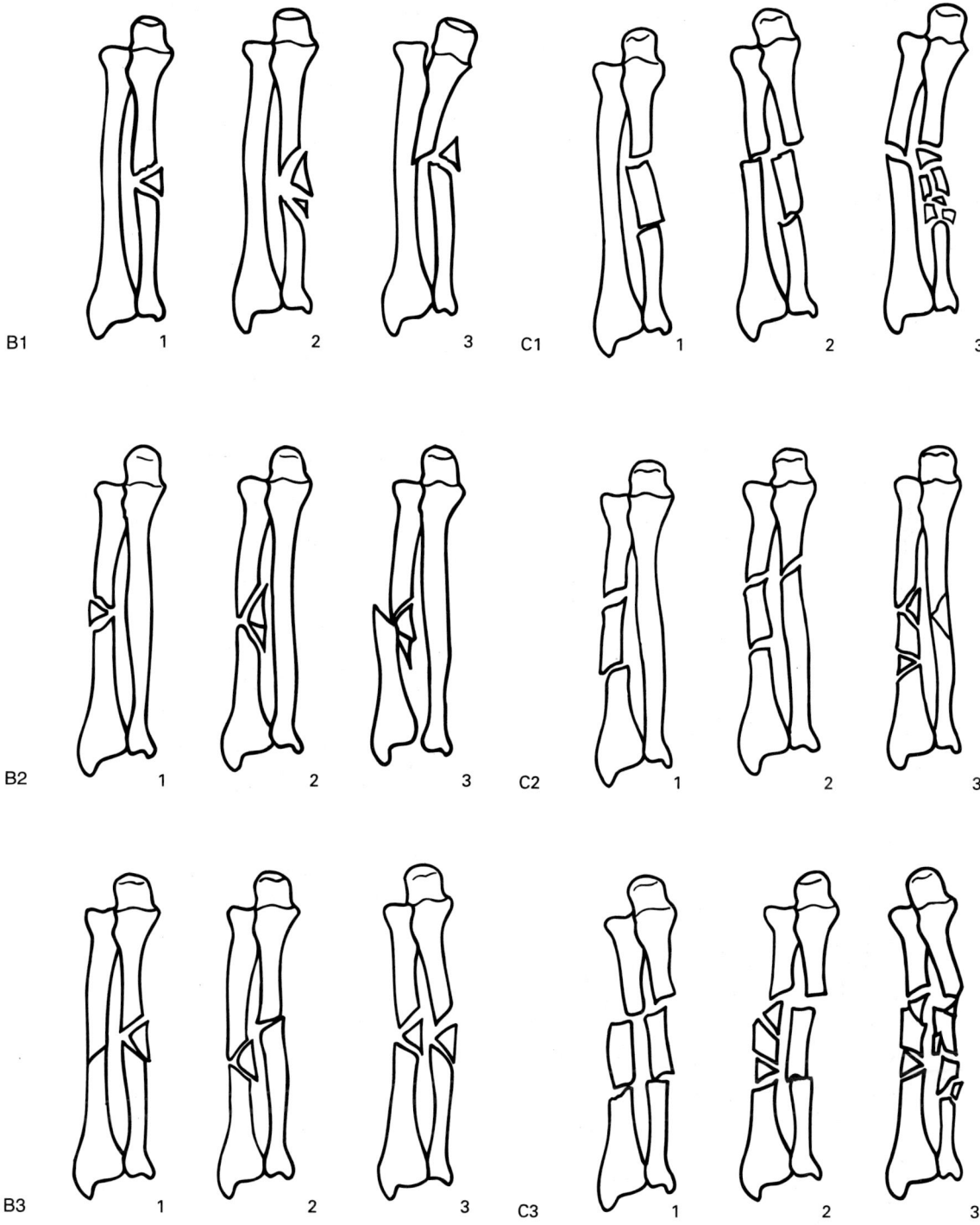

Figure 2.15 *cont.*

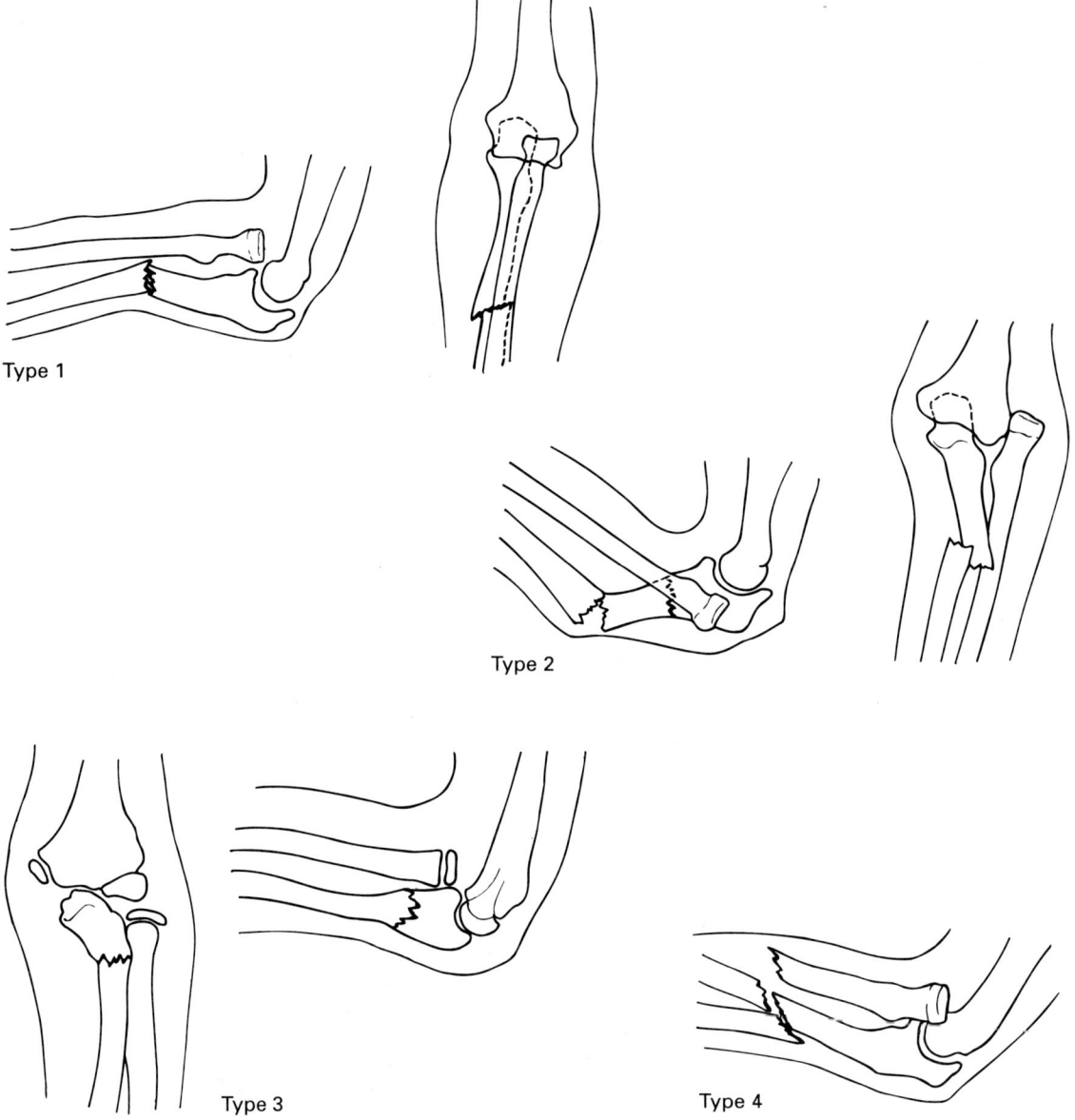

Figure 2.16 Bado's classification of Monteggia fractures

Type 1   Anterior dislocation of the radial head with fracture
         of the ulnar diaphysis with anterior angulation
Type 2   Posterior or posterolateral dislocation of the radial
         head and fracture of the ulnar diaphysis with
         posterior angulation

Type 3   Fracture of the ulnar metaphysis and lateral
         dislocation of the radial head
Type 4   Fracture of both bones with dislocation of the radial
         head

The earliest classification still in common usage today is that of Older (1965). This system is based on displacement and comminution (Figure 2.17) and has been shown to have some prognostic value (Solgaard, 1985), probably because radial shortening has an important influence on the classification.

Two years later, Frykman (1967) proposed a classification system which depended on involvement of the radiocarpal and radioulnar joints and

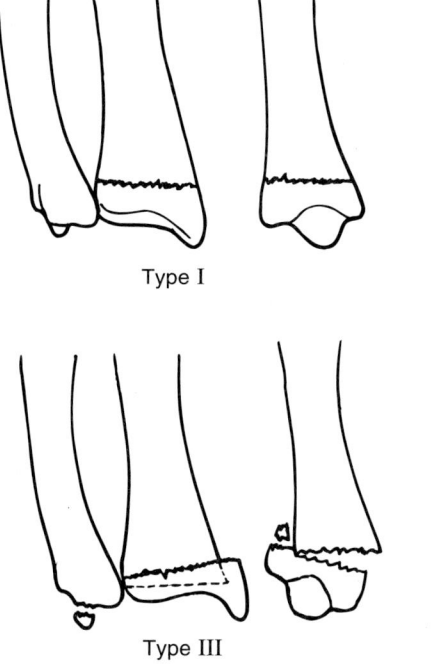

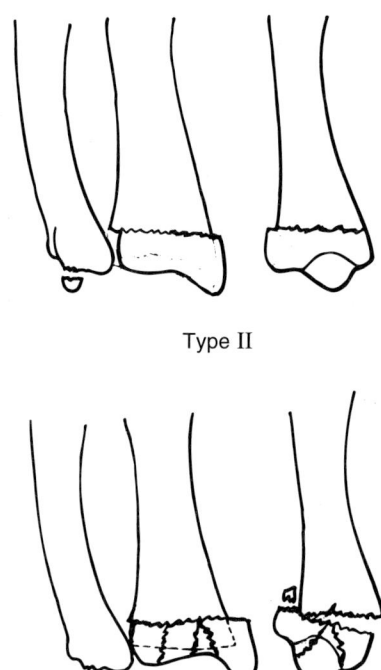

Figure 2.17   Older's classification of fractures of the distal radius

Type I   Dorsal angulation up to 5°, radial length distal to ulna at least 7 mm

Type II   Dorsal angulation, radial length 1–7 mm, no comminution

Type III   Dorsal radius comminuted, radial length less than 4 mm, distal fragment slightly comminuted

Type IV   Marked comminution, radial length usually negative

the presence of a fracture of the ulna (Table 2.5). This was probably the most commonly used fracture classification until the advent of the AO system, and it correlates loosely with function (Keating *et al.*, 1994). It ignores many features such as degree of displacement and comminution but has the advantage of being simple.

Other simple classification systems have been proposed based on the presence of comminution (Jenkins, 1989) or reducibility, stability and articular involvement (Cooney *et al.*, 1990; Rayhack, 1990). The Fernandez classification system (Fernandez and Jupiter, 1996) is based on the mechanism of injury and is intended to assist with management decisions. There are five major types (Figure 2.18). Type I is a bending fracture which is extra-articular with some comminution and equates to the traditional Colles' or Smith's fractures. Type II fractures are caused by shearing forces and are partial articular fractures such as the Barton's fracture or a radial styloid fracture. Type III fractures are caused by a compression force and are intra-articular with impaction. Type IV are avulsion fractures associated with ligamentous attachments such as the dorsal fracture dislocation. Type V fractures arise from a combination of the above mechanisms and are complex, often high-energy, injuries. Each type is accompanied by a recommendation for treatment and an assessment of the likelihood of associated lesions being present. Overall this classification is relatively simple and has the merit of relating to possible treatment, but its prognostic value has not yet been tested.

Table 2.5   **Frykman's classification of fractures of the distal radius and ulna**

Joint involvement	Ulna −	Ulna +
Extra-articular	1	2
Radiocarpal	3	4
Distal radioulnar	5	6
Both radiocarpal and distal radioulnar	7	8

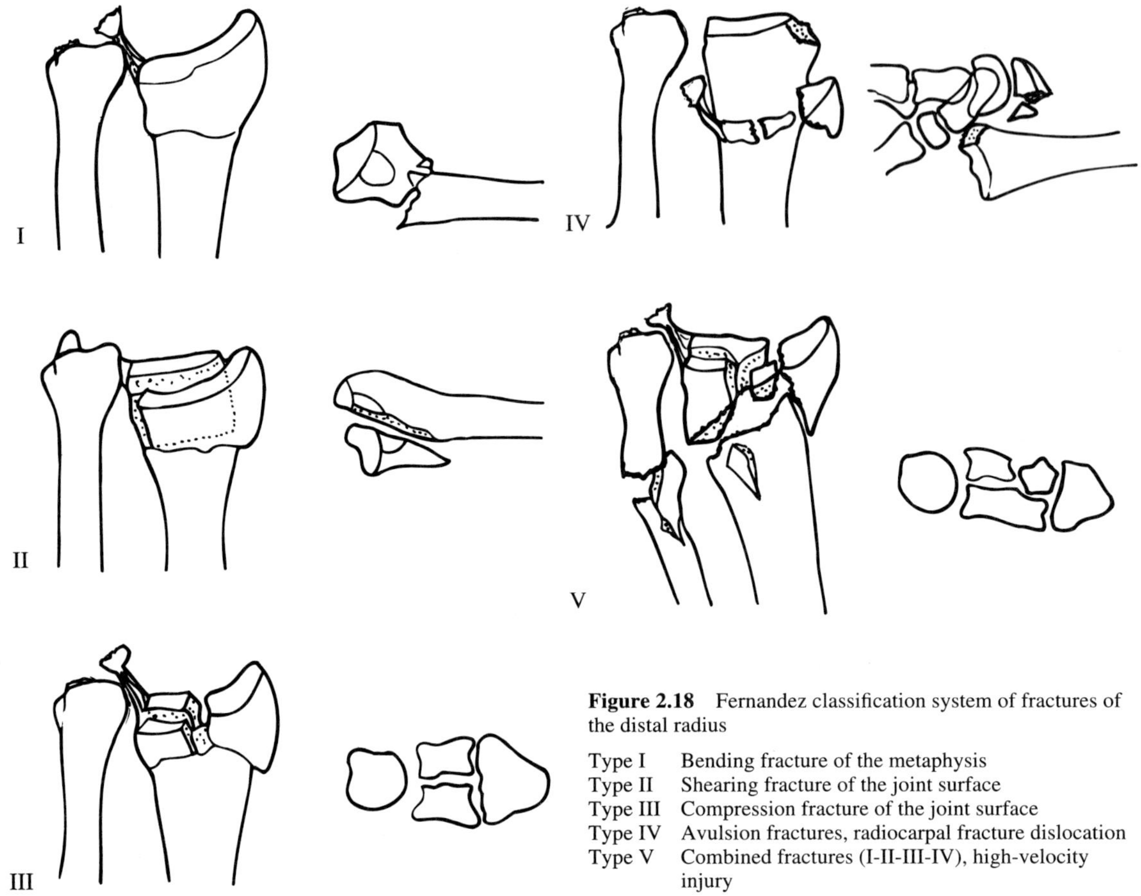

I

II

III

IV

V

Figure 2.18 Fernandez classification system of fractures of the distal radius

Type I   Bending fracture of the metaphysis
Type II   Shearing fracture of the joint surface
Type III   Compression fracture of the joint surface
Type IV   Avulsion fractures, radiocarpal fracture dislocation
Type V   Combined fractures (I-II-III-IV), high-velocity
           injury

The same authors also describe a classification of associated distal radioulnar joint injuries with particular reference to distal radioulnar joint stability. Type I distal radioulnar joint injuries are stable with a congruous distal radioulnar joint and include both avulsion fractures of the tip of the ulnar styloid and stable fractures of the ulnar neck (Figure 2.4). Type II distal radioulnar joint injuries are unstable with a significant soft tissue injury (e.g. a substance tear of the triangular fibrocartilage (TFCC) or a fracture of the base of the ulnar styloid. Type III injuries are potentially unstable with an intra-articular fracture on either side of the distal radioulnar joint (Figure 2.19). Again, each type has recommendations for treatment and can be used for assessment of prognosis although this has not been evaluated.

The most complex and comprehensive classification of distal radial fractures yet published is that of the AO group (Figure 2.20). As with other

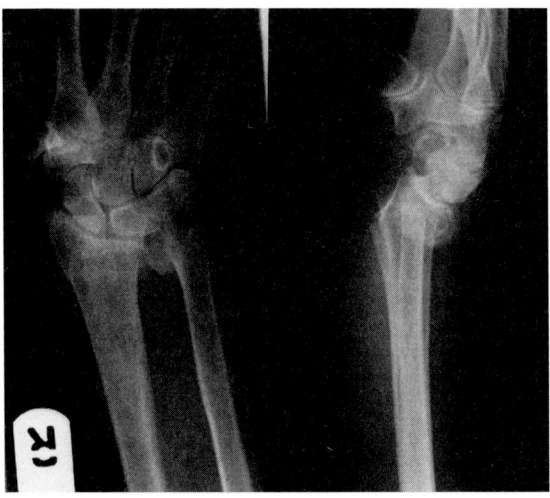

Figure 2.19 A Fernandez type III injury to the distal radioulnar joint. There is an articular injury on both sides of the joint

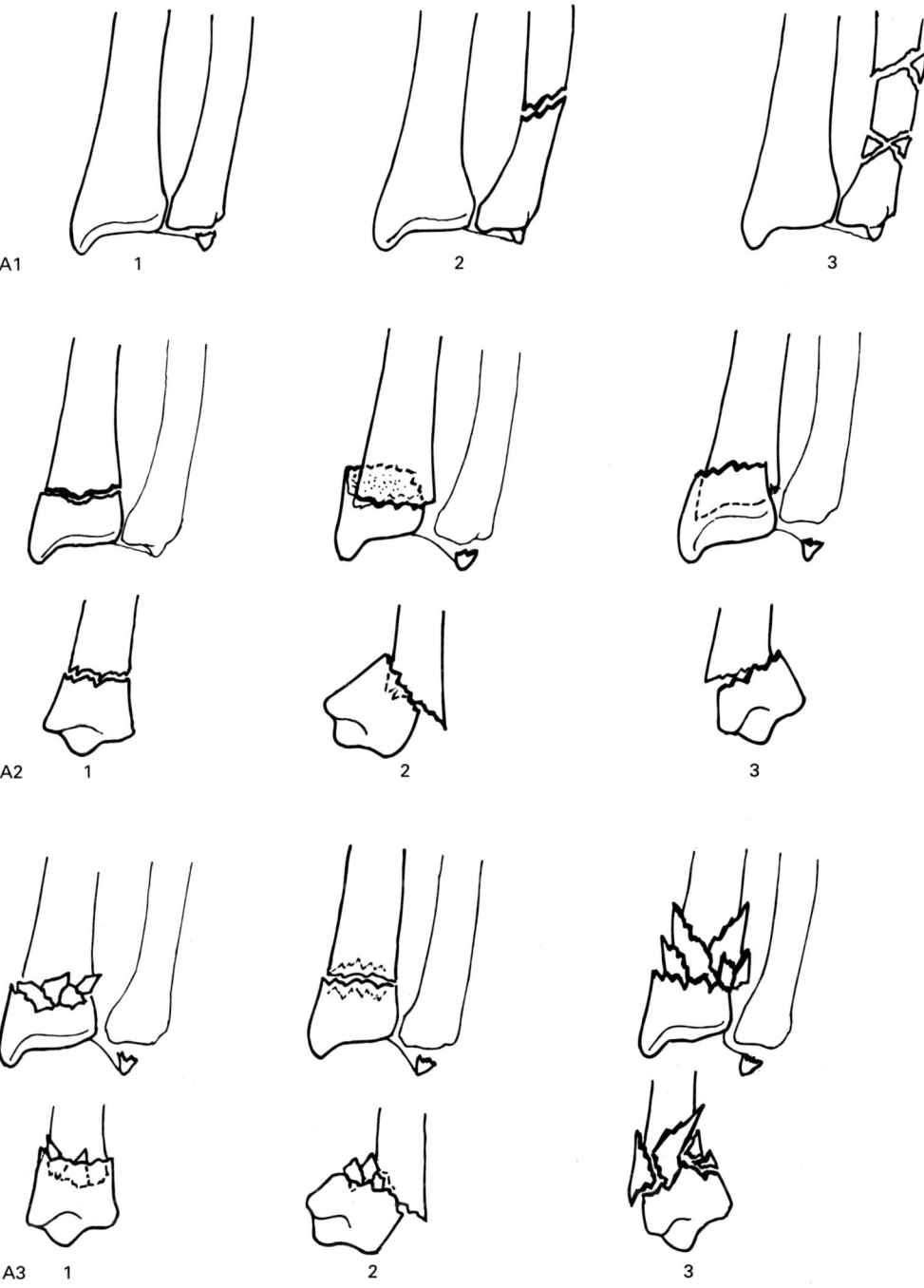

Figure 2.20   The AO classification of fractures of the distal radius

Group A1	Extra-articular fracture of the ulna, radius intact	
Subgroup	A1.1	styloid process
	A1.2	metaphyseal simple
	A1.3	metaphyseal multifragmentary
Group A2	Extra-articular fracture of the radius, simple and impacted	
Subgroup	A2.1	without any tilt

A2.2 with dorsal tilt
A2.3 with volar tilt
Group A3 Extra-articular fracture of the radius, multifragmentary
Subgroup A3.1 impacted with axial shortening
A3.2 with a wedge
A3.3 complex

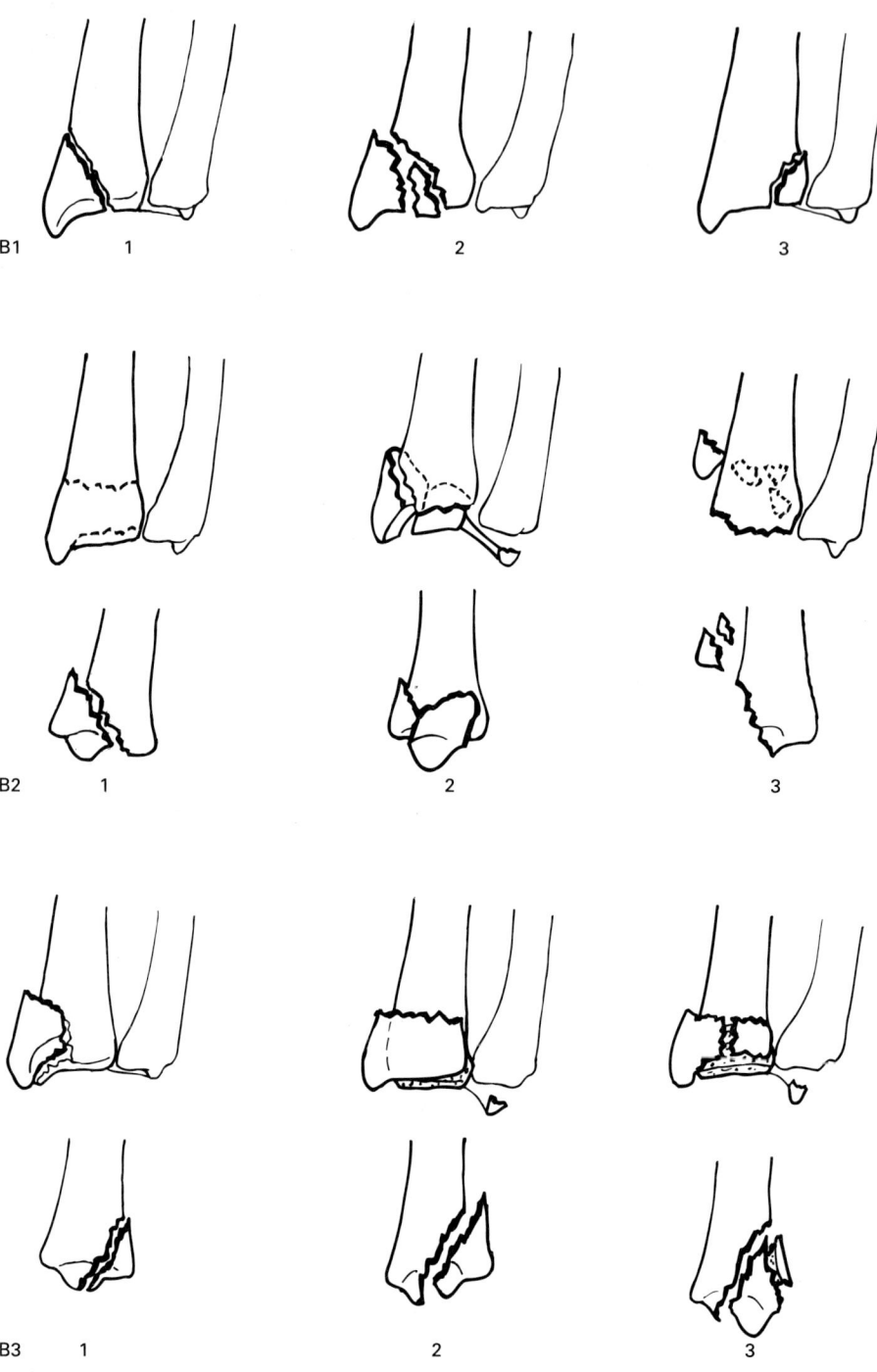

Figure 2.20 *cont.*

Group B1	Partial articular fracture of the radius, sagittal	
Subgroup	B1.1	lateral simple
	B1.2	lateral multifragmentary
	B1.3	medial
Group B2	Partial articular fracture of the radius, dorsal rim	
Subgroup	B2.1	simple

	B2.2	with lateral sagittal fracture
	B2.3	with dorsal dislocation of the carpus
Group B3	Partial articular fracture of the radius, volar rim	
Subgroup	B3.1	simple with a small fragment
	B3.2	simple with a large fragment
	B3.3	multifragmentary

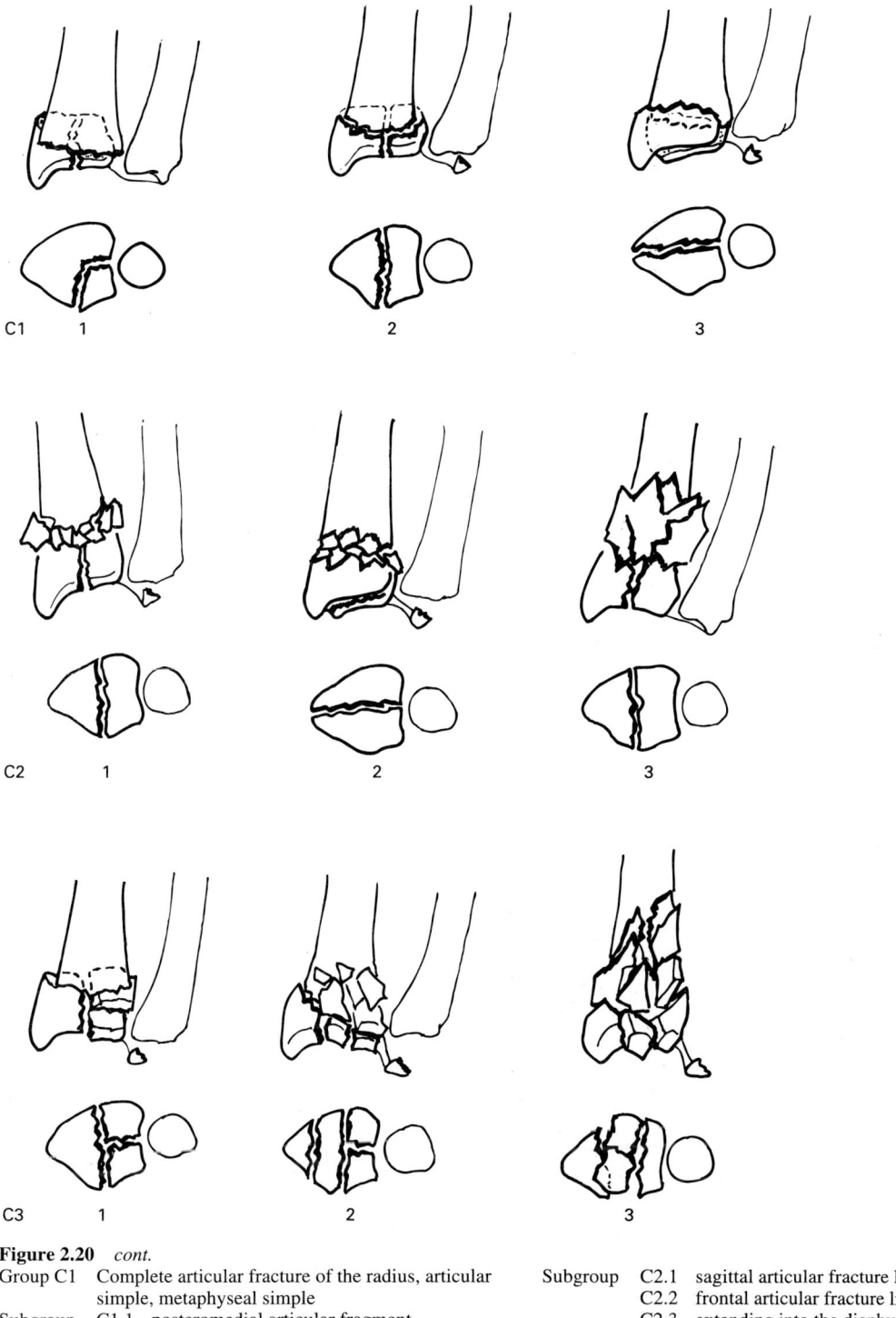

Figure 2.20 *cont.*

Group C1   Complete articular fracture of the radius, articular simple, metaphyseal simple

Subgroup   C1.1   posteromedial articular fragment
            C1.2   sagittal articular fracture line
            C1.3   frontal articular fracture line

Group C2   Complete articular fracture of the radius, articular simple, metaphyseal multifragmentary

Subgroup   C2.1   sagittal articular fracture line
            C2.2   frontal articular fracture line
            C2.3   extending into the diaphysis

Group C3   Complete articular fracture of the radius, multifragmentary

Subgroup   C3.1   metaphyseal simple
            C3.2   metaphyseal multifragmentary
            C3.3   extending into the diaphysis

metaphyseal fractures, type A fractures are extra-articular, type B partial articular and type C complete articular fractures.

The A1 fractures are extra-articular fractures of the ulna with an intact radius and they increase in severity from A1.1 to A1.3. The A2 fractures are extra-articular fractures without metaphyseal comminution, either undisplaced (A2.1), with dorsal tilt (A2.2) or with volar tilt (A2.3). Any extra-articular fracture with metaphyseal comminution falls into the A3 group with increasing comminution from A3.1 to A3.3, with A3.3 fractures being those with comminution extending into the diaphysis. The A3.2 fracture is the typical Colles' fracture with metaphyseal comminution (Figure 2.4).

The type B or partial articular fractures are those in which the fracture fragment contains a portion of the articular surface but the metaphysis remains in continuity with the epiphysis. The B1 fractures are sagittal fractures of the radius, with the best known being the radial styloid fracture. The B2 fractures are fractures of the dorsal rim of increasing complexity and the B3 fractures affect the volar rim of the distal radius (Figure 2.21).

Type C fractures are complete articular fractures where the articular surface is fractured and totally separated from the metaphysis or diaphysis. The C1 fractures are simple fractures without metaphyseal comminution. The C2 fractures are simple articular fractures with metaphyseal comminution (Figure 2.6) and the C3 fractures are the most complex, with multiple articular fracture lines and increasing metaphyseal comminution.

Varying inter- and intra-observer agreement has been reported using this system (Kreder *et al.*, 1996) with most consistency being achieved by those with experience. Its ability to predict outcome has yet to be fully assessed although it has been shown to correlate loosely to function in volar displaced fractures (Keating *et al.*, 1994).

Classification systems for intra-articular fractures of the distal radius

There are several classification systems exclusively for intra-articular fractures. Melone (1984) described four possible parts of an intra-articular fracture: the radial shaft, the radial styloid and

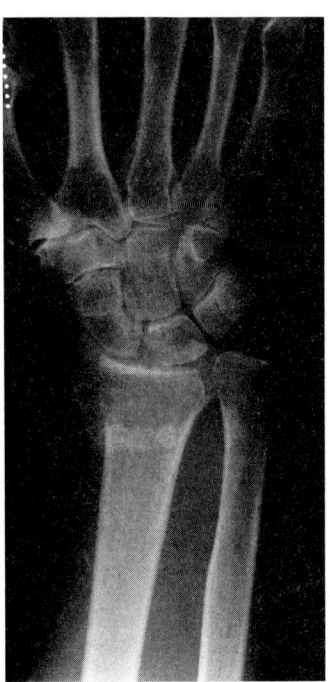
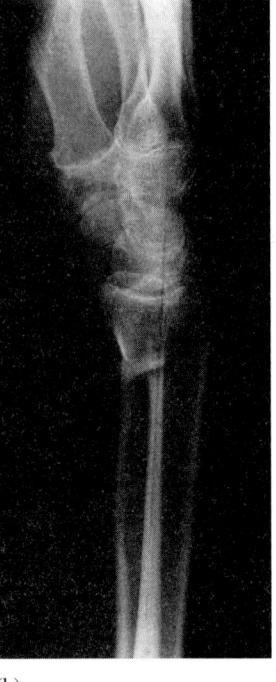

(a)   (b)

Figure 2.21   An AO type B3.2 fracture of the distal radius

dorsal and volar portions of the lunate facet, which he termed the medial complex. Type 1 injuries are undisplaced and stable. Type 2 occur when the radial styloid splits from the medial complex but the medial complex is intact (Figure 2.6). Type 3 injuries are as type 2 but with a volar spike which may damage the soft tissues. Type 4 injuries occur when, along with the other features, the dorsal and volar parts of the medial complex are split. In 1993, Melone added a type 5, which he described as an explosion injury with severe articular comminution. This classification system is intended to be morphological to aid in reduction and surgical treatment.

McMurtry (1992) recognizes partial articular fractures as two-part fractures and goes on to describe three-part, four-part and five-part fractures which are similar to Melone type 2, type 4 and type 5 injuries respectively. The Mayo classification system (Missakian *et al.*, 1992) is similar to the above two and emphasizes the individual articular surfaces of the distal radius including the sigmoid fossa of the distal radio-ulnar joint.

References

Andersen, D.J., Blair, W.F. and Steyers, C.M. *et al.* (1996) Classification of interobserver reliability and intraobserver reproducibility. *J. Hand Surg.*, **21A**, 574–82

Bado, J.L. (1967) The Monteggia lesion. *Clin. Orthop.* **50**, 71–86

Brumback, R.J. and Jones, A.L. (1994) Interobserver agreement in the classification of open fractures of the tibia. The results of a survey of two hundred and forty-five orthopaedic surgeons. *J. Bone Joint Surg.*, **76A**, 1162–6

Colton, C.L. (1974) Fractures of the olecranon in adults: classification and management. *Injury*, **5**, 121–9

Cooney, W.P., Agee, J.M. and Hastings, H. (1990) Management of intra-articular fractures of the distal radius. *Contemp. Orthop.*, **21**, 71–104

Court-Brown, C.M., Christie, J. and McQueen, M.M. (1990) Closed intramedullary nailing. Its use in closed and type 1 open fractures. *J. Bone Joint Surg.*, **72B**, 605–11

Court Brown, C.M., McQueen, M.M., Quaba, A.A. and Christie, J. (1991) Locked intramedullary nailing of open tibial fractures. *J. Bone Joint Surg.*, **73B**, 959–64

Davidson, P.A., Moseley, J.B. and Tullos, H.S. (1993) Radial head fracture. A potentially complex injury. *Clin. Orthop.*, **297**, 224–30

Fernandez, D.L. and Jupiter, J.B. (1996) Epidemiology, mechanism, classification. In *Fractures of the Distal Radius*. (D.L. Fernandez and J.B. Jupiter, eds) pp. 263–315, New York: Springer

Frykman, G. (1967) Fracture of the distal radius including sequelae – shoulder–hand–finger syndrome, disturbance in the distal radio-ulnar joint and impairment of nerve function: a clinical and experimental study. *Acta Orthop. Scand.*, suppl. 108, 1–155

Gustilo, R.B. and Anderson, J.T. (1976) Prevention of infection in the treatment of 1,025 open fractures of long bones: retrospective and prospective analysis. *J. Bone Joint Surg.*, **58A**, 453–8

Gustilo, R.B., Mendoza, R.M. and Williams, D.N. (1984) Problems in the management of type II (severe) open fractures. A new classification of type III open fractures. *J. Trauma*, **24**, 742–6

Horne, J. and Tanzer, T.L. (1981) Olecranon fractures: a review of 100 cases. *J. Trauma*, **21**, 469–72

Jenkins, N.H. (1989) The unstable Colles' fracture. *J. Hand Surg.*, **14B**, 149–54

Johnston, G.W. (1962) A follow-up of one hundred cases of fracture of the head of the radius with a review of the literature. *Ulster Med. J.*, **31**, 51–6

Johnstone, D.J., Radford, W.J. and Parnell, E.J. (1993) Interobserver variation using the AO/ASIF classification of long bone fractures. *Injury*, **24**, 163–5

Keating, J.F., Court-Brown, C.M. and McQueen, M.M. (1994) Internal fixation of volar displaced distal radial fractures. *J. Bone Joint Surg.*, **76B**, 401–5

Kreder, H.J., Hanel, D.P. and McKee, M. *et al.* (1996) Consistency of AO fracture classification for the distal radius. *J. Bone Joint Surg.*, **78B**, 726–31

Letts, M., Locht, R. and Wiens, J. (1985) Monteggia fracture dislocations in children. *J. Bone Joint Surg.*, **67B**, 724–7

Mason, M.L. (1954) Some observations on fractures of the head of the radius with a review of one hundred cases. *Br. J. Surg.*, **42**, 123–32

McMurtry, R.Y. and Jupiter, J.B. (1992) Fractures of the distal radius. In *Skeletal Trauma*. (B.D. Browner, J.B. Jupiter, A.M. Levine and P.G. Trafton, eds) pp. 1062–94

Melone, C.P. (1984) Articular fractures of the distal radius. *Orthop. Clin. N. Am.*, **15**, 217–36

Missakian, M.L., Cooney, W.P., Amadio, P.C. and Glidewell, H.L. (1992) Open reduction and internal fixation for distal radius fractures. *J. Hand Surg.*, **17A**, 745–55

Morrey, B.F. (1993) Fractures of the proximal ulna and olecranon. In *The Elbow and its Disorders*. (B.F. Morrey, ed.) p. 418, Philadelphia: Saunders

Müller, M.E., Allgower, M., Schneider, R. and Willenegger, H. (1991) Manual of internal fixation.Techniques recommended by the AO-ASIF Group. Berlin: Springer

Müller, M.E., Nazarian, S., Koch, P. and Schatzker, J. (1990) *The Comprehensive Classification of Fractures of Long Bones*. Berlin: Springer

Newman, J.H. (1977) Displaced radial neck fractures in children. *Injury*, **9**, 114–21

Oestern, H.J. and Tscherne, H. (1984) Pathophysiology and classification of soft tissue injuries associated with fractures. In *Fractures with Soft Tissue Injuries*. (H. Tscherne and L. Gotzen, eds) Berlin: Springer

Older, T.M., Stabler, E.V. and Cassebaum, W.H. (1965) Colles' fracture: evaluation and selection of therapy. *J. Trauma*, **5**, 469–76

Poland, J. (1898) *Traumatic Separation of the Epiphyses*. London: Smith, Elder & Co.

Rang, M. (1969) *The Growth Plate and its Disorders*. Baltimore: Williams & Wilkins

Rayhack, J. (1990) Symposium on distal radial fractures (W. Cooney, ed.). *Contemp. Orthop.*, **21**, 75

Reckling, F.W. (1982) Unstable fracture-dislocations of the forearm (Monteggia and Galeazzi lesions). *J. Bone Joint Surg.*, **64A**, 857–63

Salter, R.B. and Harris, W.R. (1963) Injuries involving the epiphyseal plate. *J. Bone Joint Surg.*, **45A**, 587–622

Sanders, R., Swiontkowski, M., Hurley, J. and Spiegel, P. (1993) The management of fractures with soft tissue disruption. *J. Bone Joint Surg.*, **75A**, 778–89

Solgaard, S. (1985) Classification of distal radial fractures. *Acta Orthop. Scand.*

3

Forearm anatomy

D. Ring and J.B. Jupiter

Introduction

The contributions of the forearm to upper extremity function are often underemphasized. More often than not, the proximal and distal radioulnar articulations are included with the elbow and wrist joints respectively. As such, the complexity of the anatomical structures which have evolved in order to provide rotational motion to the upper extremity may not be fully appreciated. Unique demands are made upon the forearm which must serve a dual purpose of a rotational joint as well as a bony strut between the elbow and the hand. Injury to any component of the forearm mechanism can jeopardize the function of the entire unit. Operative exposure in this region must be performed thoughtfully and carefully in order to avoid injury to the numerous neurovascular structures coursing towards the hand, minimize scarring and intertendinous adhesions, reduce the risk of radioulnar synostosis formation, and prevent iatrogenic radioulnar or ulnohumeral instability.

Evolution of the forearm

The pattern of evolution at both the elbow and the wrist reflects a transition from stability to mobility (Almquist, 1992; Larson, 1993). The development of bipedalism freed the upper extremity for enhanced manipulative function.

In conjunction with the increase in brain size and the development of the prehensile thumb, the acquisition of forearm rotation has been considered one of the three most important aspects differentiating the most highly developed hominids as these represent important factors determining the ability to manipulate one's environment

(Almquist, 1992; Linscheid, 1992). A rotational forearm provided advantages to brachiating mammals and facilitated hunting and food gathering as well as tool use (Palmer and Werner, 1984; Almquist, 1992; Linscheid, 1992). Forearm rotation enhances feeding by allowing the hand to assume the optimum position for grasping an object and for presenting it to the mouth (Kapandji, 1982). Furthermore, forearm rotation makes it possible for the hand to reach any point of the body for protection and cleaning, and it enhances tool use by allowing both precise tool positioning as well as a rotational force (Kapandji, 1982).

The evolution of forearm rotation required withdrawal of the distal ulna from the carpus and the development of synovial lined proximal and distal radioulnar articulations (Lewis, 1965; Almquist, 1992; Kauer, 1992). It is not surprising that the development of the human fetus recapitulates this process (Kauer, 1992).

Skeletal anatomy

Ulna

The ulna is frequently considered the 'straight' bone of the forearm, primarily in relation to its more curved counterpart, the radius (Figures 3.1 and 3.2). However, the ulna does have a slight, gradual bow which must be considered in the setting of forearm injury (Figure 3.1). In particular, the assessment of plastic bowing fractures of the forearm bones, which occur commonly in children and on occasion in adults, requires a familiarity with the normal ulnar bow. There is an apex posterior bow along the entire length of ulna as seen on a lateral radiograph. In the anteroposterior plane the ulna has a slight double curvature, apex

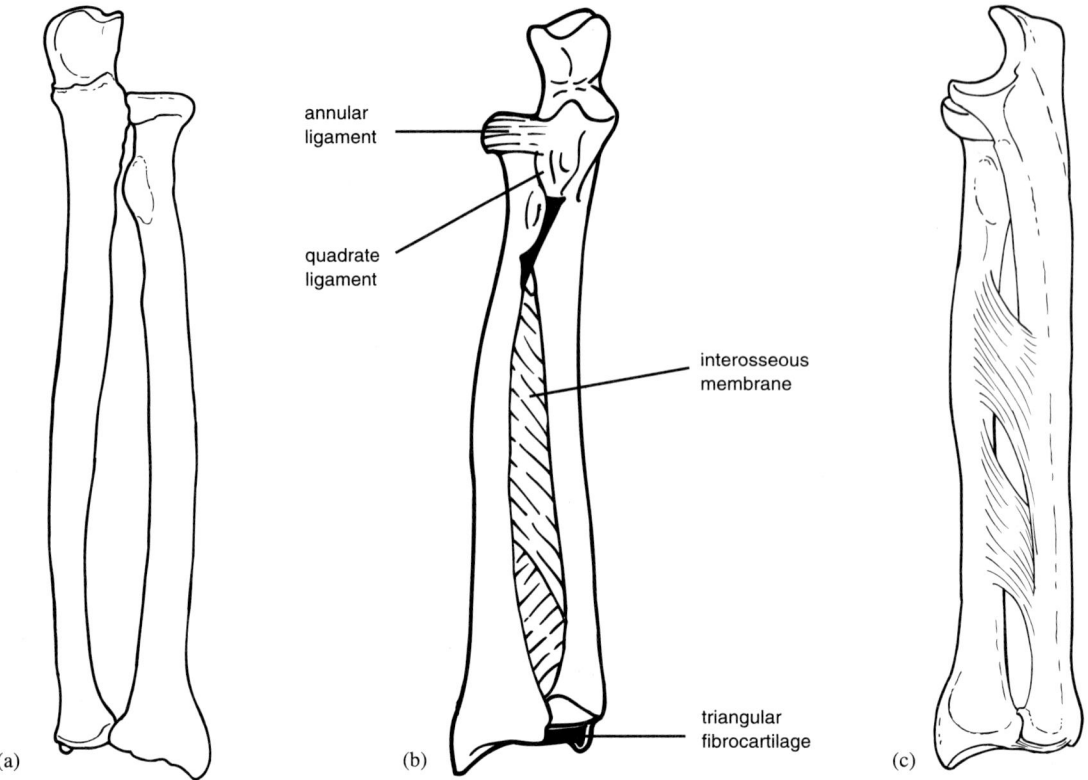

Figure 3.1(a) Anteroposterior view of the radius and ulna with the forearm in full supination. Note the double curvature of each bone in this plane. Even minor alterations in this curvature can interfere with forearm rotation. The radius and ulna articulate via synovial joints at their proximal and distal extents. (b) Radioulnar association is maintained by three groups of soft tissue structures: the annular and quadrate ligaments proximally the interosseous membrane in the mid-portion; and the triangular fibrocartilage complex distally. (c) Viewed laterally, the ulna has a gradual apex posterior bow whereas the radius again has a double curvature

lateral in the proximal half and apex medial distally (Goss, 1973; Williams and Warwick, 1980). The apex lateral curvature of the proximal half measures between 1° and 6° and may contribute to the normal carrying angle of the elbow (Keats *et al.*, 1966; Shiba *et al.*, 1988).

The ulna is triangular in cross-section through the majority of its mid-portion, becoming cylindrical distally (Figure 3.3). The laterally directed apex of the triangle corresponds with the insertion of the interosseous ligament. Proximally, this interosseous apex runs towards the posterior border of the lesser sigmoid notch, forming the supinator crest for insertion of the supinator muscle. A tubercle on this crest near the proximal limit of the supinator muscle represents the insertion point of the lateral ulnar collateral ligament complex of the elbow (Morrey and An, 1985; O'Driscoll *et al.*, 1991, 1992). The posterior border (or apex) of the triangular cross-section of

the ulnar shaft commences at the distal, posterior apex of the olecranon and follows the lateral curvature of the proximal ulna as it descends. This apex remains essentially subcutaneous as it divides the flexor and extensor musculature on the ulnar border of the forearm, and is palpable along the entire length of the bone.

The majority of the stability of the ulnohumeral joint is conferred by its bony structure. In particular, the articulation of the wide, grooved trochlear spool of the distal humerus with its corresponding ridged notch in the trochlear notch (also known as the greater sigmoid or semilunar notch) of the proximal ulna is well suited to providing both anteroposterior and varus/valgus stability. With the contributions from the proximal extensions of the coronoid and olecranon processes, the trochlear notch forms an arc of approximately 190° (Sorbie *et al.*, 1986). A 30° posterior tilt of this arc matches the corresponding anterior angu-

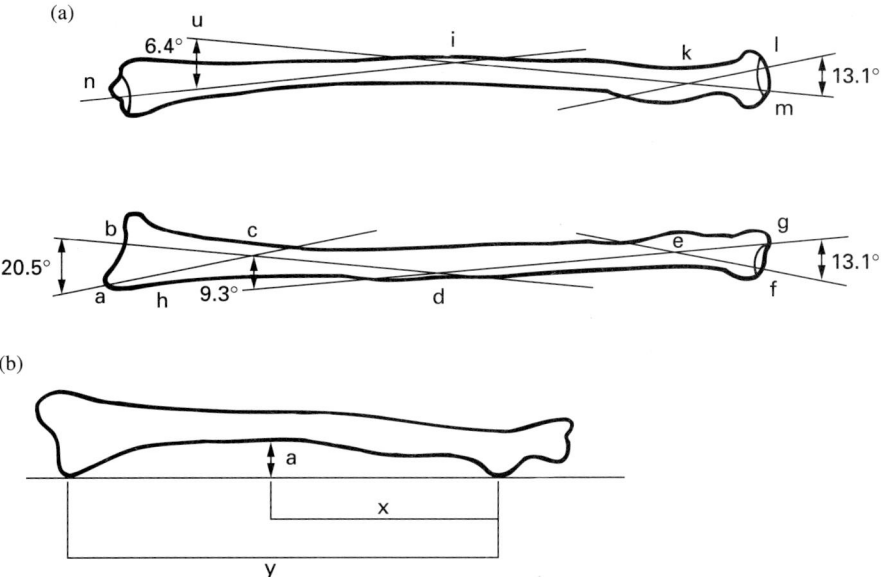

Figure 3.2 The double anatomical curvature of the radius must be preserved. (a) Sage carefully measured the angles of these curves in 50 human cadaver specimens and found that the proximal curvature averaged 13.1° apex medial and 13.1° apex anterior in the coronal and sagittal planes respectively. The distal curvature averaged 9.3° apex lateral and 6.4° apex posterior in the coronal and sagittal planes respectively. (b) Schemitch and Richards quantified the prominent distal radial bow in terms of the following measurements made from standard lateral radiographs with the forearm in neutral rotation. A reference line is drawn from the tip of the bicipital tuberosity to the ulnar-most aspect of the distal radius. The maximum radial bow (a) is measured as the number of millimetres along a perpendicular to this line from the apex of the radial bow to the reference line. The location of the maximum radial bow is measured as the percentage distance from the tip of the bicipital tuberosity to the point of the apex of maximum radial bow along the length of the reference line (i.e. x/y × 100). In their series of forearm fractures treated by plate and screw fixation, failure to restore these parameters to within approximately 4% of the opposite, uninvolved forearm was associated with a loss of 20% or more of forearm rotation

lation of the distal humeral articular surface (Morrey, 1993).

The central role that the ulnohumeral articulation plays in the overall stability of the elbow is commonly underemphasized. In fact, some authors consider the olecranon process expendable. Some have suggested that we can resect up to 80% of the olecranon in the treatment of comminuted fractures, provided that the collateral ligaments remain intact (Wainwright, 1942; McKeever and Buck, 1947; Adler *et al.*, 1962; Gartsmann *et al.*, 1981). While it may be true that satisfactory elbow function is possible following excision of olecranon fracture fragments in many patients with relatively low functional demands, it is somewhat erroneous to conclude that the contributions of the olecranon to ulnohumeral stability are expendable.

Attempts to characterize the stabilizing structures of the elbow consistently demonstrate that the ulnohumeral articulation provides the primary resistance to varus stress and is also a major stabil-

izer of the elbow under a valgus stress (Morrey and An, 1983; Sojberg *et al.*, 1987; Hotchkiss *et al.*, 1989). Rather than a threshold percentage of the total length of the olecranon process beyond which resection will lead to instability, An and co-workers have demonstrated a progressive decrease in resistance to varus and valgus force as increasing amounts of the olecranon are resected (An *et al.*, 1986). This investigation also demonstrated a greater and more rapid diminution in resistance to valgus than varus force with excision of increasing amounts of the olecranon process. In contrast, with the elbow in 90° of flexion, greater than 60% of the resistance to varus stress was preserved provided the distal portion of the trochlear notch, including the coronoid process, remained intact (An *et al.*, 1986). These findings suggest that the anterior and posterior portions of the trochlear notch (the coronoid and olecranon processes respectively) make distinct contributions to varus and valgus stability of the ulnohumeral joint.

(a)

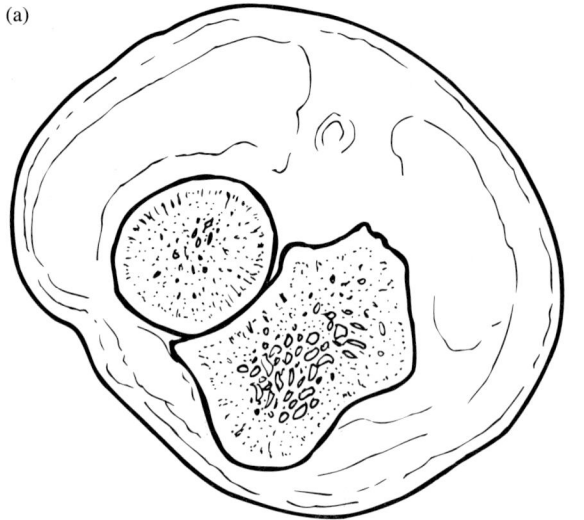

(b)

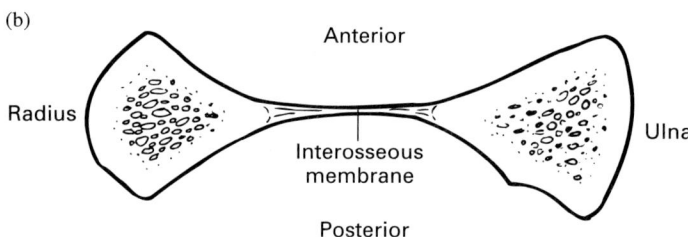

Anterior

Radius

Interosseous
membrane

Posterior

Ulna

(c)

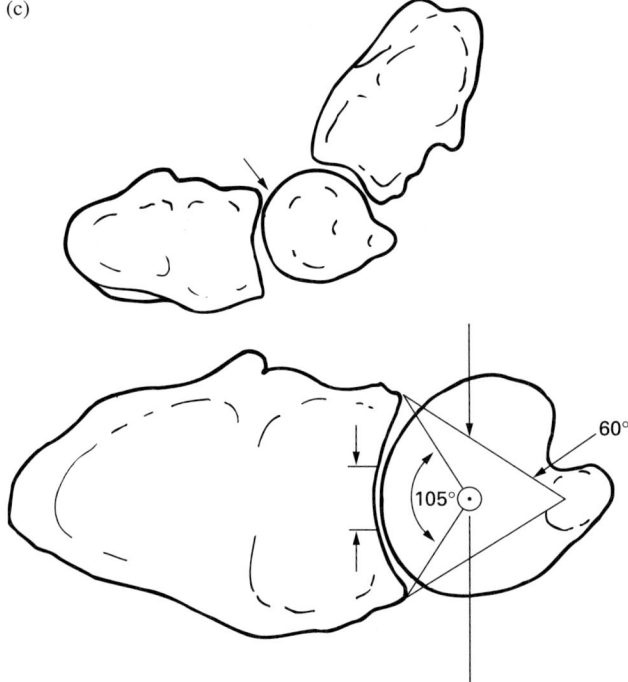

Figure 3.3   Cross-sectional anatomy of the forearm. (a) The proximal radioulnar joint. The radii of curvature of the lesser sigmoid notch of the ulna and the radial head are similar. The lesser sigmoid notch presents an arc of approximately 70° of hyaline cartilage which articulates with hyaline cartilage covering approximately 240° of the radial head circumference. The radial head is slightly ellipsoid with the result that the largest area of contact occurs in full forearm supination. (b) In the mid-portion of the forearm, the radius and ulna are triangular in cross-section with their interosseous apices representing the point of attachment of the interosseous membrane. The posterior apex (or border) of the ulna is subcutaneous and easily palpable throughout its length. (c) The distal radioulnar joint. The radius of curvature of the ulnar head is smaller than that of the sigmoid notch of the radius (8 mm compared to 15 mm). As a result, the area of contact between the radius and ulna is relatively small at this joint. This area of contact translates dorsally with pronation and volarly with supination

A relative anatomical separation of the olecranon and coronoid contributions to the trochlear notch of the ulna is defined by both a narrowing of the olecranon at its junction to the ulnar shaft as well as the dearth of articular cartilage between the articular surfaces of the two regions, which is sometimes called the transverse groove of the greater sigmoid notch (Figure 3.4a) (Tillmann, 1978; Morrey, 1993). The areas covered with articular cartilage correspond to the contact points of the ulna with the trochlea (Goodfellow and Bullough, 1967; Walker, 1977; Goel *et al.*, 1982; Stormont *et al.*, 1985). The bone in the area intervening between the olecranon and coronoid is not only narrower, but is also less dense, and is therefore more susceptible to fracture (Tillmann, 1978; Morrey, 1993).

This pattern of anterior and posterior areas of articular cartilage and joint contact is a result of the elliptical contour of the greater sigmoid notch. The roughly circular trochlear spool of the distal humerus articulates with the anterior and posterior articular surfaces provided by the coronoid and olecranon throughout the range of elbow motion, but rarely makes contact with the intervening portion of the sigmoid notch (Figure 3.4b).

The coronoid process represents an anterior extension and buttressing of the greater sigmoid notch. In addition to providing the anterior articular surface of the notch, the coronoid process serves as the point of attachment for the anterior capsule as well as the anterior portion of the medial collateral ligament (Cage *et al.*, 1995). A portion of the brachialis insertion, which extends slightly more distal on the ulna, lies on the coronoid (Cage *et al.*, 1995). Consequently, large fractures of the coronoid process result in ulnohumeral instability due to both the loss of the anterior pillar of the ulnohumeral articulation as well as the disruption of the soft tissue stabilizers of the elbow.

On the lateral aspect of the coronoid rests the radial or lesser sigmoid notch, which presents a 70° arc of articular cartilage for articulation with the radial head. Distal to this notch is a depression which provides clearance for the bicipital tuberosity during forearm rotation. The supinator crest contributes to the posterior border of this depression.

The ulna decreases in size from its proximal to its distal end. The distal terminus of the ulna expands somewhat into the ulnar head. The anterior 220° of the circumference of this head present hyaline cartilage for articulation with the sigmoid notch of the distal radius. This inferior surface, or seat, is also covered with articular cartilage and articulates with the triangular fibrocartilage. Posteromedially, the ulnar styloid process projects distally and expands somewhat the anteroposterior diameter of the head. A groove corresponding to the course of the extensor carpi ulnaris over the posterior aspect of the ulnar head is found just lateral to the styloid process. A depression know as the prestyloid recess or fovea is located radial to the base of the styloid process, towards the centre of the ulna. A number of small perforating blood vessels are found in the area of the fovea (Linscheid, 1992).

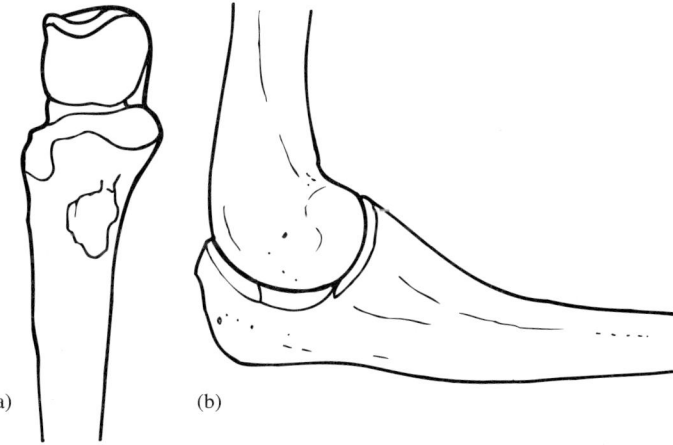

(a)   (b)

Figure 3.4 The ulnotrochlear articulation. (a) In the anteroposterior view it becomes apparent that the surface of the trochlear notch is not entirely articular. Hyaline articular cartilage is present on two distinct areas separated by the transverse groove of the greater sigmoid notch: one superiorly on the olecranon and one inferiorly on the coronoid process. The region of the transverse groove corresponds with an area of narrowing as the olecranon process attaches to the ulnar shaft. This narrowing, combined with the lack of dense subchondral bone, make this area relatively weak and subject to fracture. (b) In the lateral view, one notes that the trochlear notch is slightly ellipsoid, with the long axis directed towards the transverse groove. As a result, the contact of the relatively circular trochlear spool of the distal humerus with the trochlear notch of the ulna is limited to the anterior and posterior articular portions of the notch through a full range of ulnohumeral motion

Radius

The radius has a characteristic bow which has been demonstrated to be important to proper forearm function and must be accurately restored when this bone is fractured (Figures 3.1 and 3.2) (Kellam and Jupiter, 1991; Schemitsch and Richards, 1992; Richards, 1996). The radius has a double curvature in both the anteroposterior and lateral planes (Sage, 1982). The bicipital tuberosity, representing the insertion of the biceps brachii tendon, is at the apex of the smaller, proximal, convex medial curve, whereas the large, distal, convex lateral curve has at its apex the insertion of the pronator teres. This circumstance provides these powerful muscles with longer lever arms through which to produce rotatory torque of the radius. According to Sage's measurements, the proximal curvature of the radius averages approximately 13° apex medial in the coronal and 13° apex anterior in the sagittal anatomical plane. The distal curvature averages approximately 9.3° apex lateral in the coronal and 6.4° apex posterior in the sagittal plane (Figure 3.2a).

The large ulnar concavity of the distal curvature of the radius allows for overriding of the ulna without restriction of pronation (Kapandji, 1982). The limit of forearm pronation is reached as the flexor musculature is compressed between the forearm bones (Kapandji, 1982). This action produces a tightness in the interosseous ligament which encourages dorsal translation of the ulnar head within the sigmoid notch of the distal radius (Kapandji, 1982).

Numerous studies have demonstrated a direct relationship between the degree of forearm bone angular and rotational malalignment and restriction of rotational motion (Evans, 1945, 1951; Patrick, 1946; Matthews *et al.*, 1982; Sage, 1982; Tarr *et al.*, 1984; Sarmiento *et al.*, 1992; Schemitsch and Richards, 1992). Schemitsch and Richards found that even small deviations resulted in limitations in both forearm rotation and grip strength (Schemitsch and Richards, 1992; Richards, 1996). Their method of quantifying the larger distal radial bow consists of determining the point of the apex of the large distal curvature on a posteroanterior radiograph of the forearm. A reference line is then drawn from the apex of the bicipital tuberosity to the most ulnar aspect of the radius at the wrist. The location of the maximum radial bow is expressed as the percentage along this line, measured from the apex of the bicipital tuberosity, at which the point of the apex is found.

The distance in millimetres from the reference line to the point of the apex of the radial bow along a perpendicular to the reference line is defined as the maximum radial bow (Figure 3.2b) (Schemitsch and Richards, 1992).

The radius is cylindrical in the proximal head and neck region. The bicipital tuberosity marks the distal aspect of the neck and the junction between metaphyseal and diaphyseal bone (Sage, 1982). The superior articular surface of the radial head presents a depression for articulation with the capitellum of the humerus. The medial edge of this depression is bevelled in order to avoid impingement on the capitellotrochlear groove of the distal humeral articular surface with radial rotation (Kapandji, 1982). In addition, the superior articular surface follows the curvature of the proximal radius, forming an angle of approximately 15° with respect to the radial shaft (Evans, 1945; Sage, 1982). Approximately 240° of the outside circumference of the radial head articulates with the lesser sigmoid notch of the proximal ulna and is therefore covered with hyaline cartilage. The remaining anterolateral non-articular third is the area most often fractured, perhaps as a result of the relative weakness of this area in the absence of dense subchondral bone (Thomas, 1929).

The radial head is elliptical rather than circular in cross-section. Its long axis is approximately 15% longer than its short axis (Kapandji, 1982). The long axis of the elliptical radial head lies in the coronal plane in full pronation. Kapandji has suggested that this circumstance provides an additional 2 mm of clearance for the bicipital tuberosity by shifting the axis of rotation of the radius laterally as the bicipital tuberosity rotates towards the ulnar shaft (Kapandji, 1982).

The biceps inserts onto the roughened posterior aspect of the bicipital tuberosity. The orientation of the tuberosity can provide an indication of the rotation of the proximal radius, which may prove useful in the treatment of forearm fractures (Evans, 1945). Its apex is directed roughly opposite to that of the distal radial styloid. It points directly medial or ulnar in full supination and directly lateral in full pronation and is not visible when the proximal radius is in the neutral position (Evans, 1945, 1951).

Distal radial anatomy has received a great deal of attention due to the frequency of injury in this region and the correlation of the functional result with restoration of normal anatomy. The distal radial articular surface has a normal palmar inclination which averages between 10° and 12°

(Golden, 1963; Friberg and Lundström, 1976; Gilula, 1992; Mann *et al.*, 1992a) and an ulnar inclination averaging 22–23° (Gartland and Werley, 1951; Friberg and Lundström, 1976; Mann *et al.*, 1992b) (Figure 3.5).

The radiocarpal articular surface is divided into separate facets for articulation with the scaphoid and lunate. The ulnar portion of the lunate facet is normally at the same level or within 2 mm of the radial portion of the ulnar head. Their relative position is referred to as the ulnar variance (Figure 3.5*d*) (Hultén, 1928). Positive ulnar variance reflects a relatively long ulna and can be associated with ulnolunate impingement and degenera-

tive tears of the triangular fibrocartilage (Palmer and Werner, 1981, 1984; Palmer *et al.*, 1982). Negative ulnar variance can result in an increase in the proportion of the carpal load born by the lunate facet of the distal radius, a circumstance which may be related to the occurrence of avascular necrosis of the lunate (Hultén, 1928).

Post-traumatic alterations in ulnar variance can cause similar problems, particularly ulnocarpal abutment in the setting of residual radial shortening (Aro and Koivunen, 1991; Oskam *et al.*, 1993). Radial shortening in this setting can also cause incongruity of the distal radioulnar joint (Werner *et al.*, 1989, 1992; Adams, 1993a; Kihara

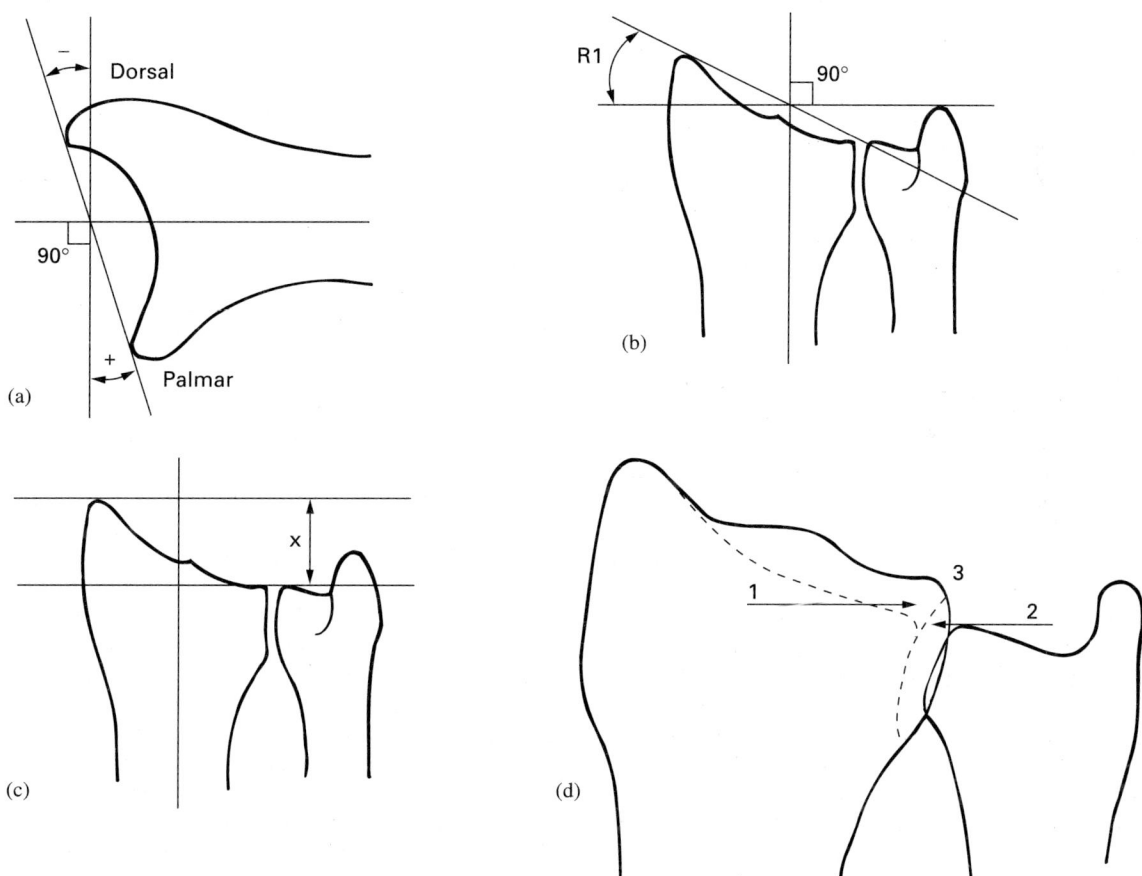

Figure 3.5   The orientation of the distal radial articular surface has been characterized in terms of a number of measurements. (a) The palmar tilt or inclination is measured from a sagittal view as the angle between a line connecting the distal-most points of the dorsal and palmar rims of the articular surface and a perpendicular to the longitudinal axis of the radius. (b) The ulnar inclination is measured from a coronal view as the angle between a line connecting the distal-most points of the radial and ulnar rims of the articular surface and a perpendicular to the longitudinal axis of the radius. (c) The radial length is measured from a coronal view as the distance in millimetres between a line drawn at the tip of the radial styloid process, perpendicular to the longitudinal axis of the radius and a second perpendicular at the level of the distal articular surface of the ulnar head. (d) The ulnar variance represents the vertical distance between the distal ends of the ulnar head and the ulnar-most aspect of the distal radial articular surface

et al., 1996). Symptoms referable to this articulation represent one of the most common residual problems following fracture of the distal radius (Stewart *et al.*, 1985; Martini, 1986; Frahm *et al.*, 1989; Aro and Koivunen, 1991; Roysam, 1992; Tsukazaki and Iwasaki, 1993).

The traditional method of assessing relative radial shortening, the so-called 'radial length', involves calculation of the distance between perpendiculars to the radial shaft drawn at the tip of the radial styloid and at the ulnar aspect of the distal radial articular surface (Figure 3.2c). The average is 11–12 mm (Gartland and Werley, 1951). This distance is less useful, however, since it reflects loss of ulnar inclination of the distal radial articular surface in addition to relative radial shortening.

Radial width, measured as the distance between the most lateral tip of the radial styloid process to the longitudinal axis through the centre of the radius on the anteroposterior radiograph (Friberg, 1976; Van der Linden *et al.*, 1981; Abbaszadegan *et al.*, 1989; Mann *et al.*, 1992b), is felt to be important in evaluating the results of the treatment of fractures of the distal radius. The difference between this measurement on the involved and uninvolved wrists is termed the radial shift.

The sigmoid notch, on the distal-most aspect of the medial surface of the distal radius, is a hyaline cartilage-covered 60–70° arc with a radius of curvature of approximately 15 mm which articulates with the ulnar head at the distal radioulnar joint (Ekenstam and Hagert, 1985a; Linscheid, 1992). Kapandji emphasized that, rather than presenting a perfectly cylindrical surface, the surface of the notch was slightly conical, with a smaller radius of curvature distally (Kapandji, 1981, 1982). Furthermore, the articular surface has been described as having a normal radial tilt of about 20° (Kapandji, 1982; Ekenstam and Hagert, 1985a).

Other investigators have observed that the shape and orientation of the sigmoid notch of the distal radius vary in relation to the ulnar variance (Forstner, 1987, 1990; De Smet and Fabry, 1993). The average radial angulation of the sigmoid notch has been measured as 9.4°. The angle was greater in ulna negative wrists (average 15.7°) and actually directed ulnarly in many ulna positive wrists (average 8.0°) (De Smet and Fabry, 1993). Ulna neutral wrists had a cylindrical sigmoid notch, whereas ulna negative and ulna positive wrists had conical or hemispherical shaped notches respectively (Forstner, 1987, 1990; De Smet and Fabry, 1993).

These findings suggest that operative procedures and traumatic injuries which relatively shorten or lengthen the ulna with respect to the radius may cause distal radioulnar incongruity. At least one investigation has documented increased contact pressures in the distal radioulnar joint following relative lengthening or shortening of the ulna (Werner *et al.*, 1989).

Lister's tubercle, a prominence about which the extensor pollicis longus tendon angulates in its course from forearm to thumb, provides a useful guide for distal radial rotational alignment and a landmark to both the division between the second and third dorsal compartments as well as the location of the radiocarpal joint, into which an arthroscopic entrance portal is typically made 1 cm distal to this tubercle.

Radioulnar articulation

The radius rotates about the relatively stationary ulna along an axis which passes roughly through the centre of the radial head proximally and the fovea of the ulnar head distally. Rotation of the radius occurs via axial rotation of the radial head at the proximal radioulnar joint, whereas distally, the motion is a combination of axial rotation and translation of the radius relative to the ulna (Gemmill, 1990; Kapandji, 1981, 1982; Cone *et al.*, 1983; Ekenstam and Hogert, 1985a; Olerud *et al.*, 1988; Schuind *et al.*, 1991; Hagert, 1992). The relatively small contact area between the sigmoid notch of the radius and the ulnar head translates dorsally within the notch in pronation and volarly in supination (Gemmill, 1900; Kapandji, 1982; Cone *et al.*, 1983; Ekenstam and Hagert, 1985a; Olerud *et al*, 1988; Schuind *et al.*, 1991).

The association of the radius and ulna is maintained by ligamentous structures at the proximal and distal radioulnar joints as well as by the interosseous ligament, a ligamentous sheet interconnecting the two bones along their mid-portion.

The proximal radioulnar joint is stabilized by the annular and quadrate ligaments and the interosseous ligament (Martin, 1958a; Spinner and Kaplan, 1970a) (Figure 3.6). The annular ligament is a strong band of tissue arising from the margins of the lesser sigmoid notch on the proximal ulna which tapers in circumference distally conforming to the transition from radial head to neck. The radial collateral ligament of the elbow inserts onto the annular ligament (Figure 3.6b). An extension of this ligament, the lateral ulnar collateral ligament, which may not be separable in anatomical

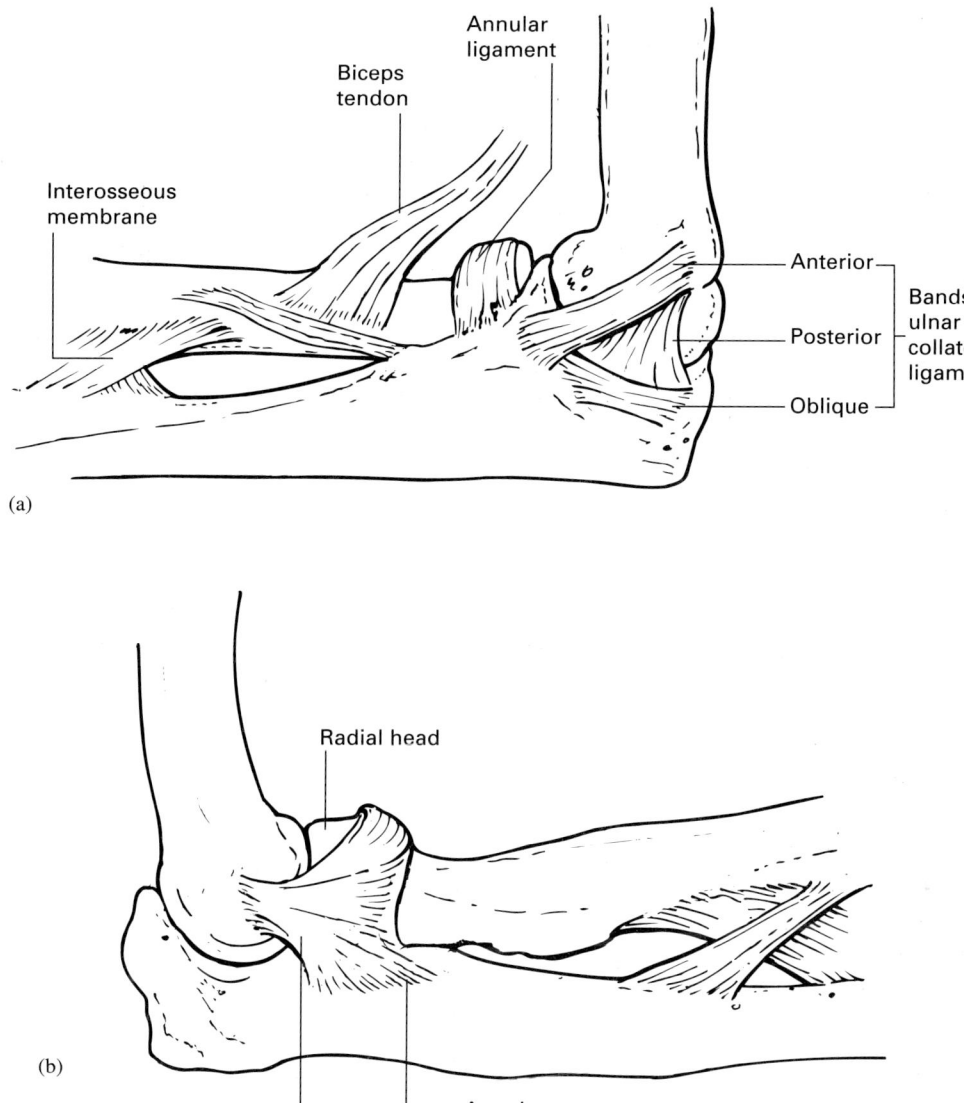

Figure 3.6 The ligamentous structures stabilizing the proximal radio-ulnar joint

dissections, runs with the annular ligament and inserts on the supinator crest of the ulna (Morrey and An, 1985; O'Driscoll *et al.*, 1991, 1992). It is important to be mindful of the fact that these structures are intimately interwoven, and that injury to the lateral ligament complex, which can occur with traumatic proximal radioulnar joint dislocations or iatrogenically during operative exposure of the radial head, could potentially contribute to postero-lateral rotatory instability of the ulnohumeral joint.

The quadrate ligament is described as a thin ligamentous structure which covers the capsule at the inferior margin of the annular ligament and attaches to the ulna (Spinner and Kaplan, 1970b; Kapandji, 1982). Its existence as a discrete entity and its contributions to the stability of the proximal radioulnar joint have been disputed by some authors (Martin, 1958a). The oblique cord represents a thickening in the fascia overlying the deep head of the supinator, which runs between the radius just below the bicipital tuberosity and the tuberosity for insertion of the brachialis on the proximal ulna (Martin, 1958b).

The anatomical structures contributing to the stability of the distal radioulnar joint have received a great deal of attention (Palmer and Werner, 1981, 1984; King *et al.*, 1986a; Palmer *et al.*, 1988; Schuind *et al.*, 1991; Adams, 1993a, b; Adams and Holley, 1993). The previously described volar and dorsal radioulnar ligaments (Ekenstam and Hagert, 1985a,b) are simply the peripheral portions of the triangular fibrocartilage, the central portion of which acts primarily as a cushion between the carpus and the ulnar head (Palmer and Werner, 1981). In fact, there is a linked group of structures which contribute not only to distal radioulnar joint stability, but also to ulnocarpal stability and cushioning. Because these structures, including the articular disc, the dorsal and volar radioulnar ligaments, the meniscus homologue, the ulnar collateral ligament and the sheath of the extensor carpi ulnaris are inseparable in anatomical dissections, it has proven more useful and accurate to consider them as a single, complex structure: the triangular fibrocartilage complex (Palmer and Werner, 1981). This is not intended to minimize the functional role of any component part, but rather to emphasize their interrelationship.

It is now generally accepted that the triangular fibrocartilage complex is the most important stabilizer of radioulnar articulation at the distal radioulnar joint. Considering the entire forearm unit, the most important soft tissue structure maintaining association of the radius and ulna is the interosseous ligament. This is a ligamentous structure running obliquely from distal-ulnar to proximal-radial between the two bones. It is thickest and strongest in its central portion.

Radiographic anatomy

In the assessment of distal radioulnar joint congruity, standard radiographs are both difficult to interpret and susceptible to even very small changes in radiographic technique and positioning (Mino *et al.*, 1983, 1985; Nakamura *et al.*, 1995). Recommendations for improving the reliability of standard radiographs focus on reproducible positioning techniques and comparison with the uninvolved limb (Nakamura *et al.*, 1995). Computed tomography scans are required to assess distal radioulnar joint congruity in patients with casts, bony deformity or pain preventing proper positioning (Mino *et al.*, 1985).

The interpretation of computed tomography scans is also difficult, and a number of distinct criteria have been described for defining subluxation. These include: (1) the radioulnar line method (Mino *et al.*, 1983, 1985) in which subluxation is diagnosed if the ulnar head does not lie entirely between two lines, one connecting the dorsal edge of the radial sigmoid notch to the dorsal edge of the osseous groove for the tendons of the first extensor compartment and the other connecting the volar edges of these grooves; (2) the congruity method (Weschler *et al.*, 1987) which assesses the congruity of the arcs of the ulnar head and the sigmoid notch; and (3) the epicentre method (Weschler *et al.*, 1987) in which a perpendicular to the line connecting the dorsal and volar edges of the sigmoid notch which originates at a point midway between the centre of the ulnar styloid and the centre of the ulnar head is expected to fall within the middle half of the sigmoid notch. Unfortunately, these methods provide somewhat discordant results, are occasionally abnormal in asymptomatic individuals and have limited correlation with clinical findings (Pirela-Cruz *et al.*, 1991; Staron *et al.*, 1994). Comparison of computed tomography scans of both distal radioulnar joints under applied stresses has been suggested (Pirela-Cruz *et al.*, 1991).

Detection of proximal radioulnar incongruity is relatively straightforward. Provided that the ulnohumeral joint remains intact, a line through the radial shaft should bisect the capitellum in any radiographic projection unless the proximal radioulnar joint is disrupted (Guistra *et al.*, 1974).

Muscle tendon units

Active forearm rotation is produced primarily by four muscles, two originating and inserting in the forearm (so-called intrinsic muscles) and two which cross the elbow joint (extrinsics). The extrinsic muscles insert on the apex of one of the radial curves and produce motion in a manner analogous to a crank shaft. The intrinsic muscles are short, flat muscles with broad insertions which wrap around the shaft of the bone; they produce rotation via an unwinding action (Kapandji, 1982). Both the intrinsic supinator and extrinsic biceps insert on the proximal radial shaft and produce supination. The extrinsic pronator teres and intrinsic pronator quadratus insert on the midshaft and distal radius respectively and produce pronation (Figure 3.7). Contraction of the brachioradialis encourages neutral forearm rotation. The power of

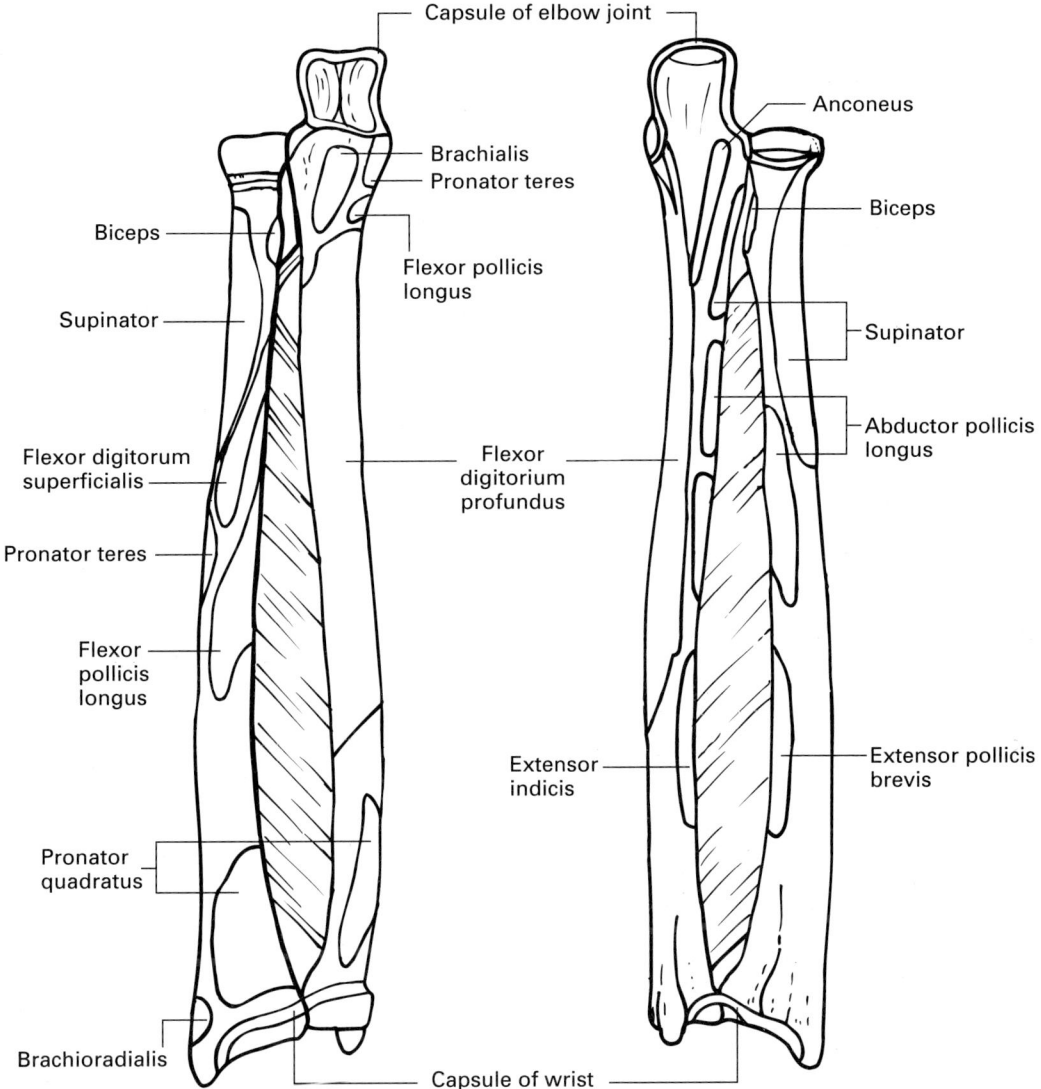

Figure 3.7 The muscle attachments of the radius and ulna

supination exceeds that of pronation by approximately 15% (Askew *et al.*, 1987). Malunion of the radius can decrease the mechanical efficiency of the muscles producing forearm rotation by shortening the lever arms (Kapandji, 1982).

The forearm musculature is commonly considered as three separate compartments based on fascial divisions and nerve supply: the volar or flexor compartment innervated by the median and ulnar nerves; the dorsal or extensor compartment innervated by the posterior interosseous nerve; and the mobile wad of Henry (the brachioradialis and the extensor carpi radialis longus and brevis)

innervated by the radial nerve. The divisions between the compartments delineate safe intervals for operative exposure (Figure 3.8).

Anatomical studies suggest that the fascial divisions between these compartments are sufficiently pliant that fascial release of one compartment usually decompresses the remaining two (Gelberman *et al.*, 1978, 1981). As a result, in the treatment of compartment syndrome of the forearm, pressures in the dorsal and mobile wad compartments rarely, but occasionally, remain elevated following release of the volar forearm musculature (Gelberman *et al.*, 1978, 1981).

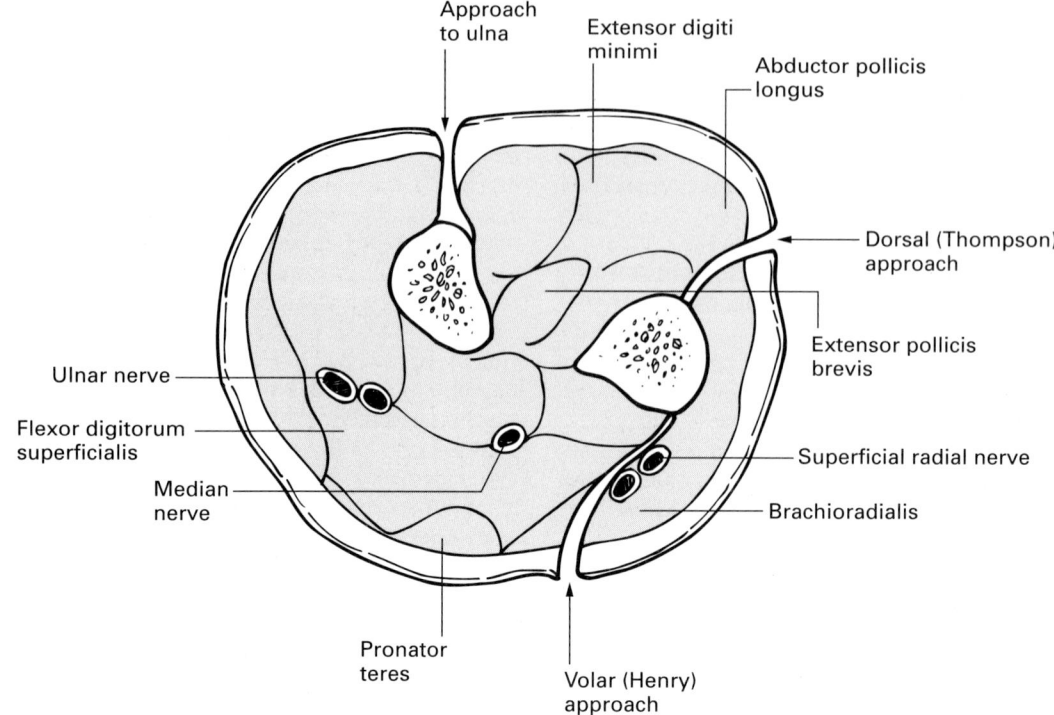

Figure 3.8   Cross-section of mid-forearm. This drawing depicts the muscular and neurovascular structures in the mid-forearm and their locations. The surgical intervals take advantage of internervous planes and represent safe approaches to the bones

Muscle tissue becomes sparse in the distal forearm, where the transition from muscle to tendon is completed. On the extensor surface of the distal radius and ulna, the tendons organize and are confined within compartments defined by the attachment of the extensor retinaculum to the dorsal radial and ulnar periosteum. Commonly referred to by number counting radial to ulnar, the first dorsal compartment contains the abductor pollicis longus and the extensor pollicis brevis; the second contains the radial wrist extensors, extensor carpi radialis brevis and longus; the third contains the extensor pollicis longus as it angles about the fulcrum provided by Lister's tubercle; the fourth contains the extensor digitorum communis and extensor indicis tendons; the fifth contains the extensor digiti quinti tendon; and the sixth contains the extensor carpi ulnaris, lying in a groove in the ulnar head, just dorsoradial to the ulnar styloid.

Neurovascular anatomy

Three large nerves enter the forearm at the elbow: the ulnar, radial and median nerves. For the ulnar

nerve, the forearm is primarily a conduit to the hand. It passes from the extensor compartment of the arm to the flexor compartment of the forearm under the medial epicondyle of the distal humerus. It then dives below the flexor carpi ulnaris, under the fascial band formed by the connection between its humeral and ulnar heads, innervating this muscle as well as the ulnar half of the flexor digitorum profundus. The remainder of the flexor musculature of the hand and wrist is innervated by the median nerve. The ulnar nerve is incorporated into the epimysium of the flexor digitorum profundus, lying between the flexor carpi ulnaris and the flexor digitorum superficialis muscles. The ulnar nerve lies just lateral to the tendon of the flexor carpi ulnaris at the wrist.

The median nerve is found medial to the brachial artery, overlying the brachialis muscle at the elbow. After entering the forearm in the cubital fossa, it passes between the humeral and ulnar heads of the pronator teres and then disappears under the superior margin of the flexor digitorum superficialis between its radial and ulnar origins. It lies between the superficial and deep digital flexor musculature, often incorporated into the epimysium of the flexor digitorum superficialis, until it reaches the wrist at which point the nerve emerges in a relatively superficial position between the flexor carpi radialis and flexor digitorum superficialis tendons. The anterior interosseous branch of the median nerve arises proximal to the elbow and is a distinct branch at the level of the superior margin of the flexor digitorum superficialis muscle. This branch supplies the flexor pollicis longus and the radial half of the flexor digitorum profundus muscle, as well as the pronator quadratus.

The radial nerve bifurcates just proximal to the elbow. Its deep branch, the posterior interosseous nerve, courses over the radial head and dives between the two heads of the supinator muscle at the arcade of Frohse, a fibrous thickening of the fascial margin of the superficial head of the supinator. The posterior interosseous nerve is typically separated from the radial shaft by the deep head of the supinator muscle, but occasionally lies in direct contact with the periosteum of the radial neck, making it particularly susceptible to damage when internal fixation devices are implanted in this region (Davies and Laird, 1948).

The superficial branch of the radial nerve runs along the undersurface of the brachioradialis with the radial artery and provides sensory branches to the dorsoradial aspect of the wrist and hand. The brachioradialis and extensor carpi radialis longus and brevis (the so-called mobile wad as described by Henry [1973]) are innervated by a branch of the radial nerve prior to its bifurcation. The extensor carpi radialis brevis is often innervated by a branch of the posterior interosseous nerve and rarely by a motor branch from the superficial radial nerve. In either case, the nerve branch enters the muscle sufficiently proximal that the plane between the extensor carpi radialis brevis and extensor digitorum communis (Thompson's interval) is usually safe in the forearm (Sunderland, 1946). The remainder of the wrist and digit extensor musculature on the dorsal aspect of the forearm is innervated by the posterior interosseous nerve. The posterior interosseous nerve terminates in an articular branch which lies in the floor of the fourth compartment of the extensor retinaculum.

As a result of the fact that the two muscles producing supination of the forearm are innervated by distinct nerves (the musculocutaneous and the posterior interosseous nerves), supination is more likely to be preserved following nerve injury. On the other hand, pronation is produced entirely by the median nerve. This circumstance may be related to the fact that loss of pronation can be accommodated by abduction of the shoulder, whereas limitation of supination is difficult to compensate for (Kapandji, 1982).

The skin of the forearm is supplied primarily by three nerves: the medial, lateral and posterior antebrachial cutaneous nerves. The lateral antebrachial cutaneous nerve is the continuation of the musculocutaneous nerve, which emerges from between the biceps brachii and the brachialis muscles on the lateral aspect of the distal arm. This nerve innervates the skin of the lateral half of the anterior aspect of the forearm as well as the direct lateral aspect of the forearm. The medial antebrachial cutaneous nerve is a branch from the medial cord of the brachial plexus which runs down the arm with the brachial artery, emerges in the middle of the arm, and divides into an anterior and a posterior branch. These large branches supply the majority of the anteromedial and posteromedial skin surface of the forearm. The posterior antebrachial cutaneous nerve of the forearm supplies the posterolateral aspect of the forearm integument.

The arterial supply to the upper extremity is characterized by extensive collateralization. Thus, while the brachial artery – which enters the forearm superficial to the brachialis muscle, lateral and adjacent to the median nerve – represents the primary blood supply of the forearm, distal branches such as the radial, ulnar and interosseous arteries are also supplied by large collaterals: the radial recurrent, anterior and posterior ulnar recurrent and interosseous recurrent arteries respectively. The radial recurrent artery represents the continuation of the radial collateral branch of the profunda brachii artery and travels with the radial nerve onto the anterior aspect of the elbow. The middle collateral branch of the profunda brachii becomes the interosseous recurrent artery on the posterolateral aspect of the elbow. The anterior and posterior ulnar recurrent arteries are named for their positions relative to the elbow joint and begin as the inferior and superior ulnar collateral arteries respectively in the arm.

Bifurcation of the brachial artery into the radial and ulnar arteries occurs at the level of the radial neck. The ulnar artery is crossed by the median nerve as it courses under the pronator teres and flexor digitorum superficialis, and eventually meets the ulnar nerve with which it continues through the forearm lying between the flexor digitorum superficialis and the flexor carpi ulnaris and overlying the flexor digitorum profundus. The radial artery passes medial to the biceps tendon and lies on the surface of the supinator muscle, meeting the superficial radial nerve, with which it courses toward the wrist on the undersurface of the brachioradialis muscle. The radial and ulnar arteries lie radial to the flexor carpi radialis tendon and the ulnar nerve (as well as the flexor carpi ulnar tendon) at the wrist respectively.

The common interosseous artery arises just below the bicipital tuberosity and almost immediately branches into anterior and posterior interosseous arteries at the superior margin of the interosseous ligament. These arteries, associated with their corresponding nerves, run along the anterior and posterior surfaces of the interosseous ligament, anastomosing in the distal forearm.

Kinesiology

It is difficult to separate the forearm and carpal contributions to the total pronation and supination of the hand (Cyriaz, 1926; King *et al.*, 1986b). Some studies have found that when carpal rotation is included, total forearm rotation approaches 260°, whereas isolated distal radioulnar (or forearm) motion is closer to 190° (King *et al.*, 1986b). Pronation is limited by compression of the flexor musculature between the radius and ulna, whereas supination is limited proximally by the restraint of the annular ligament (as reinforced by the anterior fibres of the lateral and medial collateral ligament complexes of the elbow), the quadrate ligament, the tone of the pronator quadratus and impingement of the ulnar styloid process on the posterior margin of the sigmoid notch of the distal radius (Kapandji, 1982).

Using simple goniometric measurements, pronation averages between 71° and 80° and supination between 80° and 84° (Silver, 1923; Glanville and Kreezer, 1937; Dorinson and Wagner, 1948; Boone and Azen, 1979). More sophisticated techniques demonstrate a total arc of forearm rotation less than 160° with supination (75–88°) being greater than pronation (70–71°) (Darcus and Salter, 1953; Salter and Darcus, 1953). Maximum forearm rotation decreases with extension of the ulnohumeral joint, perhaps as a result of tension in the musculature crossing the elbow (Wagner, 1977).

Forearm rotation represents the most important contribution to the rotational motion of the upper extremity. In combination with the rotational motion of the shoulder joint, forearm rotation makes it possible to position the hand through an entire 360° arc of motion when the upper extremity either hangs vertically down beside the trunk or is abducted to 90°. Flexion or extension at the shoulder joint reduces rotational motion at the shoulder, thereby limiting total upper extremity rotation to 270°. With the shoulder fully abducted, shoulder rotation is negligible and nearly all of the rotational motion of the upper extremity is provided by the forearm (Kapandji, 1982).

Most simple activities of daily living can be performed within an arc of approximately 50° each of pronation and supination (Chao *et al.*, 1980; Morrey *et al.*, 1981). On the other hand, many activities such as supporting a tray or accepting objects into the hand require near full supination, whereas many other activities such as pressing downwards, leaning upon an object or dribbling a basketball become restricted even with relatively small decreases in pronation (Kapandji, 1982).

As a result of the translational contribution to forearm rotation at the distal radioulnar joint, the axis of rotation is not constant. The so-called instant centre of rotation, or the centre of rotation at any given position of forearm rotation, translates a few millimetres about the centre of the radial and ulnar heads as the forearm courses through a full arc of rotation (Carret *et al.*, 1976; Cone *et al.*, 1983; King *et al.*, 1986b; Robbin *et al.*, 1986; Adams *et al.*, 1993a,b). Distally, the average centre of rotation is usually found near the fovea of the ulnar head (King *et al.*, 1986b; Adams 1993a,b).

Radial deformity alters the location of the average centre of rotation, a circumstance which is likely to disturb the mechanics of the forearm by creating incongruity at the radioulnar joints and altering the tension in the soft tissue structures, resulting in limited rotational motion (Adams *et al.*, 1993a). This is a reflection of the fact that the entire forearm is actually an articulation with two separated and relatively small areas of articular contact proximally and distally. Kapandji (1982) has likened the forearm articulation to a door which requires coaxial axes of rotation at both hinges in order to preserve free rotation. Radial deformity alters the alignment and axis of one or both radioulnar joints resulting in limitation of forearm motion. A door with hinges on different axes would have to be sawed in half in order to allow free rotation (Kapandji, 1982).

While it is clear that forearm rotation consists primarily of rotation of the radius and hand about the ulna, some have suggested that the distal ulna also moves in both the adduction/abduction and flexion/extension planes with forearm rotation (Heiberg, 1984; Duchenne, 1949; Ray *et al.*, 1951; Capner, 1956; Rose-Innes, 1960; Vesely, 1967). Ulnar motion is purportedly more pronounced when the forearm rotation occurs about an axis passing through a more radial digit (Ray *et al.*, 1951; Rose-Innes, 1960). Some claim that this motion of the distal ulna results from a small amount of varus/valgus motion and even axial rotation of the ulna at the ulnohumeral articulation, whereas others believe it to be an artifact of the measurement technique used (Chao and Morrey, 1978; Youm *et al.*, 1979; Morrey *et al.*, 1981). Dbjay suggests that the perceived ulnar motion is a reflection of the slight axial rotation of the humerus which occurs when the forearm rotates about a more radial axis (Dbjay, 1972; Kapandji, 1982).

Ulnar variance increases between 1 and 2 mm with pronation or grasp (Palmer and Werner, 1981, 1984; Epner *et al.*, 1982; Palmer *et al.*, 1982, 1984; Linscheid, 1992; Schuind *et al.*, 1992; Friedman *et al.*, 1993). A corresponding increase in radiocapitellar contact and force transmission has been measured during pronation (Morrey *et al.*, 1988). Tightening of the soft tissues with forearm rotation, the interosseous ligament in particular, has been offered as an explanation for these observations (Morrey *et al.*, 1988).

Whereas the radial head conforms tightly to the lesser sigmoid notch of the ulna at the proximal radioulnar joint, the radius of curvature of the ulnar head is much smaller than that of the sigmoid notch of the distal radius (approximately 8 mm compared to 15 mm respectively), with the result that there exists a relatively small area of contact between the two bones at the distal radioulnar joint in any particular position (Gemmill, 1990; Cone *et al.*, 1983; Olerud *et al.*, 1988; Schuind *et al.*, 1992). This arrangement allows for the radius to translate with respect to the ulnar head so that in full pronation the ulnar head rests at the dorsal-most aspect of the sigmoid notch and in full supination it rests at the volar-most aspect.

The volar radioulnar ligament is in maximum tension in full supination and the dorsal ligament in full pronation, representing the limit of axial rotational motion (Schuind *et al.*, 1991; Adams and Holley, 1993). However, it is the restraint of translational rather than rotational motion which determines the stability of the distal radioulnar

joint. This explains the seemingly paradoxical observation that instability of the distal radioulnar articulation is dependent on the integrity of the ligament opposite to the direction of dislocation: dorsal dislocation of the ulnar head reflects rupture of the volar radioulnar ligament and vice versa. Similar to the situation at the proximal radioulnar joint (Spinner and Kaplan, 1970b), when the soft tissue stabilizers of the distal radioulnar joint are damaged, the most stable forearm position is full supination, perhaps reflecting the tautness of the interosseous ligament in this position (King *et al.*, 1986a).

The radial head helps to transmit load to the humerus, and when it remains intact minimal shortening of the radius with respect to the ulna is observed under an axial load (Morrey *et al.*, 1988). Many studies have documented relative radial shortening with development of problems at the distal radioulnar joint following radial head resection for fracture (Taylor and O'Connor, 1964; Mikic and Vukadinovic, 1983; Carn *et al.*, 1986; Postacchini and Morace, 1992). Investigations that have attempted to isolate the contributions of the soft tissue stabilizers to the axial stiffness of the forearm have produced slightly varied numbers, but are consistent in the conclusion that when the interosseous ligament remains intact, substantial proximal migration of the radius with respect to the ulna will not occur (Hotchkiss *et al.*, 1989; Reardon *et al.*, 1991; Schuind *et al.*, 1991; Rabinowitz *et al.*, 1994).

Considering two studies which measured the relative proximal migration of the radius with respect to the ulna following radial head resection and serial sectioning of the soft tissue stabilizing structures, one found little (0.4 mm) and the other substantial (7 mm) relative shortening with radial head excision alone (Reardon *et al.*, 1991; Rabinowitz *et al.*, 1994). Both investigations demonstrated that loss of both the triangular fibrocartilage complex and the interosseous ligament were required for marked (greater than 10 mm) proximal migration to occur. Sectioning of only one of these structures resulted in shortening between 4 and 10 mm under axial load (Reardon *et al.*, 1991; Rabinowitz *et al.*, 1994).

The relative contribution of the triangular fibrocartilage and the interosseous ligament to the stiffness of the forearm under axial load was investigated in a study by Hotchkiss and co-workers (1989) in which stiffness rather than displacement was measured directly following radial head excision and serial sectioning of the soft tissue structures. They found that section of the triangular fibrocartilage complex decreased the axial stiffness by only 8%, whereas isolated sectioning of the central band of the interosseous ligament resulted in a 71% decrease in axial stiffness (Hotchkiss *et al.*, 1989). This study also emphasized the importance of this central band as compared to the proximal and distal portions of the interosseous ligament, noting that their experiments demonstrated a relatively small 11% decrease in axial stiffness following combined sectioning of the interosseous ligament proximal to the central band and the triangular fibrocartilage.

One should not, however, conclude from these data that the proximal and distal portions of the interosseous ligament are irrelevant. Another study investigating simulated isolated distal third fractures of the radius (so-called Galleazzi fractures) followed by serial sectioning of the soft tissue structures demonstrated up to 5 mm of shortening with osteotomy alone, and marked shortening (greater than 10 mm) only after both the triangular fibrocartilage and the distal portion of the interosseous ligament had been sectioned (Moore *et al.*, 1985).

These anatomical studies demonstrate that the interosseous ligament throughout its length is a major stabilizer of the forearm under axial load. In the clinical setting, provided that the normal curvature of the radius and ulna are maintained, the integrity of the interosseous ligament is usually sufficient to maintain proximal radioulnar joint congruity despite rupture of the annular and quadrate ligaments as occurs in Monteggia-type fracture dislocations of the forearm (Spinner and Kaplan, 1970b). Similarly, provided that the interosseous ligament remains intact, substantial proximal migration of the radius following radial head fracture or excision is unusual (Morrey *et al.*, 1979). Complete dislocation of the distal radioulnar joint cannot occur with disruption of the triangular fibrocartilage complex alone. The interosseous ligament must also be at least partially disrupted (King *et al.*, 1986a). Radioulnar diastasis also indicates damage to the interosseous ligament (Palmer and Werner, 1984; King *et al.*, 1986a).

Operative exposures

Skin incision

The skin of the upper extremity is well vascularized owing to the large number of distinct angio-

somes and the excellent collateralization (Taylor *et al.*, 1990). As a result, large flaps can be raised with little risk to skin of adequate quality. It is important to avoid transecting a large cutaneous neural branch and causing a painful neuroma. Since the blood supply to the skin tends to follow the cutaneous nerves, incisions between regions supplied by different nerves have the added benefit of improving flap vascularity (Taylor *et al.*, 1990). There is little need for curved incisions in the forearm since the ability to raise large flaps provides broad access to the underlying anatomical structures.

In the elbow region, a straight posterior incision proves useful for the majority of exposures. This straight posterior incision represents an incision of great utility in the elbow region comparable to the straight anterior incision which has become popular for exposure of the knee. The incision falls between innervating cutaneous nerves and allows for the creation of extensive medial and lateral skin flaps, allowing access to nearly the entire elbow. We have found that curving to avoid the prominence of the olecranon, as recommended by others, is unnecessary as scar sensitivity is uncommon.

The skin over the extensor surfaces of the hand, wrist and elbow is sufficiently lax and elastic that the relaxed skin tension lines can be crossed without causing scar contracture or hypertrophy. On the other hand, incisions should cross the flexor creases of the wrist or elbow obliquely (Kraissl, 1951).

It is preferable to avoid incising the skin directly over neurovascular structures in the setting of acute trauma. If excessive swelling prevents closure of the wound, these structures will remain exposed. For forearm compartment release we incise the skin on the ulnar aspect of the wrist, creating a radially based flap to ensure coverage of the median nerve. Preservation of subcutaneous venous structures should help limit oedema formation.

Indirect reduction

It is important to attempt to preserve the vascularity of the bone and surrounding soft tissue envelope during operative exposure. An abundant blood supply is important not only to enhance and accelerate the healing of traumatic and elective discontinuities of bone, but also to avoid infection, wound healing problems and other complications.

The reward to the surgeon who cultivates skills that effect reduction of skeletal deformity without a broad devascularizing exposure of the bone and excluding the need for circumferentially placed bone clamps will be more rapid and predictable bone healing. Most of these so-called indirect reduction techniques are based upon the utilization of continuous longitudinal traction as applied through an external fixator type of distraction device or directly through an applied plate using a plate-tensioning device (Mast *et al.*, 1979). The operative exposure of a fractured bone which has been realigned using a distractor can be limited to the area required for placement of the plate. Periosteum that is preserved in this region through careful extraperiosteal dissection of the bone can be expected to at least partially survive when a limited contact plate is used.

Ulna

Operative exposure of the ulna is straightforward as a result of the subcutaneous location of the entire length of the posterior apex of its triangular shaft, the absence of major neurovascular structures in the plane and its location between both cutaneous and muscular nerve supplies. A straight longitudinal skin incision over the ulna falls in the plane between the medial and lateral posterior cutaneous nerves of the forearm. The dorsal cutaneous branch of the ulnar nerve crosses the extreme distal end of the ulna running from volar to dorsal, and care must be taken to avoid injuring its branches. The posterior apex of the ulnar shaft defines the plane between the extensor forearm musculature innervated by the radial nerve and the flexor musculature innervated by the ulnar nerve.

In the mid-forearm it is preferable to expose the volar (flexor or medial) rather than the dorsal (extensor or posterior) surface of the ulna in order to avoid violation of the interosseous ligament which can contribute to the formation of a radio-ulnar synostosis. The ulnar nerve and artery lie underneath the flexor carpi ulnaris on top of the flexor digitorum profundus and are easily avoided provided that elevation of the flexor carpi ulnaris is performed close to the bone and does not stray into its substance.

We expose the proximal ulna through a straight dorsal incision. Proximally, plate fixation should be applied to the broad dorsal surface of the ole-cranon, which represents the tension aspect of the bone. Exposure of the tip of the olecranon is often necessary in order to obtain adequate fixation of a small proximal ulnar fracture. This requires partial elevation of the triceps insertion via a longitudinal incision of its most distal aspect.

Occasionally, it may be necessary to expose the dorsal aspect of the ulna. This can be performed safely between the fifth and sixth extensor compartments, taking care to avoid injury to the dorsal sensory branch of the ulnar nerve.

Radius

Dorsal or Thompson exposure

The radial shaft can be exposed through either a dorsal or volar dissection. The dorsal or posterior approach is commonly referred to by the eponym Thompson after the surgeon who popularized the approach (Thompson, 1918). The popularity of this approach has waned as a result of the potential of injury to the posterior interosseous nerve, which must be dissected from the substance of the supinator and protected, as well as the narrow skin bridge which remains between the incision used for exposure of the radius and that used to expose the ulna when both bones require exposure (Richards, 1996).

A straight longitudinal skin incision of appropriate length is made along the line connecting the lateral epicondyle at the elbow with Lister's tubercle at the wrist while the elbow is at 90° of flexion and the forearm is in neutral rotation. Following elevation of skin flaps, the internervous interval between the extensor digitorum communis (supplied by the posterior interosseous nerve) and the extensor carpi radialis brevis (supplied by the radial nerve) is most easily identified by locating the point at which the abductor pollicis longus and extensor pollicis brevis emerge from between the mobile wad and dorsal compartment musculature in the distal half of the forearm. The deep fascia is incised directly adjacent to this interval and the muscles are separated in a distal to proximal direction until their common aponeurosis is encountered (Figure 3.9). The supinator muscle covering the proximal radius is thereby exposed.

Utilization of the proximal portion of the dorsal surface of the radius for plate fixation requires identification and mobilization of the posterior interosseous nerve, as this nerve may lie almost directly adjacent to the bone at this level and could potentially be trapped beneath a plate (Davies and Laird, 1948). The posterior interosseous nerve emerges from beneath the superficial and deep heads of the supinator muscle approximately 1 cm proximal to the distal limit of this muscle. It can

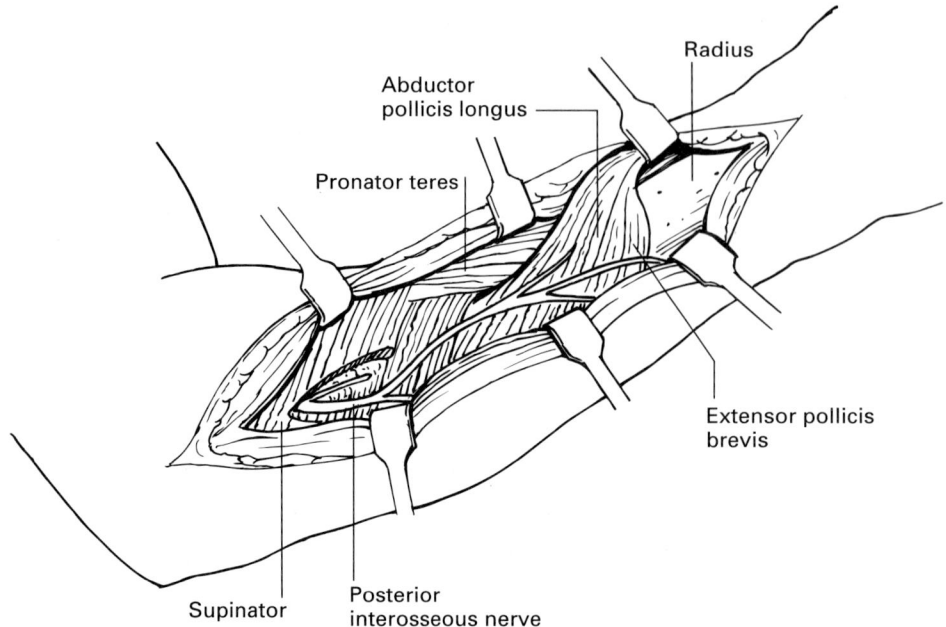

Figure 3.9   The dorsal, or Thompson, approach to the radius utilizes the plane between the radial nerve-innervated mobile wad of Henry (brachioradialis and extensor carpi radialis longus and brevis) and the extrinsic hand extensor musculature innervated by the posterior interosseous nerve. The interval between the extensor carpi radialis brevis and extensor digitorum communis muscles is located distally where the abductor pollicis longus and extensor pollicis brevis emerge from between the two muscles. The interval is developed proximally, exposing the underlying supinator. The posterior interosseous nerve must be mobilized from its location between the superficial and deep heads of the supinator in order to ensure that it is not injured during proximal dissection close to the bone. After isolation and protection of this nerve is complete, elevation of the supinator from the radius is undertaken

be identified at this point and then dissected free from the muscle, with care taken to preserve its muscular branches (Figure 3.9). Following sufficiently proximal mobilization of the nerve, exposure of the radial shaft can be performed by rotating the radius into full supination and detaching the insertion of the supinator from the anterior aspect of the radius.

Exposure of the mid-portion of the bone is facilitated by mobilization and retraction of the crossing abductor pollicis longus and extensor pollicis brevis muscles. Exposure of the radius distal to the extensor pollicis brevis is performed in the interval between the radial wrist extensors (extensor carpi radialis brevis and longus muscles) and the extensor pollicis longus muscle which ultimately produce the tendons occupying the third and second dorsal extensor compartments respectively.

Anterior or Henry exposure

Exposure of the anterior surface of the radius is both safer and more extensile than a dorsal exposure (Henry, 1973). A straight longitudinal inci-

sion along a line between the lateral margin of the biceps tendon at the elbow and the radial styloid process at the wrist will afford access to the plane between the mobile wad and the flexor musculature of the forearm. This incision falls roughly between the regions of the medial and lateral antebrachial cutaneous nerves. The deep fascia is incised adjacent to the medial border of the brachioradialis and a plane is developed between this radial nerve-innervated muscle and the median nerve-innervated flexor carpi radialis and pronator teres muscles (Figure 3.10). Dissection is initiated distally and proceeds proximally following the course of the radial artery. Arterial branches to the brachioradialis and the recurrent radial artery arising near the elbow are ligated and the radial artery is mobilized and retracted medially with the flexor carpi radialis muscle. The superficial radial nerve is encountered on the undersurface of the brachioradialis and remains lateral with this muscle.

Deep dissection is initiated proximally where the biceps tendon is followed towards its insertion on the bicipital tuberosity of the radius. Full

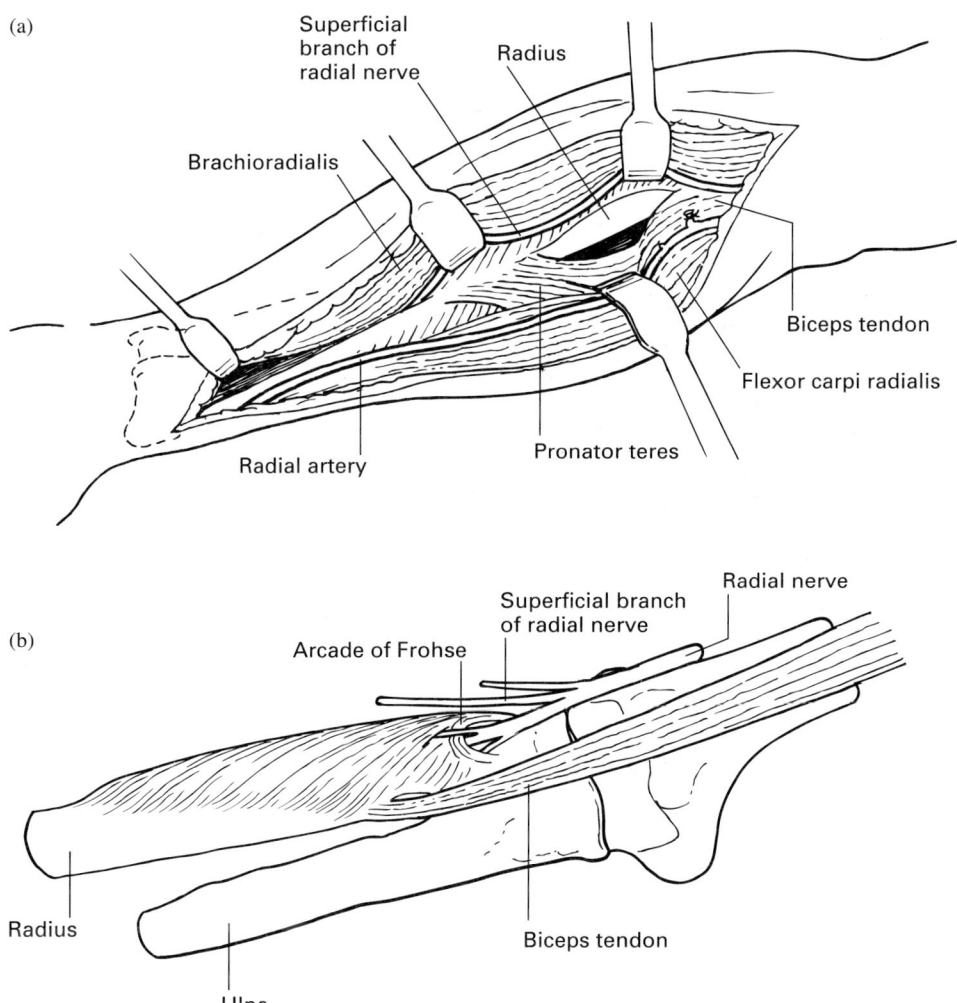

(a)

Superficial branch of radial nerve

Radius

Brachioradialis

Biceps tendon

Flexor carpi radialis

Radial artery

Pronator teres

(b)

Radial nerve

Superficial branch of radial nerve

Arcade of Frohse

Radius

Biceps tendon

Ulna

Figure 3.10 The volar, or Henry, approach to the radius exploits the interval between the radially innervated brachioradialis and the median nerve-innervated pronator teres and flexor carpi radialis. (a) The superficial radial nerve is retracted laterally with the brachioradialis. The radial artery is retracted medially by ligating small branches coursing laterally. (b) Full supination exposes the insertion of the distal biceps tendon onto the radial tuberosity. Dissection between the supinator and this tendon is often facilitated by entering a small bursa which is common in this area. Once the margin of the supinator insertion is clearly identified, elevation of this muscle from the radius can proceed safely as the posterior interosseous nerve lies well lateral between the two heads of the supinator

supination of the forearm displaces the posterior interosseous nerve laterally and brings the insertion of the supinator muscle anterior. The insertion of the supinator muscle is identified by deepening the muscular plane along the lateral aspect of the biceps tendon (Figure 3.10*b*). Here one may encounter a bursa between the biceps tendon and the supinator which further facilitates this dissection. The posterior interosseous nerve remains well protected within the substance of the supinator muscle during elevation of its insertion from

the radius, provided that excessive lateral traction is not applied.

The insertion of the pronator teres must be detached and the body of the flexor digitorum superficialis elevated in order to expose the midportion of the radius. This is performed by pronating the arm in order to bring the lateral limit of these structures into view.

The distal portion of Henry's anterior approach to the radius is often utilized to obtain volar exposure of fractures of the distal radius. Distally, dissection proceeds through the sheath of the flexor carpi radialis tendon between the tendon and the radial artery. The lateral border of the pronator quadratus is exposed via supination of the radius. This lateral border is divided, leaving a cuff for later suture repair, and subperiosteal mobilization of the muscle proceeds medially towards the distal radioulnar joint. The flexor pollicis longus is also elevated, from its lateral side to protect its innervation, as needed for more proximal exposure of the distal portion of the volar surface of the radius.

The important volar carpal ligaments should not be violated. Partial visualization of the articular surface in a fracture of the distal radius can sometimes be obtained through the fracture itself.

Distal radius

Alternative volar exposure

Exposure to the ulnar-most aspect of the distal radius can be difficult through Henry's approach. Furthermore, neurovascular decompression is often indicated in the setting of high-energy trauma with elevated pressures in the carpal tunnel, Guyon's canal and the forearm musculature. In these circumstances, an alternative approach to the volar surface of the distal radius between the ulnar neurovascular bundle and the flexor tendons often proves useful.

In this approach, an incision which begins distally as a standard incision for carpal tunnel decompression crosses the volar wrist creases obliquely, creating a radially based flap of skin for coverage of the median nerve in case swelling prevents closure of the wound, and extends towards the midline proximally. The interval between the ulnar nerve and artery and the contents of the carpal canal is developed and the carpal contents are mobilized radially. The pronator quadratus is partially divided at its distal, ulnar aspect over the distal radioulnar joint. The proximal attachment of this muscle can usually be left intact. In younger individuals the transverse retinacular ligament can be repaired by Z-lengthening.

Dorsal exposures

The most commonly used exposure is between the second and fourth dorsal extensor compartments with the tendon of the third extensor compartment, the extensor pollicis longus, transposed dorso-radially out of its compartment (Figure 3.11). A straight longitudinal skin incision over Lister's tubercle and crossing the wrist joint is used. The extensor retinaculum is divided between the second and third dorsal compartments in the interval defined by Lister's tubercle. When using dorsal plate fixation, an ulnarly based flap of extensor retinaculum can be developed and preserved for later suture between the extensor tendons and the plate extending onto the radial aspect of the dorsal surface of the radius. The tendon of the extensor pollicis longus is then freed along its length and left dorsally and radially transposed in the subcutaneous tissues. Replacement in its compartment would risk irritation, attrition and rupture.

The integrity of the second and fourth compartments should remain undisturbed in order to minimize the risk of intratendinous adhesions, tendinitis and tendon rupture. Subperiosteal elevation of these compartments exposes the entire dorsal distal radial surface (Figure 3.11*b*). Longitudinal incision of the wrist capsule in line with the periosteal incision is safe and provides exposure of the distal radial articular surface, the carpal bones and the intrinsic intercarpal ligaments. Suture repair of the capsulotomy is performed prior to closure.

Dorsal exposure can also be performed more radially between the first and second extensor compartments, or more ulnarly between the fourth and fifth compartments. When performing the more radial exposure, one must take care to protect the radial sensory nerve and dorsal radial artery. The distal radioulnar joint can be visualized via capsulotomy through the incision between the fourth and fifth extensor compartments (Alexander and Lichtman, 1984; Menon *et al.*, 1984). Alternatively, the distal radioulnar joint can be approached through the floor of the fifth extensor compartment (Wehbe, 1986). The tendon should be replaced in its sheath at closure.

Radial head exposure

The simplest and safest exposure of the radial head is between the extensor carpi ulnaris and the anconeus. This exposure represents a portion of the commonly used extensile lateral approach to the elbow described by Kocher (1911). The interval utilized by the Thompson approach to the radius between the extensor digitorum communis and the extensor carpi radialis brevis places the posterior interosseous nerve at risk. Kocher's interval is posterior to the nerve, which is protected by the extensor carpi ulnaris and the extensor digitorum communis muscles, and anterior to

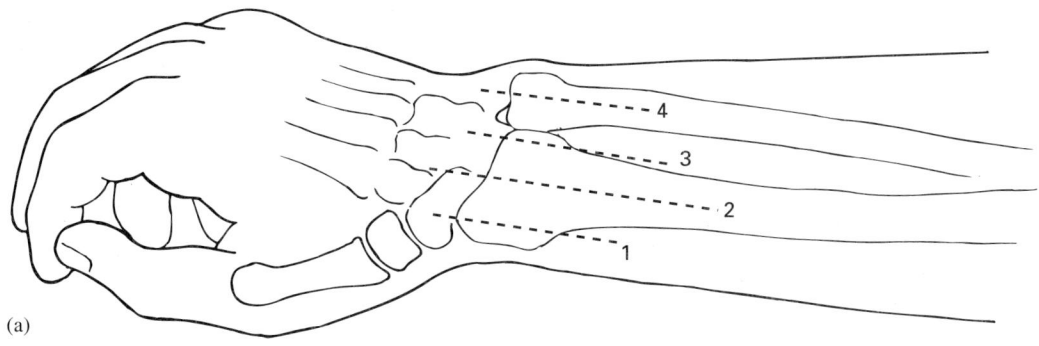

(a)

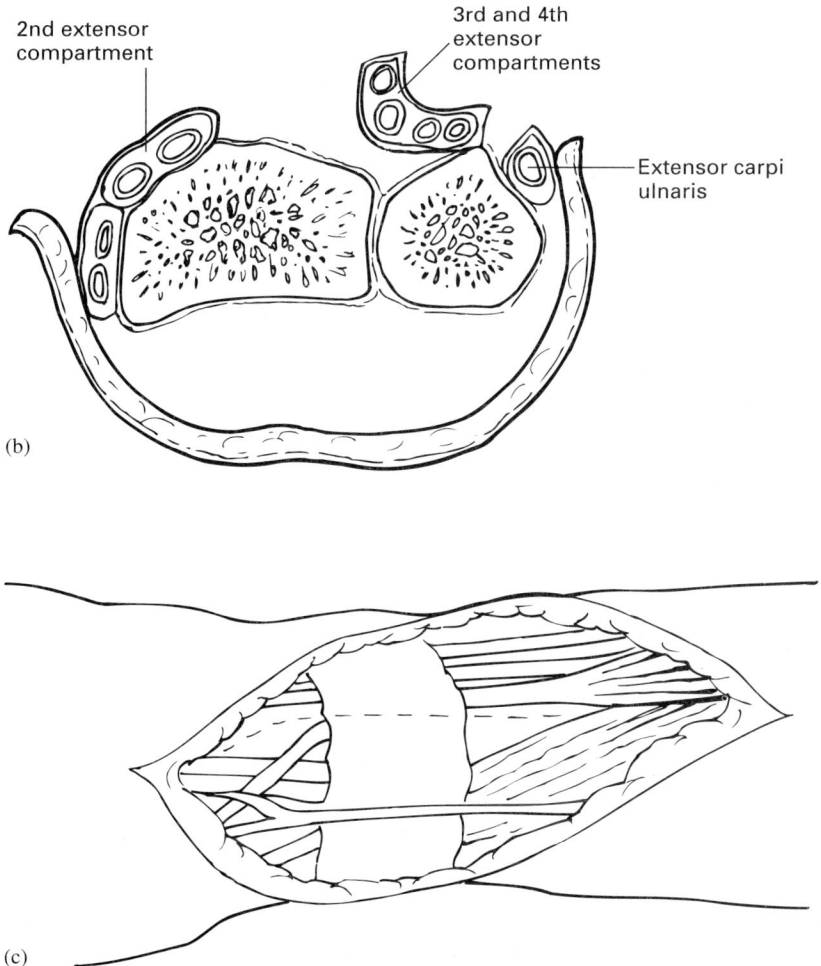

2nd extensor
compartment

3rd and 4th
extensor
compartments

Extensor carpi
ulnaris

(b)

(c)

Figure 3.11 Dorsal approaches to the distal radius. (a) Various intervals between extensor tendon compartments can be chosen based upon the area of exposure required. (b, c) The most common exposure, between the second and fourth extensor compartments, is best performed by incising the extensor retinaculum over Lister's tubercle between the second and third compartments, mobilizing the extensor pollicis longus from the third compartment and dissecting the second and fourth dorsal compartments subperiosteally from the dorsal surface of the distal radius. Maintaining the integrity of these compartments will help to limit adhesions and extensor tendon irritation

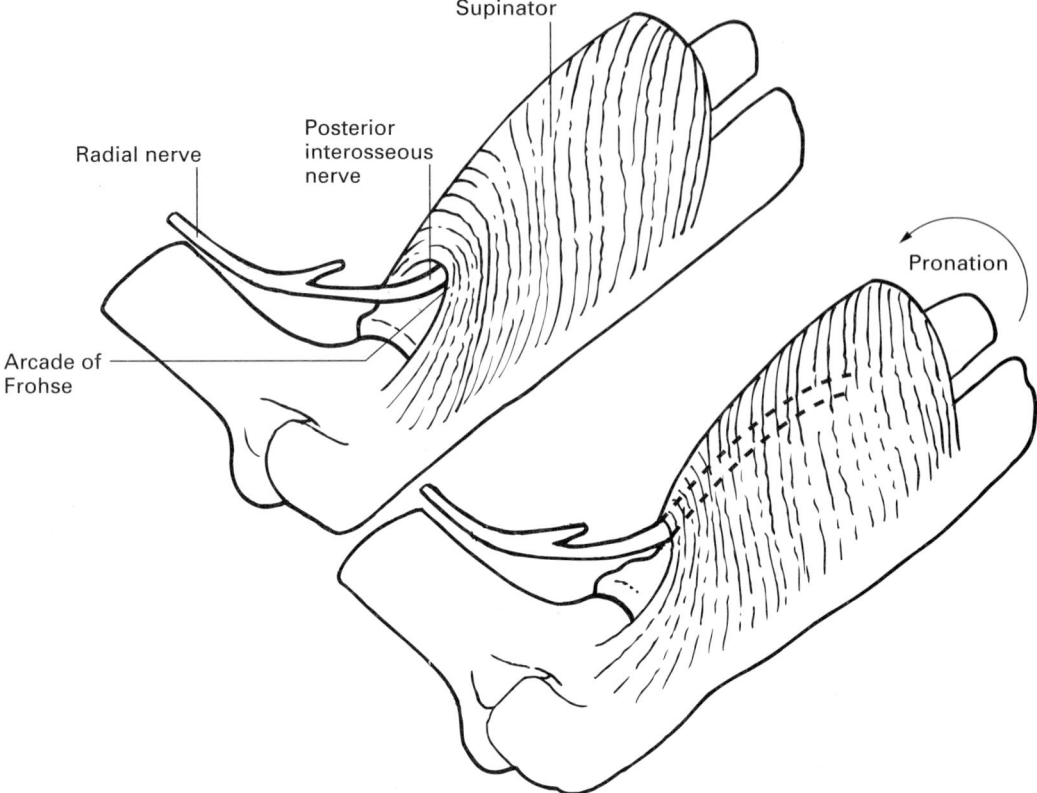

Figure 3.12   During radial exposure, an increased margin of safety can be obtained by fully pronating the arm. This results in medial translation of the posterior interosseous nerve of approximately 1 cm

the lateral ulnar collateral ligament of the elbow. Forearm pronation can provide an added measure of safety by translating the posterior interosseous nerve approximately 1 cm medially (Figure 3.12) (Kaplan, 1941; Strachan and Ellis, 1971).

If an isolated procedure on the radial head is intended, an oblique incision is made running from the lateral epicondyle of the humerus over the interval of the anconeus and extensor carpi ulnaris muscles, terminating near the posterior apex of the ulna. When a more complex operative procedure is undertaken, Kocher's interval can be located through the standard straight posterior incision for exposure of the elbow following elevation of the lateral skin flap. Through a small incision, the interval may be easier to find after the forearm fascia is incised. The anconeus and extensor carpi ulnaris are then elevated from the lateral elbow capsule. Exposure may be facilitated by releasing the insertion of the anconeus from the humerus subperiosteally. The proximal radius and radiocapitellar joint are then exposed through a

longitudinal capsular incision which must be carefully repaired upon closure of the wound.

Simultaneous exposure of the radius and ulna

Simultaneous exposure of the proximal radius and ulna was proposed by Boyd (1940) as a useful method of exposure for Monteggia fractures. There is concern, however, that such a simultaneous exposure could increase the risk for the formation of a proximal radioulnar synostosis. Annular ligament repair is almost never necessary in Monteggia fractures as concentric reduction of the proximal radioulnar joint and appropriate healing of the annular, quadrate and interosseous ligaments is dependent upon anatomical reduction of the ulnar fracture.

Radial head fractures are common in posterior Monteggia injuries. When a radial head fracture must be addressed in this setting, we prefer to use Kocher's interval between the anconeus and extensor carpi ulnaris. Unlike Gordon (1967), who treated fractures of the ulna and radial head through a single lateral incision, we approach both the ulna and the radial head through a straight posterior incision by elevating a lateral skin flap. An attempt is made to avoid stripping the anconeus from the ulnar fracture fragments by using a distractor to facilitate reduction and by limiting exposure of the dorsal surface of the ulna to that which is required for plate fixation.

Boyd's approach is useful in the resection of a proximal radioulnar synostosis. Through a straight posterior incision with medial and lateral skin flaps, the anconeus and extensor carpi ulnaris are elevated from the lateral aspect of the ulna. The insertion of the supinator is then detached from the supinator crest and reflected anteriorly, being careful to preserve the attachment of the lateral ulnar collateral ligament to the tubercle of the supinator crest proximally as well as to avoid excessive traction on the posterior interosseous nerve, which lies within the substance of the supinator. In the setting of a proximal radioulnar synostosis, elevation of the supinator from the synostosis will continue directly onto the radius until adequate exposure is achieved. If the dissection extends distally along the radius, the recurrent interosseous artery may be encountered and can be ligated to facilitate further exposure (Pankovich, 1977).

Forearm compartment release

Fascial release for compartment syndrome can be performed through the standard volar Henry-type exposure in the setting of a forearm fracture, or through a straight ulnar McConnell-type incision when exposure of the bones is not required (McConnell, 1920; Henry, 1973). This ulnar incision provides a flap for median nerve coverage at the wrist and allows exposure of the median and ulnar nerves under the superficial flexor muscles when required. Access to the deep flexor musculature is obtained without dissection through the superficial flexors.

References

Abbaszadegan, H., Jonsson, U. and Von Sivers, K. (1989) Prediction of instability of Colles' fractures. *Acta Orthop. Scand.*, **60**, 646–50

Adams, B.D. and Holley, K.A. (1993) Strains in the articular disk of the triangular fibrocartilage complex: a biomechanical study. *J. Hand Surg.*, **18A**, 919–25

Adams, B.D. (1993a) Effects of radial deformity on distal radioulnar joint mechanics. *J. Hand Surg.*, **18A**, 492–8

Adams, B.D. (1993b) Partial excision of the triangular fibrocartilage complex articular disk: a biomechanical study. *J. Hand Surg.*, **18A**, 334–40

Adler, S., Fay, G.D. and MacAusland, W.R. Jr. (1962) Treatment of olecranon fractures. Indictations for excision of the olecranon fragment and repair of the triceps tendon. *J. Trauma*, **2**, 597–602

Alexander, C.E. and Lichtman, D.M. (1984) Ulnar carpal instabilities. *Orthop. Clin. N. Am.*, **15**, 307–20

Almquist, E.A. (1992) Evolution of the distal radioulnar joint. *Clin Orthop.*, **275**, 5–13

An, K.N., Morrey, B.F. and Chao, E.Y.S. (1986) The effect of partial removal of proximal ulna on elbow constraint. *Clin. Orthop.*, **209**, 270–9

Aro, H.T. and Koivunen, T. (1991) Minor axial shortening of the radius affects outcome of Colles' fracture treatment. *J. Hand Surg.*, **16A**, 392–8

Askew, L.J., An, K.N., Morrey, B.F. and Chao, E.Y.S. (1987) Isometric elbow strength in normal individuals. *Clin. Orthop.*, **222**, 261–6

Boone, D.C. and Azen, S.P. (1979) Normal range of motion of joints in male subjects. *J. Bone Joint Surg.*, **61A**, 756–9

Boyd, H.B. (1940) Surgical exposure of the ulna and proximal third of the radius through one incision. *Surg. Gynecol. Obstet.*, 71, 86–8

Cage, D.J.N, Abrams, R.A., Callahan, J.J. and Botte, M.J. (1995) Soft tissue attachments of the ulnar coronoid process. *Clin. Orthop.*, **320**, 154–8

Capner, N. (1956) The hand in surgery. *J. Bone Joint Surg.*, **38B**, 128–51

Carn, R.M., Medige, J., Curtain, D. and Koenig, A. (1986) Silicone rubber replacement of the severely fractured radial head. *Clin Orthop.*, **209**, 259–69

Carret, J.P., Fischer, L.P., Gonon, G.P. and Dimnet, J. (1976) Etude cinematique de la prosupination au niveau des articulations radiocubitales (radio ulnaris). *Bull. Assoc. Anat.*, **60**, 279–95

Chao, E.Y., An, K.N., Askew, L.J. and Morrey, B.F. (1980) Electrogoniometer for the measurement of human elbow rotation. *J. Biomech. Eng.*, **102**, 301–10

Chao, E.Y.S. and Morrey, B.F. (1978) Three-dimensional rotation of the elbow. *J. Biomech.*, **11**, 57–73

Cone, R.O., Szabo, R., Resnick, D. *et al.* (1983) Computed tomography of the normal radioulnar joints. *Invest. Radiol.*, **18**, 541–5

Cyriaz, E.F. (1926) On the rotatory movements of the wrist. *J. Anat.*, **60**, 199–201

Darcus, H.D. and Salter, N. (1953) The amplitude of supination and pronation with the elbow flexed to a right angle. *J. Anat.*, **87**, 169–84

Davies, F. and Laird, M. (1948) The supinator muscle and the deep radial (posterior interosseous) nerve. *Anat. Rec.*, **101**, 243–50

Dbjay, H.C. (1972) L'humerus dans la pronosupination. *Rev. Med. Limoges*, **3**, 147–50

De Smet, L. and Fabry, G. (1993) Ortientation of the sigmoid notch of the distal radius: determination of different types of the distal radioulnar joint. *Acta Orthop. Belg.*, **59**, 269–72

Dorinson, S.M. and Wagner, M.L. (1948) An exact technique for clinically measuring and recording joint motion. *Arch. Phys. Med. Rehabil.*, **29**, 468–75

Duchenne, G.B.A. (194) The physiology of motion, demonstrated by means of electrical stimulation and clinical observation and applied to the study of paralysis and deformities. (E.B. Kaplan, transl.). Philadelphia: Lippincott

Ekenstam, F. af and Hagert, C.G. (1985a) Anatomical studies on the geometry and stability of the distal radioulnar joint. *Scand. J. Plast. Reconstr. Surg.*, **19**, 17–25

Ekenstam, F. and Hagert, C.G. (1985b) The distal radio- ulnar joint. The influence of geometry and ligament on simulated Colles' fracture. An experimental study. *Scand. J. Plast. Reconstr. Surg.*, **19**, 27–31

Epner, R.A., Bowers, W.H. and Guilford, W.B. (1982) Ulnar variance – the effect of wrist positioning and roentgen filming technique. *J. Hand Surg.*, **7A**, 298–305

Evans, E.M. (1945) Rotational deformities in the treatment of fractures of both bones of the forearm. *J. Bone Joint Surg.*, **27B**, 373–9

Evans, E.M. (1951) Fractures of the radius and ulna. *J. Bone Joint Surg.*, **33B**, 548–61

Forstner, H. (1987) Das distale Radio Ulnar Gelenk. *Unfallchirurg*, **90**, 512–7

Forstner, H. (1990) The morphology of the distal radioulnar joint: aspects and implications for orthopaedic surgery. *Handchir. Mikrochir. Plast. Chir.*, **22**, 296–303

Frahm, R., Saul, O. and Drescher, E. (1989) CT-diagnostik bei fehlstellungen nach distaler radiusfrakktur. *Radiologe*, **29**, 68–72

Friberg, R. and Lundström, B. (1976) Radiographic measurements of the radiocarpal joint in normal adults. *Acta Radiol.*, **17**, 249–56

Friedman, S.L., Palmer, A.K., Short, W.H. *et al.* (1993). The change in ulnar variance with grip. *J. Hand Surg.*, **18A**, 713–6

Gartland, J.J. and Werley, C.W. (1951) Evaluation of healed Colles' fractures. *J. Bone Joint Surg.*, **33A**, 895–907

Gartsman, G.M., Scales, J.C. and Otis, J.C. (1981) Operative treatment of olecranon fractures. *J.Bone Joint Surg.*, **63A**, 718–21

Gelberman, R.H., Garfin, S.R., Hergenroeder, P.T. *et al.* (1981) Compartment syndromes of the forearm: diagnosis and treatment. *Clin. Orthop.*, **161**, 252–61

Gelberman, R.H., Zakaib, G.S., Mubarak, S.J. *et al.* (1978) Decompression of forearm compartment syndromes. *Clin. Orthop.*, **134**, 225–9

Gemmill, F. (1900) On the movement of the lower end of the radius in pronation and supination, and on the interosseous membrane. *J. Anat. Physiol.*, **35**, 101–9

Gilula, L.A. (1992) *The Traumatized Hand and Wrist: Radiographic and Anatomic Correlation*. Philadelphia: Saunders

Giustra, P.E., Killoran, P.J., Furman, R.S. and Root, J.A. (1974) The missed Monteggia fracture. *Radiology*, **10**, 45–7

Glanville, A.D. and Kreezer, G. (1937) The maximum amplitude and velocity of joint movements in normal male human adults. *Hum. Biol.*, **9**, 197–211

Goel, V.K., Singh, D. and Bijlani, V. (1982) Contact areas in human elbow joint. *J. Biomech. Eng.*, **104**, 169–75

Golden, G.N. (1963) Treatment and prognosis of Colles' fracture. *Lancet*, **1**, 511–5

Goodfellow, J.W. and Bullough, P.G. (1967) The pattern of aging of the articular cartilage of the elbow joint. *J. Bone Joint Surg.*, **49B**, 175–81

Gordon, M.L. (1967) Monteggia fracture. A combined surgical approach employing a single lateral incision. *Clin. Orthop.*, **50**, 87–93

Goss, C.M. (ed.) (1973) *Anatomy of the Human Body* (29th American edn of *Gray's Anatomy*) pp. 208–16, Philadelphia: Lea & Febiger

Hagert, C.G. (1992) The distal radioulnar joint in relation to the whole forearm. *Clin. Orthop.*, **275**, 56–64

Heiberg, J. (1884) The movement of the ulna in rotation of the forearm. *J. Anat. Physiol.*, **19**, 237–40

Henry, A.K. (1973) *Extensile Exposure*, 2nd edn. Edinburgh: Churchill Livingstone

Hotchkiss, R.N., An, K.N., Sowa, D.T. *et al.* (1989) An anatomic and mechanical study of the interosseous membrane of the forearm: pathomechanics of proximal migration of the radius. *J. Hand Surg.*, **14A**, 256-261, 1989.

Hultén, O. (1928) Über anatomishe variationen der handgelenkknochen: Ein beitrag zur kenntnis der genese zwei vershiedener mondbeinveränderungen. *Acta Radiol.*, **9**, 155–68

Kapandji, I.A. (1981) The inferior radioulnar joint and pronosupination. In *The Hand*, vol. I. (R. Tubiana, ed.) pp. 121–9, Philadelphia: Saunders

Kapandji, I.A. (1982) *The Physiology of the Joints*, 5th edn LH. Honore, (transl.) Edinburgh: Churchill Livingstone

Kaplan, E.B. (1941) Surgical approaches to the proximal end of the radius and its use in fractures of the head and neck of the radius. *J. Bone Joint Surg.*, **23**, 86–92

Kauer, J.M.G. (1992) The distal radioulnar joint: anatomic and functional considerations. *Clin. Orthop.*, **275**, 37–45

Keats, T.E., Teeslink, R., Diamond, A.E. and Williams, J.H. (1996) Normal axial relationships of the major joints. *Radiology*, **87**, 904–7

Kellam, J.F. and Jupiter, J.B. (1991) Fractures of the forearm. In *Skeletal Trauma*, 1st edn (B.D. Browner, J.B. Jupiter, A.M. Levine and P.G. Trafton, eds) pp. 1095–124, Philadelphia: W.B. Saunders

Kihara, H., Palmer, A.K., Werner, F.W. *et al.* (1996) The effect of dorsally angulated distal radius fractures on distal radioulnar joint congruency and forearm rotation. *J. Hand Surg.*, **21A**, 40–7

King, G.J., McMurty, R.Y., Rubenstein, J.D. and Ogston, N.G. (1986a) Computerized tomography of the distal radioulnar joint: correlation with ligamentous pathology in a cadaveric model. *J. Hand Surg.*, **11A**, 711–7

King, G.J., McMurty, R.Y., Rubenstein, J.D. and Gertzbein, S.D. (1986b) Kinematics of the distal radioulnar joint. *J. Hand Surg.*, **11A**, 798–804

Kocher, T. (1911) *Textbook of Operative Surgery*, 3rd edn (H.J. Stiled and C.B. Paul, transl.) London: Adam & Charles Black

Kraissl, C.J. (1951) The selection of appropriate lines for elective surgical incisions. *Plast. Reconstr. Surg.*, **8**, 1–28

Larson, S.G. (1993) *Phylogeny. The Elbow and its Disorders*, 2nd edn (B.F. Morrey, ed.) pp. 6–15, Philadelphia: Saunders

Lewis, O.J. (1965) Evolutionary change in the primate wrist and inferior radioulnar joints. *Anat. Rec.*, **151**, 275–85

Linscheid, R.L. (1992) Biomechanics of the distal radioulnar joint. *Clin. Orthop.*, **275**, 46–55

Mann, F.A., Kane, S.W. and Gilula, L.A. (1992a) Normal palmar tilt: Is dorsal tilting really normal? *J. Hand Surg.*, **17B**, 315–7

Mann, F.A., Wilson, A.J. and Gilula, L.A. (1992b) Radiographic evaluation of the wrist. What does the hand surgeon want to know? *Radiology*, **184**, 15–24

Martin, B.F. (1958a) The annular ligament of the superior radial ulnar joint. *J. Anat.*, **52**, 473–81

Martin, B.F. (1958b) The obliqe cord of the forearm. *J. Anat.*, **52**, 609–15

Martini, A.K. (1986) Die sekundare arthrose des handgelenkes bei der in fehlstellung verheilten und nicht korrigierten distalen radius fraktur. *Aktuel. Traumatol.*, **16**, 143–8

Mast, J., Jakob, R.P. and Ganz, R. (1979) *Planning and Reduction Techniques in Fracture Surgery*. Heidelberg: Springer

Matthews, L.S, Kaufer, H., Garver, D.F. and Sonstegard, D.A. (1982) The effect on supination–pronation of angular malalignment of fractures of bones of the forearm. An experimental study. *J. Bone Joint Surg.*, **64A**, 14–7

McConnell, A.A. (1920) Approach to the median nerve in the forearm. *Dublin J. Med. Sci.*, **149**, 90–2

McKeever, F.M. and Buck, R.M. (1947) Fractures of the olecranon process of the ulna. *J.A.M.A.*, **135**, 1–5

Menon, J., Wood, V.E., Schoene, H.R. *et al.* (1984) Isolated tears of the triangular fibrocartilage of the wrist. *J. Hand Surg.*, **9A**, 527–30

Mikic, Z.D. and Vukadinovic, S.M. (1983) Late results in fractures of the radial head treated by excision. *Clin. Orthop.*, **181**, 220–8

Mino, D.E., Palmer, A.K. and Levinsohn, E.M. (1983) The role of radiography and computerized tomography in the diagnosis of subluxation and dislocation of the distal radioulnar joint. *J. Hand Surg.*, **8**, 23–31

Mino, D.E., Palmer, A.K. and Levinsohn, E.M. (1985) Radiography and computerized tomography in the diagnosis of incongruity of the distal radio-ulnar joint. *J. Bone Joint Surg.*, **67A**, 247–52

Moore, T.M., Lester, D.K. and Sarmiento, A. (1985) The stabilizing effect of soft-tissue constraints in artificial Galeazzi fractures. *Clin. Orthop.*, **194**, 189–94

Morrey, B.F., An, K.N. and Stormont, T.J. (1988) Force transmission through the radial head. *J. Bone Joint Surg.*, **70A**, 250–6

Morrey, B.F. and An, K.N. (1983) Articular and ligamentous contributions to the stability of the elbow joint. *Am. J. Sports Med.*, **11**, 315–20

Morrey, B.F. and An, K.N. Functional anatomy of the elbow ligaments. *Clin. Orthop.*, **201**, 84–99

Morrey, B.F., Askew, L.J., An, K.N. and Chao, E.Y.S. (1981) A biomechanical study of normal functional elbow motion. *J. Bone Joint Surg.*, **63A**, 872–7

Morrey, B.F., Chao, E.Y. and Hui, F.C. (1979) Biomechanical study of the elbow following excision of the radial head. *J. Bone Joint Surg.*, **61A**, 63–8

Morrey, B.F. (1993) Anatomy of the elbow joint. In: *The Elbow and its Disorders*, 2nd edn (B.F. Morrey, ed.) pp. 16–52, Philadelphia: Saunders

Nakamura, R., Horii, E., Imaeda, T. *et al.* (1995) Distal radioulnar joint subluxation and dislocation diagnosed by standard roentgenography. *Skeletal Radiol.*, **24**, 91–4

O'Driscoll, S.W., Bell, D.F. and Morrey, B.F. (1991) Posterolateral rotatory instability of the elbow. *J. Bone Joint Surg.*, **73A**, 440–6

O'Driscoll, S.W., Horii, E., Morrey, B.F. and Carmichael, S.W. (1992) Anatomy of the ulnar part of the lateral collateral ligament of the elbow. *Clin. Anat.*, **5**, 296–304

Olerud, C., Kongsholm, J. and Thuomas, K.A. (1988) The congruence of the distal radioulnar joint: a magnetic resonance imaging study. *Acta Orthop. Scand.*, **59**, 183–5

Oskam, J., Kingma, J. and Klasen, H.J. (1993) Ulnar-shortening osteotomy after fracture of the distal radius. *Arch. Orthop. Trauma Surg.*, **112**, 198–200

Palmar, A.K., Glisson, R.R. and Werner, F.W. (1984) Relationship between ulnar variance and triangular fibrocartilage complex thickness. *J. Hand Surg.*, **9A**, 681–3

Palmar, A.K., Werner, F.W., Glisson, R.R. and Murphy, D.J. (1988) Partial excision of the triangular fibrocartilage complex. *J. Hand Surg.*, **13A**, 403–6

Palmar, A.K., Glisson, R.R. and Werner, F.W. (1982) Ulnar variance determination. *J. Hand Surg.*, **7A**, 376–9

Palmer, A.K. and Werner, F.W. (1981) The triangular fibrocartilage of the wrist – Anatomy and function. *J. Hand Surg.*, **6A**, 153–62

Palmer, A.K. and Werner, F.W. (1984) Biomechanics of the distal radioulnar joint. *Clin Orthop.*, **187**, 26–35

Pankovich, A.M. (1977) Anconeus approach to the elbow joint and the proximal part of the radius and ulna. *J. Bone Joint Surg.*, **59A**, 124–6

Patrick, J. (1964) A study of supination and pronation, with especial reference to the treatment of forearm fractures. *J. Bone Joint Surg.*, **28B**, 737–48

Pirela-Cruz, M.A., Goll, S.R., Klug, M. and Windler, D. (1991) Stress computed tomography analysis of the distal radioulnar joint: a diagnostic tool for determining translational motion. *J. Hand Surg.*, **16A**, 75–82

Postacchini, F. and Morace, G.B. (1992) Radial head fracture treated by resection. Long-term results. *Ital. J. Orthop. Traumatol.*, **18**, 323–30

Rabinowitz, R.S., Light, T.R., Havey, R.M. *et al.* The role of the interosseous membrane and triangular fibrocartilage complex in forearm stability. *J. Hand Surg.*, **19A**, 385–93

Ray, R.D., Johnson, R.J. and Jameson, R.M. (1951) Rotation of the forearm – an experimental study of pronation and supination. *J. Bone Joint Surg.*, **33A**, 993–6

Reardon, J.P., Lafferty, M., Kamaric, E. *et al.* (1991) Structures influencing axial stability to the forearm: the role of the radial head, interosseous membrane, and distal radioulnar joint. *Orthop. Trans.*, **15**, 436–7

Richards, R.R. (1996) Chronic disorders of the forearm. *J. Bone Joint Surg.*, **78A**, 916–30

Robbin, M.L., An, K.N., Linscheid, R.L. and Ritman, E.L. (1986) Anatomic and kinematic analysis of the human forearm using high-speed computed tomography. *Med. Biol. Eng. Comput.*, **24**, 164–8

Rose-Innes, A.P. (1960) Anterior dislocation of the ulna at the inferior radioulnar joint. *J. Bone Joint Surg.*, **42B**, 515–21

Roysam, G.S. (1992) The distal radioulnar joint in Colles' fractures. *J. Bone Joint Surg.*, **75B**, 58–60

Sage, F.P. (1982) Medullary fixation of fractures of the forearm: a study of the medullary canal of the radius and a report on 50 fractures of the radius treated with a pre-bent triangular nail. *J. Bone Joint Surg.*, **64A**, 857–63

Salter, N. and Darcus, H.D. (1953) The amplitude of forearm and of humeral rotation. *J. Anat.*, **87**, 407–17

Sarmiento, A., Ebramzadeh, E., Brys, D. and Tarr, R. (1992) Angular deformities and forearm function. *J Orthop. Res.*, **10**, 121–33

Schemitsch, E.H. and Richards, R.R. (1992)The effect of malunion on functional outcome after plate fixation of fractures of both bones of the forearm in adults. *J. Bone Joint Surg.*, **74A**, 1068–78

Schuind, F., An, K.N., Berglund, L. *et al.* (1991) The distal radioulnar ligaments: a biomechanical study. *J. Hand Surg.*, **16A**, 1106–14

Schuind, F., Linscheid, R.L., An, K.N. and Chao, E.Y.S. (1992) Changes in wrist and forearm configuration with grasp and isometric contraction of elbow flexors. *J. Hand Surg.*, **17A**, 698–703

Shiba, R., Siu. D. and Sorbie, C. (1988) Geometirc analysis of the elbow joint. *J. Orthop. Res.*, **6**, 897–906

Silver, D. (1923) Measurement of the range of motion in joints. *J. Bone Joint Surg.*, **5**, 569–78

Sojberg, J.O., Ovesen, J. and Nielsen, S. (1987) Experimental elbow stability after transection of the medial collateral ligament. *Clin. Orthop.*, **218**, 186–90

Sorbie, C., Shiba, R., Siu, D. *et al.* (1986) The development of a surface arthroplasty for the elbow. *Clin. Orthop.*, **208**, 100–3

Spinner, M. and Kaplan, E.B. (1970a) Extensor carpi ulnaris: its relationship to the stability of the distal radioulnar joint. *Clin. Orthop.*, **68**, 124–9

Spinner, M. and Kaplan, E.B. (1970b) The quadrate ligament of the elbow – its relationship to the stability of the proximal radioulnar joint. *Acta Orthop. Scand.*, **41**, 632

Staron, R.B., Feldman, F., Haramti, N. *et al.* (1994) Abnormal geometry of the distal radioulnar joint: MR findings. *Skeletal Radiol.*, **23**, 369–72

Stewart, H.D., Innes, A.R. and Burke, F.D. (1985) Factors affecting the outcome of Colles' fracture: an anatomical and functional study. *Injury*, **16**, 289–95

Stormont, T.J., An, K.N., Morrey, B.F. and Chao, E.Y. (1985) Elbow joint contact study: comparison of techniques. *J. Biomech.*, **18**, 329–36

Strachan, J.H. and Ellis, B.W. (1971) Vulnerability of the posterior interosseous nerve during radial head resection. *J. Bone Joint Surg.*, **53B**, 320–3

Sunderland, S. (1946) Metrical and nonmetrical features of the muscular branches of the radial nerve. *J. Comp Neurol.*, **85**, 93–7

Tarr, R.R., Garfinkel, A.I. and Sarmiento, A. (1984) The effects of angular and rotational deformities of both bones of the forearm. *J. Bone Joint Surg.*, **66A**, 65–70

Taylor, G.I., Palmer, J.H. and McManamny, D. (1990) The vascular territories of the body (angiosomes) and their clinical applications. In: *Plastic Surgery* (J.G. McCarthy, ed.) pp. 329–78, Philadelphia: Saunders

Taylor, T.F.K. and O'Connor, B.T. (1964) The effect upon the inferior radio-ulnar joint of excision of the head of the radius in adults. *J. Bone Joint Surg.*, **46B**, 83–8

Thomas, T.T. (1929) A contribution to the mechanism of fractures and dislocations in the elbow region. *Ann. Surg.*, **89**, 108–21

Thompson, J.E. (1918) Anatomical methods of approach in operations on the long bones of the extremities. *Ann. Surg.*, **68**, 309–30

Tillmann, B. (1978) *A Contribution to the Functional Morphology of Articular Surface* (G. Konorza, transl.) Stuttgart: Georg Thieme, P.S.G. Publishing

Tsukazaki, T. and Iwasaki, K. (1993) Ulnar wrist pain after Colles' fracture. *Acta Orthop. Scand.*, **64**, 462–4

Van der Linden, W. and Ericson, R. (1981) Colles' fracture. How should its displacement be measured and how should it be immobilized? *J. Bone Joint Surg.*, **63A**, 1285–8

Vesely, D.G. (1967) The distal radioulnar joint. *Clin. Orthop.*, **51**, 75–87

Wagner, C. (1977) Determination of the rotatory flexibility of the elbow joint. *Eur. J. Appl. Physiol.*, **37**, 47–59

Wainwright, D. (1942) Fractures of the olecranon process. *Br. J. Surg.*, **29**, 403–6

Walker, P.S. (1977) *Human Joints and their Artificial Replacements.* Springfield: Charles C. Thomas

Wechsler, R.J., Wehbe, M.A., Rifkin, M.D. *et al.* (1987) Computed tomography diagnosis of distal radioulnar subluxation. *Skeletal Radiol.*, **16**, 1–5

Wehbe, M.A. (1986) Surgical approach to the ulnar wrist. *J. Hand Surg.*, **11A**, 509–12

Werner, F.W., Murphy, D.J. and Palmer, A.K. (1989) Pressures in the distal radioulnar joint: effect of surgical procedures used for Kienböck's disease. *J. Orthop. Res.*, **7**, 445–50

Werner, F.W., Palmer, A.K., Fortino, M.D. and Short, W.H. (1992) Force transmission through the distal ulna: effect of ulnar variance, lunate fossa angulation, and radial and palmar tilt of the distal radius. *J. Hand Surg.*, **17A**, 423–8

Williams, P.L. and Warwik R. (eds) (1980) *Gray's Anatomy*, 36th British edn. pp. 364–70, Philadelphia: Saunders

Youm, Y., Dryer, R.F., Thambyrajah, K. *et al.* (1979) Biomechanical analyses of forearm pronation–supination and elbow flexion–extension. *J. Biomech.*, **12**, 245–55

4

Fractures of the distal radius

M.M. McQueen

Introduction

Fractures of the distal radius are a common type of injury and are increasing in incidence because of the increasing age of the population. Coincident with the population ageing, middle-aged to elderly patients are becoming fitter and more active than a generation or two ago. Treatment of distal radial fractures must therefore adapt to increased demands from patients who are no longer prepared to accept disability, deformity or pain after this injury.

History

The study of the history of any subject is illuminating. Despite the separation of the centuries, much of the theory has remained basically the same and treatment methods may be very similar, although some are influenced by the advent of antisepsis, X-rays or anaesthesia.

The history of the fracture of the distal radius is unusual in that the injury was only recognized as a fracture during the eighteenth century. For over 2000 years the distal radial fracture was believed to be a dislocation, usually of the distal radioulnar joint.

The first surgeon to recognize that these injuries were fractures was Pouteau in 1783 who described the fracture at the distal end of the radius with dorsal displacement. His work was not widely publicized, and in 1814 Abraham Colles, unaware of Pouteau's work, described the fracture which now bears his name. He correctly stated that fracture was the commonest injury to the distal radius and remarked on the rarity of dislocation in that area.

Colles' description, although now famous, of a fracture 'about an inch and a half above the carpal extremity of the radius....the carpal surface of the radius being directly slightly backwards', received little attention at the time. It was not until Dupuytren, one of the greatest French surgeons of all time, brought the world's attention to the injury that it was universally acknowledged as a fracture rather than a dislocation. Clearly Dupuytren's reputedly forthright manner and forceful personality were a persuasive influence on his colleagues and students!

Following acceptance of the lesion as a fracture, surgeons concentrated on describing the different types of fractures. Goyrand (1832) differentiated between dorsal and volar displacement and described the associated fracture of the styloid process of the ulna. Barton (1838) described 'a subluxation of the wrist consequent to fracture through the articular surface of the carpal extremity of the radius' which could be either dorsal or volar. In 1847 Robert Smith, Professor of Surgery in Dublin, described the fracture which bears his name as 'a fracture of the lower end of the radius with displacement of the lower fragment along with the carpus forwards'.

Treatment at this time was understandably non-operative. Closed manipulation was performed, presumably without anaesthetic, and many different types of bandaging and splints were devised. Some were palmar, some volar, some extended into the hand, others to the distal radius. Clearly these splints and bandages caused significant complications. Dupuytren noted that 'the surgeons have been so intent on preserving fractures in their proper position that the extreme constriction employed has actually caused destruction of the parts'. Amputation was at times the result of distal radial fracture.

Because of the problems of immobilization, Lucas-Champonnière advocated what is now termed early mobilization (Gibbon, 1926). He treated his patients in a revolutionary way with simple and temporary immobilization, massage and passive mobilization. He taught that restoration of function is the most important objective in the treatment of fractures.

The advent of X-rays at the end of the nineteenth century contributed much to the understanding of the different patterns of injury, although many had been described without the benefit of imaging. This led the understanding and management of distal radial fractures to present practice.

Clinical features

The clinical symptoms and signs of a distal radial fracture are straightforward. The history is usually of a fall from standing height onto the affected hand with immediate pain, swelling and sometimes deformity. Direct enquiry should be made about sensation in the hand to exclude nerve damage. Assessment of the patient's general health and fitness is crucial, as this will significantly affect management decisions. Clearly different treatment options will be required for a frail, unfit, dependent patient compared to a fit and independent subject.

Physical examination is equally simple. There is tenderness to palpation over the distal radius with a variable degree of swelling. If the fracture is displaced, the classic 'dinner fork' or 'silver fork' deformity will be seen (Figure 4.1). This is produced by the dorsally displaced carpus in a

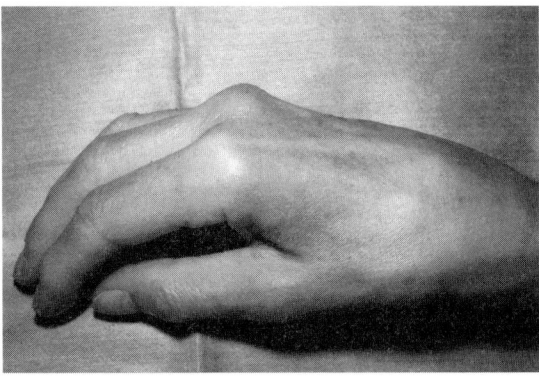

Figure 4.1   A dinner fork deformity after distal radial fracture

dorsally displaced fracture or by the prominence of the proximal surface of the fracture and the distal ulna in a volar displaced fracture. Often the hand is deviated radially. A full neurological examination of the hand and an assessment of tendon function should be performed to exclude injury to nerves and tendons or an acute compartment syndrome. A careful inspection of the soft tissues is necessary to exclude an open fracture.

Investigation

Since X-rays were first introduced in the late nineteenth century, they have remained the mainstay of investigation of the distal radial fracture. One anteroposterior and one lateral view provide sufficient information to make a diagnosis and formulate a treatment plan in the majority of cases. Assessment of displacement, the amount of comminution and articular involvement will influence the treatment chosen.

There are many methods of measuring displacement in the distal radius. The main measurements are those of dorsal/volar angle, radial shortening, ulnar variance, radial deviation or shift and radial angle.

Most authors agree on the method of measuring dorsal/volar angle (Gartland and Werley, 1951; Older *et al.*, 1965; Van der Linden and Ericson, 1981; McQueen *et al.*, 1986; Dias *et al.*, 1987a; Porter and Stockley, 1987). This is achieved by measurement of the angle between a line connecting the most distal point of the volar and dorsal cortex and a line drawn perpendicular to the long axis of the radius on the lateral X-ray (Figure 4.2). The normal inclination is volar with some variation in the normal amount from 4° to 22° (Friberg and Lundström, 1976).

Radial shortening is measured on the anteroposterior view. Many authors measure the distance between two lines perpendicular to the long axis of the radius: one at the tip of the radial styloid and one at the level of the distal articular surface of the ulna (Gartland and Werley, 1951; Dowling and Sawyer, 1961; Van der Linden and Ericson, 1981). Some authors believe that shortening is best measured by the distance between the medial corner of the distal radius and the ulnar articular surface on the anteroposterior view, sometimes called the ulnar variance (Melone, 1984; Warwick *et al.*, 1993; McBirnie *et al.*, 1995). The latter (Figure 4.2) is a more exact reflection of the disruption of the distal radioulnar joint and correlates

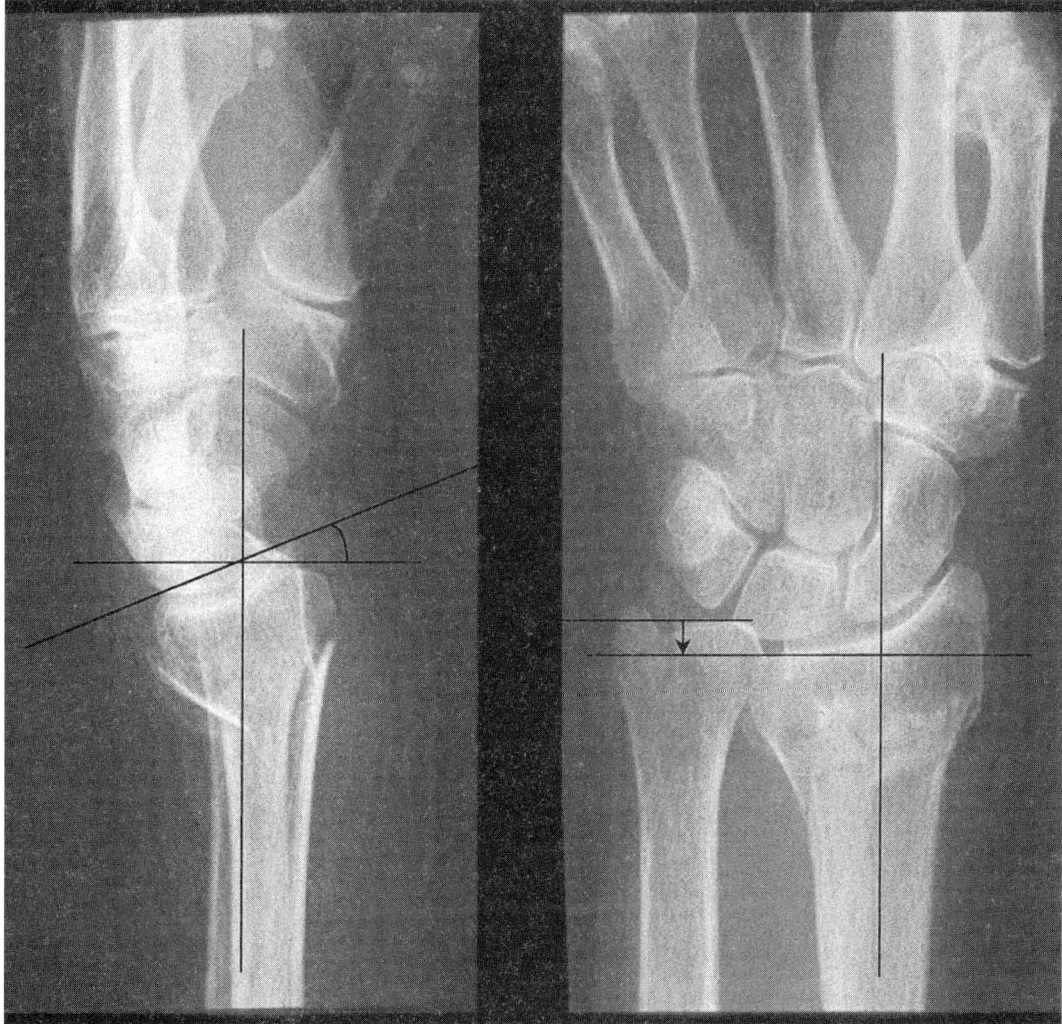

Figure 4.2   Methods of measuring dorsal angle and radial shortening (ulnar variance). There is 20° of dorsal angulation (measured from neutral). Radial shortening is also present

with long-term function (Warwick *et al.*, 1993; McQueen *et al.*, 1996). The former measurement will be affected by radial tilt, which may exaggerate the amount of shortening.

Radial shift is the increase in the distance from the long axis of the radius to the most radial point of the styloid process (Van der Linden and Ericson, 1981). This measurement and the dorsal angle are the only two measurements in the wrist that are independent of each other (Van der Linden and Ericson, 1981), although the authors did not examine ulnar variance.

The radial angle is the angle between a line perpendicular to the long axis of the radius and a line joining the radial and ulnar margins of the articular surface. This measurement has not been correlated with long-term function.

One of the most important assessments of radiological deformity is carpal alignment. Taleisnik and Watson (1984) pointed out that loss of the normal palmar tilt predisposes the carpus to a dorsal collapse, defined as dorsal displacement of the long axis of the capitate relative to the long axis of the radius on the lateral view (Figure 4.3). This is the single most significant deformity to affect wrist and hand function (McQueen *et al.*, 1996).

More sophisticated imaging techniques are infrequently required. Computerized tomography

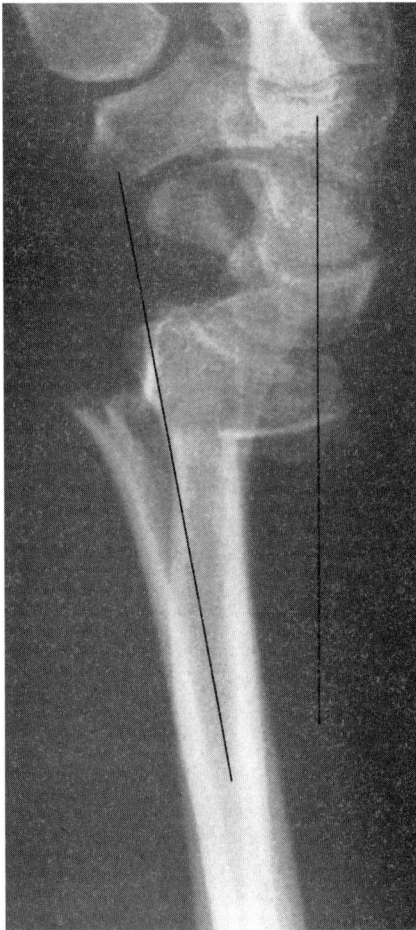

Figure 4.3 Dorsal carpal malalignment is obvious. In more subtle cases, a line drawn down the long axis of the capitate will intersect outwith the carpus with a line down the radius

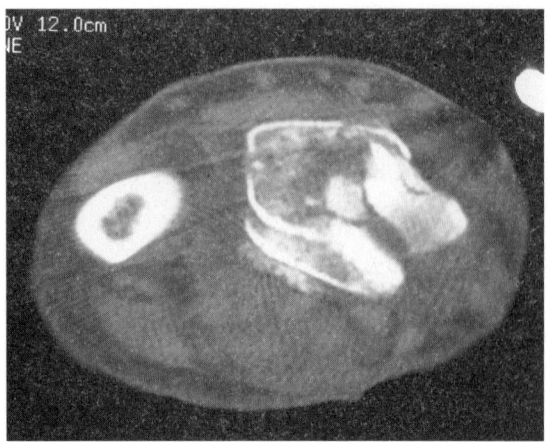

Figure 4.4 A CT scan of a distal radial fracture. There is dorsal comminution with significant damage to a detached radial styloid

(CT) can be useful in the delineation of severe articular injuries as an aid to internal fixation (Figure 4.4). A CT scan is also the most effective method of imaging damage to the distal radio-ulnar joint.

Initial management

After the initial examination, a decision must be taken about further management. Many fractures will be undisplaced and will require minimal treatment. If a fracture is displaced, a decision must be made about the need for manipulation or operative treatment based on the amount of displacement, the possibility of instability and the functional demands of the patient.

There is no absolute rule about the amount of displacement that requires reduction, although functional deficit is more likely to ensue if there is more than 10° of dorsal angulation (McQueen and Caspers, 1988; Solgaard, 1988). Radial shortening of more than a few millimetres also has a negative effect on function (Villar *et al.*, 1987; McQueen *et al.*, 1996). However, it is likely that the most significant influence on function is carpal malalignment. In a series of 120 unstable distal radial fractures, the most significant negative influence on outcome at 1 year was carpal malalignment (McQueen *et al.*, 1996). Taleisnik and Watson (1984) believe that carpal malalignment after distal radial fractures is due to loss of the normal palmar tilt. It is recommended therefore that reduction be carried out where there is loss of carpal alignment, depending on the patient's functional needs.

After reduction of the fracture, its stability must be assessed. Stability may be defined as the ability to maintain the reduced position in a cast. Prediction of instability with any degree of certainty is difficult because of its multifactorial causes. Increasing age, severe initial displacement, metaphyseal comminution, intra-articular extension and ulnar fracture all contribute to instability (Abbaszadegan *et al.*, 1989; Jenkins, 1989; Lafontaine *et al.*, 1989). Ideally, instability should be recognized at the time of injury, but in many cases it can only be diagnosed with any degree of certainty by observing the behaviour of the fracture (Figure 4.5).

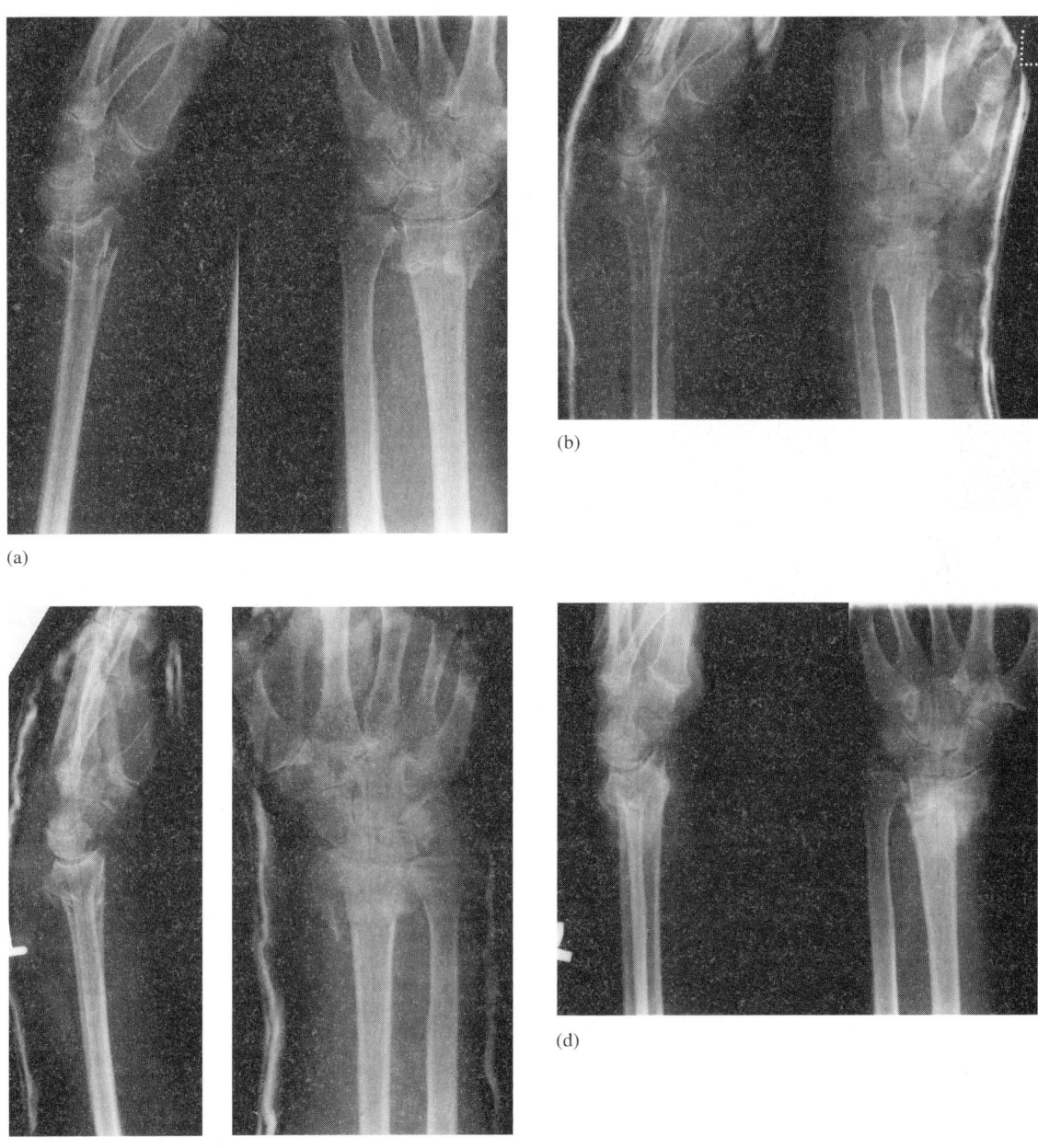

(a)

(b)

(c)

(d)

Figure 4.5　A distal radial fracture showing instability. (a) Radiographs after injury showing dorsal angulation, radial shortening and radial deviation. There is marked dorsal metaphyseal comminution. (b) The fracture has been manipulated. There is some residual dorsal angulation with minimal shortening. Inability to fully reduce a fracture is often an indication of significant instability. (c) One week after injury the original deformity has recurred. No action was taken. (d) Six weeks after injury the fracture has united; it has collapsed further into dorsal angulation and radial deviation

If the fracture is sufficiently displaced then closed reduction may be employed. This is a simple technique and is successful in approximately 95% of cases (McQueen *et al.*, 1994). It is

unlikely to be successful in displaced articular fractures with rotation of the articular fragments.

Closed reduction is usually performed using regional anaesthesia (Biers or axillary block) or local anaesthetic infiltration of the fracture haematoma. Regional anaesthesia provides better pain relief and improved reduction (Abbaszadegan and Jonsson, 1990) but it does have the disadvantage of requiring the presence of a second doctor. General anaesthetic may also be used.

Closed reduction is performed initially by applying longitudinal traction to the wrist. This may not regain volar tilt since the volar capsular ligaments tighten first, thus preventing further reduction of the dorsal displaced fracture (Bartosh and Saldaña, 1990). Palmar tilt may be obtained in some cases by palmar translation of the hand (Agee, 1993). In cases with 'bayoneting' of the volar cortex, increasing the deformity first usually allows end-to-end reduction of the volar cortex.

Non-operative management

Non-operative management is indicated for stable distal radial fractures and is usually straightforward since by definition the radiological position is maintained in a cast. There is debate about several aspects of non-operative management, including the type of plaster required, whether braces are advantageous and the length of time required for immobilization.

Type of plaster

Over the years there has been considerable debate as to the best type of plaster for a distal radial fracture. Traditionally a forearm cast is used for immobilization. Sarmiento (1965) considered the brachioradialis to be a significant deforming force in distal radial fractures even when the elbow is immobilized, and that its effect should be minimized by the use of a short-arm cast applied with the forearm in supination. Pool in 1973 published a randomized prospective study of the outcome after distal radial fracture treated by forearm cast and long-arm plaster. He found no anatomical advantage of either, but the patients treated in a long-arm plaster had an increased risk of long-term restriction of rotation. Other authors have demonstrated no differences in outcome between full casts and back slabs (Van der Linden and Ericson, 1981).

In 1975 Sarmiento *et al.* introduced a forearm brace for the treatment of distal radial fractures which allowed early movement and reported improved early results but not in a randomized

study. Since then there have been conflicting reports of the efficacy of braces. On balance it would seem that there is little difference in the long-term outcome (Bunger *et al.*, 1984; Stewart *et al.*, 1984; Moir *et al.*, 1995), but braces may well improve the speed of rehabilitation in the first few weeks (Gibson and Bannister, 1983; Moir *et al.*, 1995). Their disadvantages are the expertise required to apply them correctly and their increased cost.

Position of immobilization

Whether treated in plaster or brace, there has been debate about the position in which the wrist should be immobilized. Conventionally the wrist is placed in slight flexion and ulnar deviation, but extremes of positioning should be avoided as they do not confer an anatomical advantage, and complications, especially carpal tunnel syndrome, are likely to ensue. Despite the conventional view, Gupta (1991) showed that immobilization in dorsiflexion gave slightly improved anatomical and functional results.

The accumulated evidence would seem to suggest that if a fracture is stable there is little significant difference between treatment with plasters and braces, but the elbow should not be included as this risks loss of rotation. The position of the wrist does not significantly affect the final outcome provided extreme flexion is avoided. All the above series show some loss of position, indicating that if a fracture is inherently unstable no adjustment of plasters or positions will prevent redisplacement.

Length of immobilization

In view of the suggested improvement of early outcome with mobilization of the wrist, it may be unnecessary to immobilize the wrist at all. More rapid recovery occurs in undisplaced or minimally displaced fractures treated without plaster without significant disadvantages in terms of pain and maintenance of radiological position (Davis and Buchanan, 1987; Dias *et al.*, 1987a). Even in displaced fractures requiring reduction, functional improvement without anatomical disadvantage is apparent with 3 weeks rather than 5 weeks of immobilization (McAuliffe *et al.*, 1987).

Immobilization should be minimized or even abandoned in undisplaced fractures and need be no longer than 3–4 weeks in displaced stable fractures, depending on levels of discomfort.

Operative management

Surgical management of distal radial fractures is required when a satisfactory position cannot be obtained or maintained in a fit and active patient. Age is not a contraindication to surgery. The aims of surgical management are to improve long-term function and early recovery with minimal complications.

These indications encompass many different fracture types and many different operative techniques, but knowledge of all is necessary for managing wrist fractures so that treatment may be adapted to the needs of each patient and fracture.

External fixation

Bridging external fixation

In 1944, a new technique for the treatment of comminuted fractures of the distal radius was described (Anderson and O'Neil, 1944) which allowed 'castless fixation' and full function of the fingers using Kirschner wires in the second metacarpal and in the proximal radius. The principle was to obtain and maintain traction across the wrist until healing occurred.

Few further reports appeared until Vidal and his co-authors (1977) described the principle of tension on the ligaments and capsule allowing reduction of the fracture, and coined the term 'ligamentotaxis'.

Over the ensuing time, many authors have reported their results with external fixation bridging the wrist joint. In recent years some external fixators have been designed incorporating ball joints to allow mobilization of the wrist during the period of fixation.

Bridging external fixation may be used for dorsally displaced unstable fractures, severe intra-articular injuries or open fractures.

Technique

The upper limb is prepared and draped in a standard fashion and a tourniquet is used. If reduction is performed prior to application of the fixator then a sterile traction apparatus must be used, although most devices are sufficiently versatile to allow reduction of the fracture after application of the fixator.

Open pin placement techniques should be used to minimize inadvertent damage to the underlying soft tissues. Two pins are placed in the second metacarpal from laterally to medially or dorsolaterally to a volar-medial direction, and two pins are placed in the radius approximately 7–10 cm proximal to the fracture in a medial to lateral direction (Figure 4.6). Reduction can then be performed after assembly but before tightening of the fixator.

There is no clinical evidence of superior results from the use of any particular type of fixator. The type of fixator used should be one that is light and versatile. Although there is laboratory work which shows different mechanical characteristics of wrist external fixators (Simpson *et al.*, 1994), there is no evidence that this makes any difference to the clinical and radiological results.

Outcome

There are conflicting opinions about the effectiveness of bridging external fixation employing ligamentotaxis, and the results are difficult to interpret because of the heterogeneity of the fractures and varying outcome measures.

Anatomical results are generally considered acceptable with mean values for dorsal angulation less than 10° (Cooney *et al.*, 1979; Jonsson, 1983; Weber and Szabo, 1986; Prince and Worlock, 1988; Howard *et al.*, 1989, McQueen *et al.*, 1992, 1996; Sommerkamp *et al.*, 1994). It is unusual, however, for volar tilt to be regained probably because of the anatomy of the wrist ligaments. The volar ligaments are short and tighten first when the wrist is distracted, thus making it difficult to tilt the radius in a volar direction (Bartosh and Saldaña, 1990). Because of this, bridging fixation does not consistently achieve realignment of the carpus (McQueen *et al.*, 1996). Collapse can occur during the period of fixation, probably because the wrist ligaments lose their elasticity with prolonged traction (Figure 4.6). For the same reason and also because of resorption at the fracture site during healing, there is some loss of radial length during the period of fixation or afterwards (Cooney *et al.*, 1979; Prince and Worlock, 1988; McQueen *et al.*, 1992, 1996; Sommerkamp *et al.*, 1994).

In comparative studies, bridging external fixation consistently achieves better anatomical results than remanipulation and cast management (Howard *et al.*, 1989; Kongsholm and Olerud, 1989; Roumen *et al.*, 1991; McQueen *et al.*, 1996). Despite this, however, functional results are disappointing and in some series show no improvement over cast management (Roumen *et al.*, 1991; McQueen *et al.*,

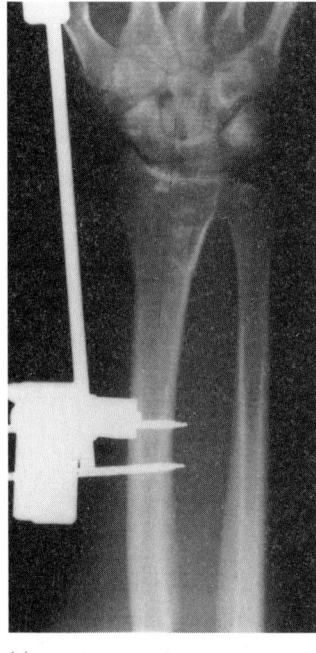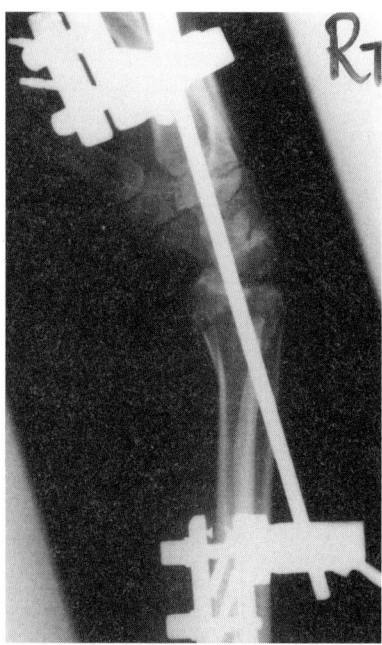

(a)

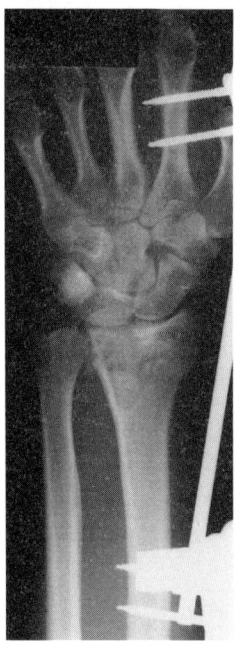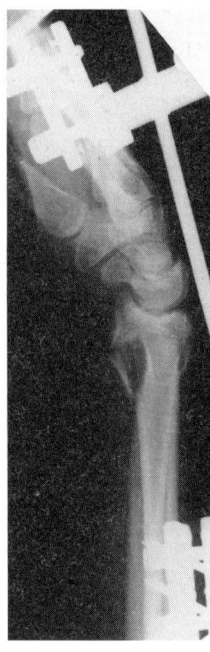

(b)

Figure 4.6   (a) Bridging external fixation of a comminuted fracture of the distal radius. A good reduction has been obtained. (b) Five days later, the fracture has collapsed into dorsal angulation and there is some radial shortening

1996) although they may be improved in younger patients (Howard *et al.*, 1989).

Overall stiffness is not a long-term problem after external fixation, with a majority of patients regaining over 80% of their normal range (Cooney *et al.*, 1979; McQueen *et al.*, 1992, 1996).

Weakness of grip strength is common, with reported values less than 75% of normal (Cooney *et al.*, 1979; Jonsson, 1983; McQueen *et al.*, 1992, 1996; Sommerkamp *et al.*, 1994).

Dynamic external fixators have achieved some popularity over recent years. They were designed with a moving part to allow the wrist to mobilize at some stage during the period that the fixator is in place, in the hope that this would improve the outcome. This has not, however, proved to be the case, and the outcome with this method of fixation is no different from the outcome following static external fixation (Sommerkamp *et al.*, 1994; McQueen *et al.*, 1996).

Non-bridging external fixation

External fixation of the distal radius can be achieved without the need to cross the wrist joint. Pins are placed into the distal radial fragment rather than the second metacarpal (Figure 4.7). Until recently, only a few reports of this method had been published (Jenkins *et al.*, 1987; Melendez *et al.*, 1989). These authors indicated that reduction was achieved with reasonable wrist movement. Recently, however, non-bridging external fixation of distal radial fractures has been shown to be significantly superior to bridging fixation, both anatomically and functionally (McQueen, 1998). It is indicated in unstable distal radial fractures which are either extra-articular or have an undisplaced articular extension and in which the distal fragment has more than 1 cm of intact volar cortex. This technique can be used to treat displaced articular fractures after reassembly of the joint surface, provided that there is sufficient space in the distal fragment to insert the pins.

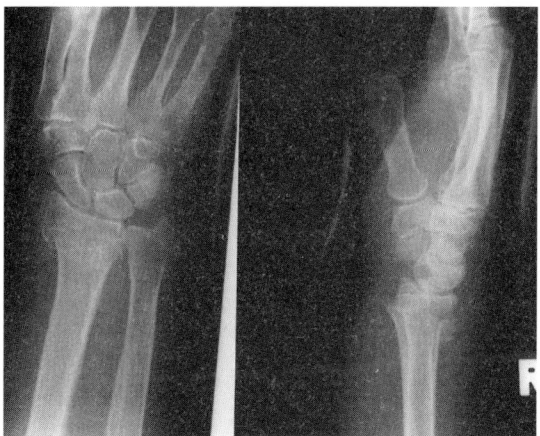

(a)

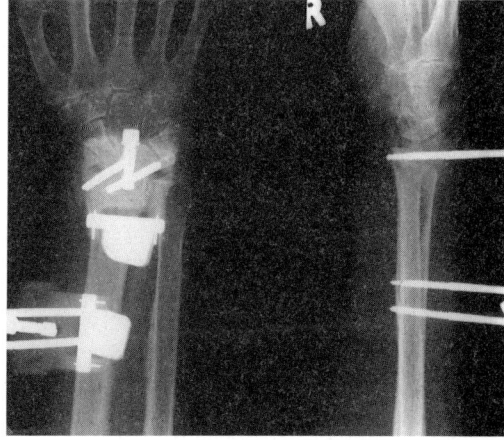

(b)

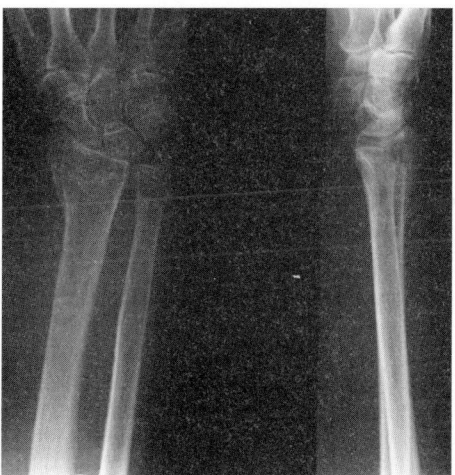

(c)

Figure 4.7 Non-bridging external fixation. (a) A redisplaced distal radial fracture 10 days after initial manipulation and plaster management. (b) Application of a non-bridging external fixator. The normal volar tilt and consequently the carpal alignment have been regained, as has radial length. (c) The same fracture 1 year after injury. The volar tilt and radial length have been maintained

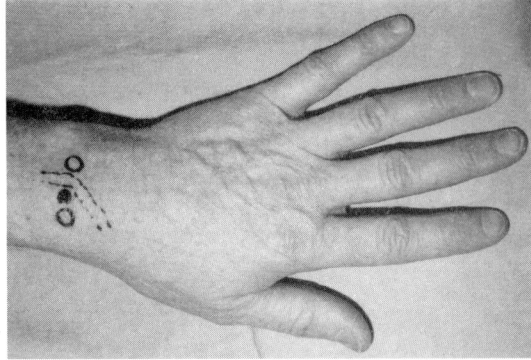

Figure 4.8 Placement of pins in the distal fragment. The sites of pin insertion are indicated by open circles, and Lister's tubercle by the black dot. The dotted lines indicate the position of the extensor pollicis longus tendon

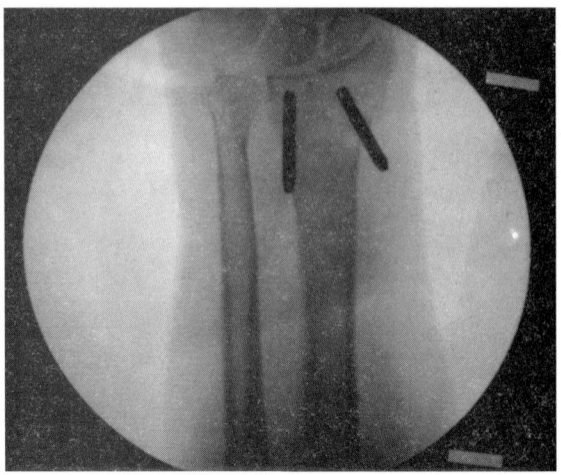

Figure 4.9 Placement of pins in the distal fragment on either side of Lister's tubercle. Note the undisplaced articular extension of the fracture in the sagittal plane

Technique

The upper limb is prepared and draped allowing access to the forearm and hand, and a tourniquet is used. The distal pins are inserted from dorsal to volar using an open technique, placing one pin on either side of the extensor pollicis longus tendon and Lister's tubercle (Figures 4.8 and 4.9). Two parallel pins are inserted using image intensifier control; they should be parallel to the wrist joint on the lateral view (Figure 4.7*b*) and should engage the volar cortex. Gentle pressure with the thumb on the pins will then reduce the dorsal fragment (Figure 4.10). Two pins are inserted proxi-mal to the fracture in the radial shaft in a similar manner to bridging fixation, and the fixator is assembled (Figure 4.11).

Outcome

There is currently only one prospective randomized study comparing bridging and non-bridging external fixation (McQueen, 1998). This compared the anatomical and functional outcome of bridging and non-bridging fixation. Carpal alignment,

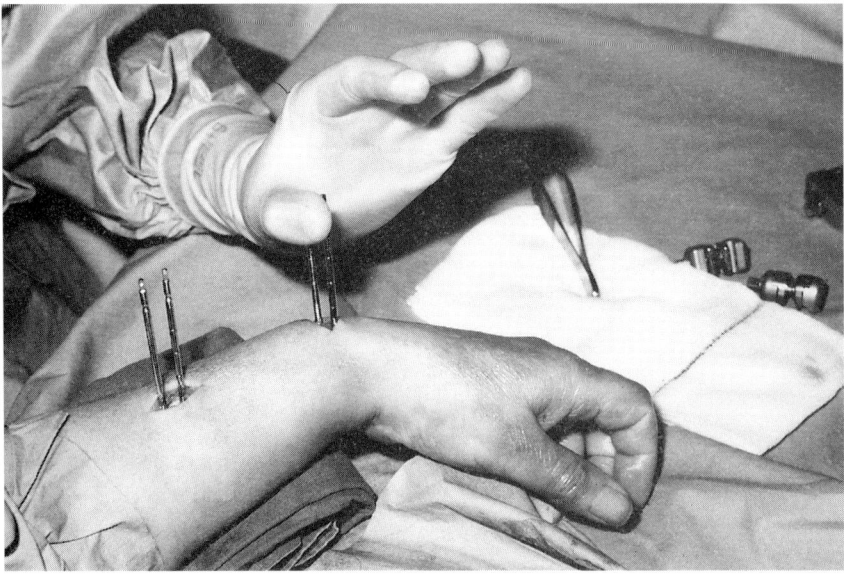

Figure 4.10 Little force is required to reduce the fracture, using the pins as a 'joystick'

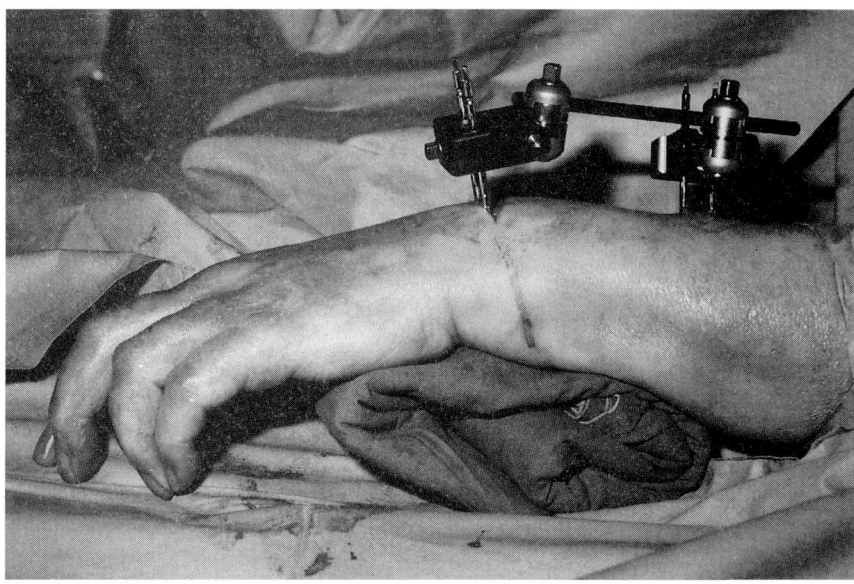

Figure 4.11 The assembled non-bridging external fixator

normal volar tilt and radial length were restored and maintained until final review by non-bridging external fixation but not by bridging fixation. These anatomical differences were statistically significant and were echoed in the functional results. Grip strength was significantly better after non-bridging fixation at all stages of review, as was the range of flexion. Other movements showed significant improvement at early stages, probably related to the absence of any restriction of wrist movement during the treatment period (Figure 4.12). Non-bridging external fixation is therefore the treatment of choice for unstable distal radial fractures, where it is technically possible.

Internal fixation

Internal fixation of the distal radius is used widely, in particular for volar displaced fractures and for displaced intra-articular fractures. Both of these usually require open reduction, although in combination with imaging techniques or arthroscopy, articular fractures can sometimes be reduced by closed means. Closed reduction and stabilization of extra-articular unstable fractures can be performed using multiple Kirschner wiring techniques, a method particularly popular in continental Europe. Plating is most often employed for volar displaced fractures, although dorsal plates can be used for dorsally displaced injury.

Plating

Plating is most commonly used for volar displaced fractures, when a small T-shaped plate is applied to the volar surface of the distal radius (Figure 4.13). This technique is popular because the volar surface of the distal radius is easily accessible and smooth, lending itself to plate application. Dorsal plating is less popular because of the difficulty of contouring a plate to the irregular surface of the dorsum of the distal radius and because of the risk of tendon rupture (Lugger and Pechlaner, 1984).

Technique

For volar plating of the distal radius it is usual to use the distal portion of Henry's approach (see Chapter 3). This allows good exposure, particularly of the radial side of the bone.

Reduction of the fracture is usually fairly simple by extension of the wrist and manipulation of the fracture fragments. A small T plate may then be applied to the volar surface of the radius. It is important to realize that the commercially available plates are not specifically designed for the distal radius and should be bent to fit the curve of the bone after fracture reduction. If there is dorsal comminution and the plate is not bent, then the fracture may be pushed into dorsal angulation (Figure 4.14). If possible, screws should be placed in the distal fragment through the transverse limb of the plate to support its buttressing effect.

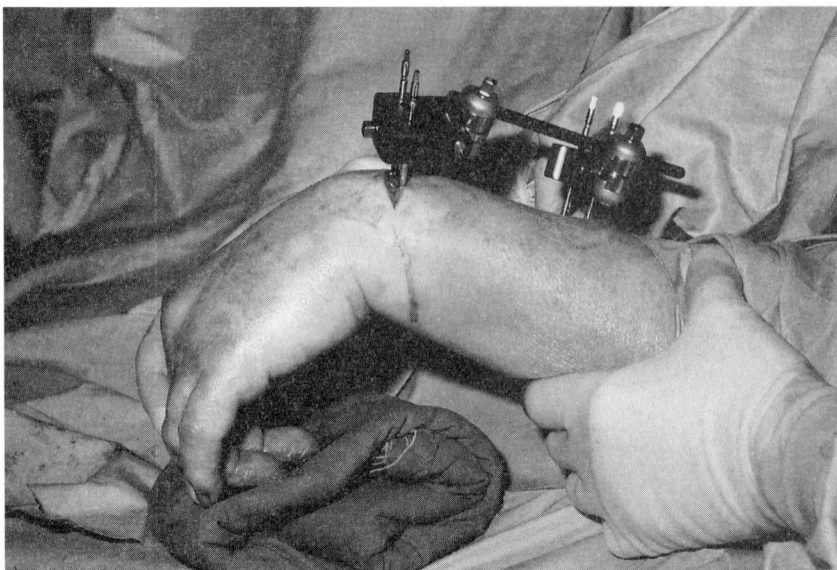

Figure 4.12 Unrestricted wrist movement is possible with non-bridging fixation

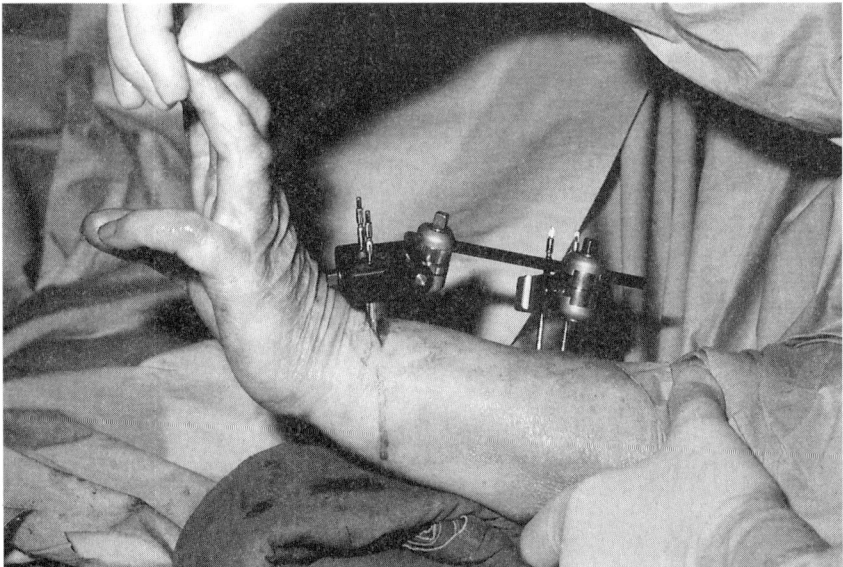

Wound closure is then effected and the fixation supported by a forearm cast for several weeks.

Dorsal plating is more complex. The approach is through a straight dorsal incision between the second and third or third and fourth extensor compartments (Chapter 3). Standard T plates are difficult to contour on to the irregular surface of the distal radius and cannot be separated from the extensor tendons, thus placing the tendons at risk of rupture. Because of the extent of exposure and technical difficulties, the use of dorsal plates is rarely indicated in the majority of dorsally displaced fractures. However, there may be a role for the use of 2 mm plates in displaced articular fractures (Rikli and Regazzoni, 1996).

Outcome of volar plating

Until recently there were few reports on outcome after volar plating of the distal radius. In 1994

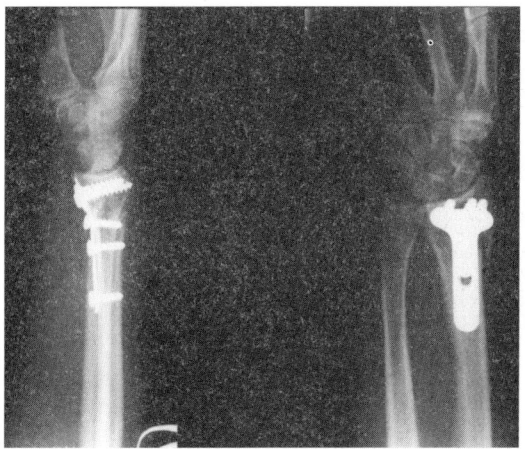

Figure 4.13 A plate applied to the volar surface of the distal radius. It has been well contoured and the fracture is reduced

Keating and his co-authors reported 79 patients treated with volar buttress plating, of whom 57 maintained an anatomical reduction. Malunion was cited as the main reason for a poor functional outcome due either to poor contouring of the plate or to a combination of volar and dorsal comminution allowing radial shortening to occur. Return of grip strength was disappointing, possibly related to a relatively high number of soft tissue complications of the fracture. These findings were confirmed by Jupiter and his colleagues (1996) who added radiological evidence of osteoarthrosis to the factors predisposing to a poor outcome.

Outcome of dorsal plating

Dorsal plating has enjoyed sporadic popularity in the management of fresh distal radial fractures although it is more commonly used for the treatment of distal radial malunions (Fernandez, 1988; Hove and Mölster, 1994; Jupiter and Ring, 1996). The usual implant used is a 3.5 mm T plate which was designed for use on the volar side of the distal radius. It is difficult to contour to the irregular surface of the dorsum of the distal radius and may result in attrition and rupture of extensor tendons (Lugger and Pechlaner, 1984; Mannerfelt *et al.*, 1990) although subtendinous placement of the extensor retinaculum at the end of the procedure may reduce this complication. Further disadvantages are the fairly extensive approach required and the necessity to support the fixation with plaster, thus immobilizing the wrist. Rikli and Regazzoni (1996) recently reported on a technique using two 2 mm dorsal plates placed dorsally on the lateral and intermediate columns of the wrist. The smaller plates may reduce the problems with tendon attrition, although a fairly extensive approach is still required. Good functional and anatomical results were reported.

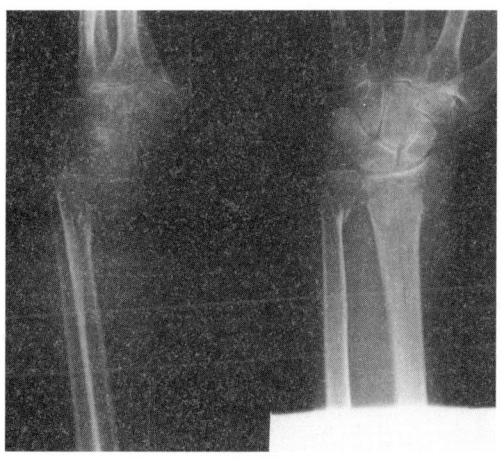

(a)

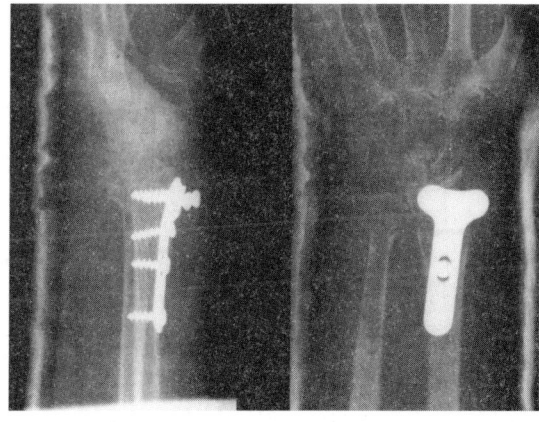

(b)

Figure 4.14 (a) A volar displaced fracture of the distal radius with volar comminution. The presence of dorsal comminution is not immediately obvious. (b) The same fracture with a plate applied to the volar surface. The plate is too straight and dorsal comminution is now evident. There is a dorsal angulation deformity with dorsal carpal malalignment and the screws have not held in the distal fragment

Percutaneous pinning

Percutaneous pinning of distal radial fractures was first described by Lambotte in the early twentieth century and since then has been used fairly extensively, mainly in continental Europe. It is indicated for unstable distal radial fractures, which are reducible by closed means.

Technique

There are many different descriptions of pin placement for percutaneous pinning, but all employ closed reduction and X-ray control under sterile conditions. Pins may be introduced through the radial styloid (Mah and Atkinson, 1992), crossing across the fracture (Stein and Katz, 1975; Clancey, 1984), ulna to radius with or without transfixion of the distal radioulnar joint (De Palma, 1952; Dowling and Sawyer, 1961) or through the fracture site, the intrafocal pinning or Kapandji technique (Kapandji, 1976). With the exception of the last technique, all require protection with a forearm cast.

Outcome

Reports of percutaneous pinning techniques lack detail of functional outcome and comparative studies. Kapandji's initial indications (1976) were primarily for fractures in younger patients with good bone stock. Attempts to extend the technique to older patients with osteoporosis can result in difficulty in maintaining fracture reduction, and most series with older patients have a significant secondary instability rate after pinning (Clancey, 1984; Mah and Atkinson, 1992; Lenoble *et al.*, 1995; Dowdy *et al.*, 1996).

Functional results are also poorly reported. The only prospective randomized comparison between Kapandji pinning and closed reduction with plaster immobilization was carried out by Stoffelen (1997). The average age of his two groups was 55 and 60 years. He found no significant differences in functional outcome between the two groups despite the fact that the closed reduction and cast group had a higher proportion of males with high-energy injury, which should have been disadvantageous to their outcome.

Bone substitutes

Charnley (1970) first used methylmethacrylate cement to fill the defect in the metaphyseal bone

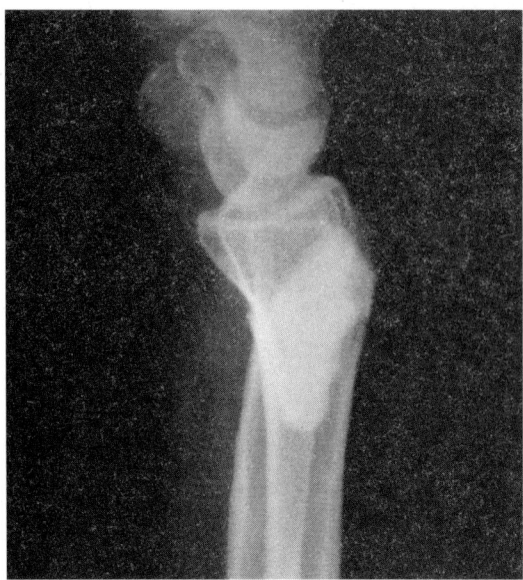

Figure 4.15   Methylmethacrylate cement used to fill the defect in the metaphysis of the distal radius after fracture

which allows redisplacement of a distal radial fracture after reduction (Figure 4.15). Since then there have been several reports detailing short-term functional advantages with the use of bone cement (Schmalholz, 1990; Kyoshige, 1995). Concern remains, however, about the possibility of thermal damage to surrounding soft tissues or joints if extrusion of cement from bone was to occur (Mjoberg *et al.*, 1984). This technique is always performed open.

Recent developments of bone mineral substitutes which are injectable, are non-exothermic and harden rapidly *in situ* to support compromised cancellous bone (Constantz *et al.*, 1995) are a significant improvement over bone cement. Preliminary clinical studies show early functional advantages using this form of treatment (Kopylov *et al.*, 1996; Husband *et al.*, 1997) (Figure 4.16).

Displaced intra-articular fractures

Displaced intra-articular fractures are severe injuries, although fortunately they constitute less than 1% of the total population of distal radial fractures. Their management requires expertise in all the treatment methods applicable to distal radial fractures and knowledge of the management of their complications.

Some displaced articular fractures reduce with closed traction and may be treated with closed

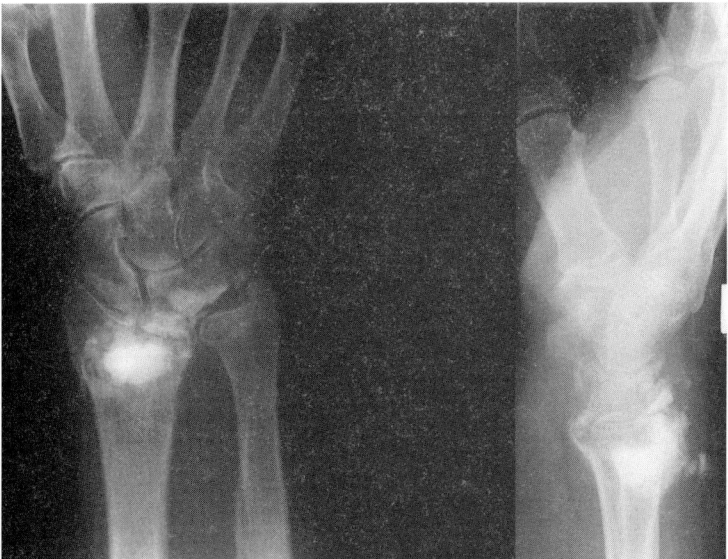

Figure 4.16   A bone mineral substitute filling the metaphyseal defect. There has been some extension into the radiocarpal joint. This has not been a clinical problem with this technique

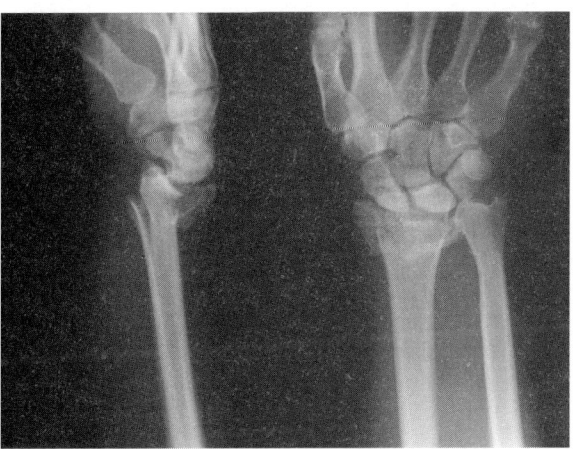

(a)

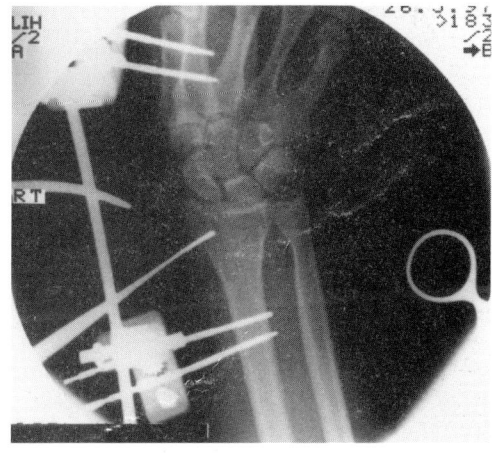

(b)

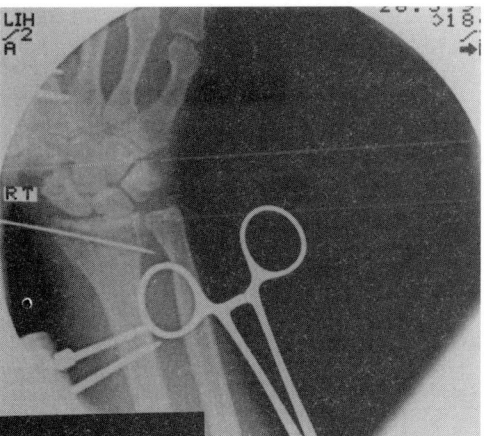

(c)

Figure 4.17   Percutaneous reduction and fixation of a displaced intra-articular fracture of the distal radius.
(a) The initial fracture with multiple intra-articular fracture lines.
(b) A bridging external fixator has been applied, improving the overall alignment but leaving a portion of the radial styloid depressed. A small lever has been inserted percutaneously to elevate the fragment. (c) The fragment is reduced and supported with a percutaneous wire

external fixation. The partial articular fracture is usually well suited to internal fixation. However it is the complete articular fracture that is irreducible closed which presents the greatest challenge to surgeons.

Techniques

The specific technique for reduction and fixation of a displaced articular fracture of the distal radius should be tailored to the patient and fracture type, although some general indications may be applied.

The simple fractures may be reduced with manipulation and traction and can then be immobilized by percutaneous pinning (Fernandez and Geissler, 1991; Seitz *et al.*, 1991). If there is significant impaction of a joint fragment, percutaneous reduction may be necessary with or without adjunctive bone grafting (Figure 4.17). Usually there is comminution of the metaphysis in association with these injuries, and bridging external fixation is required to stabilize this element of the fracture.

More complex fractures which are irreducible closed will require open reduction (Figure 4.18). Preoperative CT scanning can be helpful in planning the approach, although in some cases combined dorsal and volar approaches are required. Fixation may be by Kirschner wires, screws, plates, external fixation or a combination of these techniques (Fernandez and Geissler, 1991; Jupiter and Lipton, 1993; Rikli and Reggazoni, 1996; Hove *et al.*, 1997). Bone grafting is required where there are defects in metaphyseal bone to support the fixation. Soft tissue injuries are commonly associated with these severe fractures and the need for soft tissue procedures should always be considered.

Outcome

It is generally agreed that significant displacement of the articular surface will lead to post-traumatic arthritis (Figure 4.19). Knirk and Jupiter (1986) found late arthritis in all wrists with an articular step-off of 2 mm or more but also in 91% of wrists with any degree of articular step-off. This is supported by more recent cadaver work measuring the effect of intra-articular displacement on contact stresses in the wrist, showing that displacement of 1 mm or more significantly increases contact stresses in the wrist joint (Baratz *et al.*, 1996; Wagner *et al.*, 1996).

Provided that articular displacement is corrected, good medium-term results are reported, with around 60–80% of movement and grip strength being regained (Porter and Tillman, 1992; Jupiter and Lipton, 1993) but the long-term incidence of osteoarthritis is unknown.

Complications

Malunion

Malunion of distal radial fractures is the commonest complication of the injury and is often the underlying cause for other complications such as median neuropathies and distal radioulnar joint problems. In a prospectively documented series of 4025 patients with distal radial fractures treated in the Edinburgh Orthopaedic Trauma Unit, the malunion rate was 35%. Other authors have reported similar figures (Cooney *et al.*, 1980; Altissimi *et al.*, 1986; Porter and Stockley, 1987; Solgaard, 1988) although rates vary depending on the radiological definition of malunion. It is now established that deformity after wrist fracture is likely to lead to functional impairment (Stewart *et al.*, 1985a; Villar *et al.*, 1987; McQueen and Caspers, 1988; Solgaard, 1988) and this is most closely related to carpal alignment (McQueen *et al.*, 1996) and radial shortening (Villar *et al.*, 1987; McQueen *et al.*, 1996). Intra-articular malunion of more than 2 mm can also lead to post-traumatic arthrosis (Knirk and Jupiter, 1986).

Symptomatic malunion is correctable by radial osteotomy with or without ulnar procedures. Restoration of volar tilt will correct carpal alignment (Talcisnik and Watson, 1984) and may improve radial length. Osteotomy combined with ulnar procedures is recommended in cases of limited forearm rotation or ulnocarpal impingement (Fernandez and Jupiter, 1996). Intra-articular osteotomy may be indicated in patients with intra-articular malunion before arthrosis has developed (Marx and Axelrod, 1996). Despite the available techniques, however, early intervention with prevention of malunion is a preferable course.

Compression neuropathy

Compression neuropathy of the median, ulnar or radial nerves is reported as occurring in 8–17% of distal radial fractures, with the median nerve being most common (Cooney *et al.*, 1980; Stewart *et al.*, 1985b; Altissimi *et al.*, 1986; Porter and Stockley,

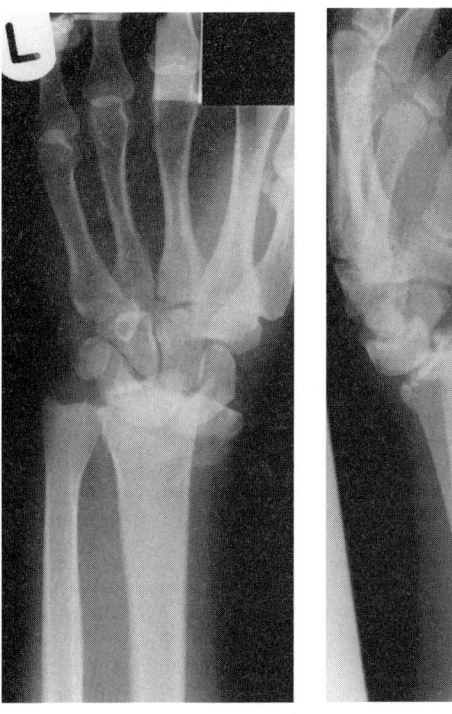

(a)

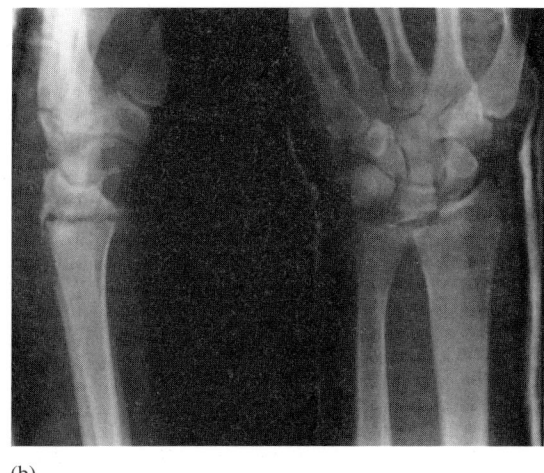

(b)

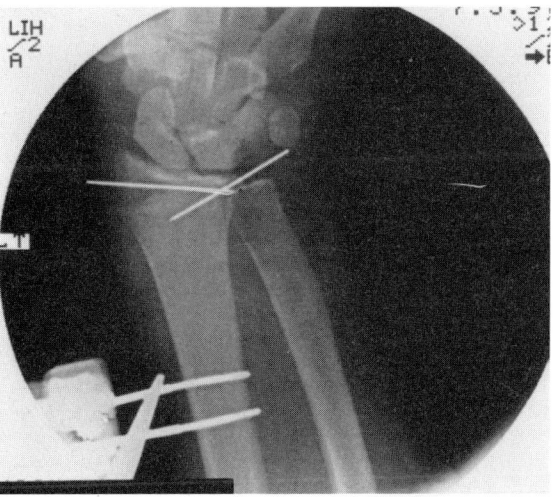

(c)

Figure 4.18   (a) A severely displaced intra-articular fracture of the distal radius in a young man who fell from a height of 30 feet. (b) Closed reduction has achieved overall alignment but with a rotated displaced lunate facet. (c) Open reduction revealed that the lunate facet had rotated through 180°, leaving the articular surface facing the elbow. This was reduced and held with Kirschner wires, and a bridging external fixator was applied

1987; Aro *et al.*, 1988). Median neuropathy may be related to the original injury, particularly in displaced fractures (Porter and Stockley, 1987), to immobilization in extreme flexion causing increased carpal tunnel pressures (Cooney *et al.*, 1980; Kongsholm and Olerud, 1986), to fracture fragments compressing the nerve (Cooney *et al.*, 1980) or to malunion of the distal radius (Stewart *et al.*, 1985; Aro *et al.*, 1988). Guidelines for the management of median neuropathy associated with distal radial fracture should follow the principles established by McCarroll (1984) who recommends decompression if a complete lesion persists after reduction of a fracture, if an incomplete lesion deteriorates at any stage or persists unchanged for longer than 7 days, or if the fracture

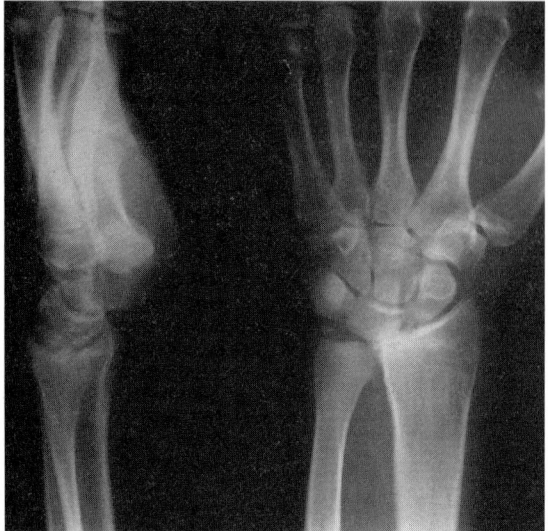

Figure 4.19 Radiographic osteoarthrosis 6 months after the same fracture as in Figure 4.18 despite good articular alignment. The wrist remains painful

requires operative intervention. In cases with malunion, osteotomy should be considered in conjunction with median decompression. Ulnar and radial neuropathy are less common and are usually treatment-related, either from cast compression or fixator pins (Cooney *et al.*, 1980).

Reflex sympathetic dystrophy

The reported incidence of reflex sympathetic dystrophy varies considerably from 1.4% to 42% (Bacorn and Kurtzke, 1953; Atkins *et al.*, 1990), probably related to different diagnostic criteria. Its underlying cause is obscure, although carpal tunnel syndrome has been implicated in its aetiology (Jupiter *et al.*, 1994).

Diagnosis is based on several criteria (Veldman *et al.*, 1993). At least four of the features of unexplained diffuse pain, diffuse swelling, difference in skin colour and temperature relative to the opposite side and a limited active range of movement should be present. In addition, these symptoms and signs should be present in an area larger than the area of primary injury and should increase with use.

Treatment should be as early as possible and may be by sympathetic blockade, intravenous guanethidine, corticosteroids, hydroxy-radical scavengers and intensive physiotherapy. Despite active treatment, the outcome is often poor.

Tendon rupture

Both flexor and extensor tendon ruptures occur after distal radial fracture, although the former are extremely rare. Extensor pollicis longus rupture is by far the commonest, although its incidence is usually less than 1% (Cooney *et al.*, 1980; Hove, 1994). Various mechanisms of injury have been proposed, although the most popular is probably a combination of attrition and impaired blood supply (Hirasawa *et al.*, 1990).

Most tendon ruptures occur several weeks to months after injury (Cooney *et al.*, 1980; Hove, 1994) and may be associated with minimally displaced or undisplaced fractures. Direct tendon repair is not usually possible because the abnormality extends over several centimetres. Treatment is most often by extensor indicis proprius transfer which yields good functional results (Hove, 1994).

Treatment-related complications

Complications related to treatment of distal radial fractures are unfortunately common occurrences. Some of the complications discussed above may be treatment related, such as carpal tunnel syndrome induced by a cast. It must be remembered that cast complications are as frequent as operative complications (Cooney *et al.*, 1980; Altissimi *et al.*, 1986) and that the permanent disability caused by poor cast application may prove more serious than that caused by skeletal deformity.

Tightness of a cast can lead to swelling of the hand and fingers, which if not relieved may lead to intrinsic contractures and finger stiffness. This must be prevented by elevation, splitting or removing the cast, and early finger movements. A cast must not be applied over the metacarpophalangeal joints, as this will contribute to finger stiffness.

Pin-related problems, usually infection or radial neuritis, may be caused by external fixation or percutaneous pinning. They are rarely serious and are usually preventable with good techniques such as open pin placement and meticulous pin track care. Other surgical complications are discussed in the relevant sections.

References

Abbaszadegan, H. and Jonsson, U. (1990) Regional anaesthesia preferable for Colles' fractures. *Acta Orthop. Scand.*, **61**, 348–9

Abbaszadegan, H., Jonsson, U. and von Swers, K. (1989) Prediction of instability of Colles' fractures. *Acta Orthop. Scand.*, **60**, 646–50

Agee, J.M. (1993) Distal radius fracture. Multiplanar ligamentotaxis. *Hand Clin.*, **9**, 577–86

Altissimi, M., Antenucci, R., Fiacca, C. and Mancini, G. (1986) Long term results of conservative treatment of fractures of the distal radius. *Clin. Orthop.*, **206**, 202–10

Anderson, R. and O'Neil, G. (1944) Comminuted fractures of the distal end of the radius. *Surg. Gynaecol. Obstet.*, **78**, 434–40

Aro, H., Kourinen, T., Katevuo, K. *et al.* (1988) Late compression neuropathies after Colles' fractures. *Clin. Orthop.*, **233**, 217–25

Atkins, R.M., Duckworth, T. and Kanis, J.A. (1990) Features of algodystrophy after Colles' fracture. *J. Bone Joint Surg.*, **72B**, 105–10

Bacorn, R.W. and Kurtzke, J.F. (1953) Colles' fracture: a study of 2,000 cases from the New York State Workers' Compensation Board. *J. Bone Joint Surg.*, **35A**, 643

Baratz, M.E., Des Jardins, J.D., Anderson, D.D. and Imbriglio, J.E. (1996) Displaced intra-articular fractures of the distal radius: the effect of fracture displacement on contact stresses in a cadaver model. *J. Hand Surg.*, **21A**, 183–8

Barton, J.R. (1938) Views and treatment of an important injury of the wrist. *Med. Examiner*, **1**, 365–8

Bartosh, R.A. and Saldaña, M.J. (1990) Intraarticular fractures of the distal radius. A cadaveric study to determine whether ligamentotaxis restores palmar tilt. In *Fractures of the Distal Radius* (P. Saffar and W.P. Cooney, eds) pp. 37–40, London: Martin Dunitz

Bunger, C., Solund, K. and Rasmussen, P. (1984) Early results after Colles' fractures. Functional bracing in supination v dorsal plaster immobilisation. *Arch. Orthop. Trauma Surg.*, **103**, 251–6

Charnley, J. (1970) *Acrylic Cement in Orthopaedic Surgery.* pp. 67–71, Edinburgh: Churchill Livingstone

Clancey, G.J. (1984) Percutaneous Kirschner wire fixation of Colles' fractures. A prospective study of thirty cases. *J. Bone Joint Surg.*, **66A**, 1008–14

Colles, A. (1814) On the fracture of the carpal extremity of the radius. *Edin. Med. Surg. J.* **10**, 182–6

Constantz, B.R., Ison, I.C., Fulmer, M.T. *et al.* (1995) Skeletal repair by in situ formation of the mineral phase of bone. *Science*, **267**, 1796–9

Cooney, W.P. III, Linscheid, R.L. and Dobyns, J.H. (1979) External pin fixation for unstable Colles' fractures. *J. Bone Joint Surg.*, **61A**, 840–5

Cooney, W.P., Dobyns, J.H. and Linscheid, R.L. (1980) Complications of Colles' fractures. *J. Bone Joint Surg.*, **62A**, 613–9

Davis, T.R. and Buchanan, J.M. (1987) A controlled prospective study of early mobilisation of minimally displaced fractures of the distal radial metaphysis. *Injury*, **18**, 283–5

De Palma, A.F. (1952) Comminuted fractures of the distal end of the radius treated with ulnar pinning. *J. Bone Joint Surg.*, **34A**, 615–62

Dias, J.J., Wray, C.C., Jones, J.M. and Gregg, P.J. (1987a) The value of early mobilisation in the treatment of Colles' fractures. *J. Bone Joint Surg.*, **69B**, 463–7

Dias, J.J., Wray, C.C. and Jones, J.M. (1987b) The radiological deformity of Colles' fractures. *Injury*, **18**, 304–8

Dowdy, P.A., Patterson, S.D., King, G.J. *et al.* (1996) Intrafocal (Kapandji) pinning of unstable distal radius fractures: a preliminary report. *J. Trauma*, **40**, 194–8

Dowling, J.J. and Sawyer, B. (1961) Comminuted Colles' fractures. Evaluation of a method of treatment. *J. Bone Joint Surg.*, **43A**, 657–68

Fernandez, D.L. (1988) Radial osteotomy and Bower arthroplasty for malunited fractures of the lower end of the radius. *J. Bone Joint Surg.*, **70A**, 1538–51

Fernandez, D.L. and Geissler, W.B. (1991) Treatment of displaced articular fractures of the distal radius. *J. Hand Surg.*, **16A**, 375–84

Fernandez, D.L. and Jupiter, J.B. (1996) Malunion of the distal end of the radius. In *Fractures of the Distal Radius* (D.L. Fernandez and J.B. Jupiter, eds) pp. 263–315, New York: Springer

Friberg, S. and Lundstrom, B. (1976) Radiographic measurement of the radio-carpal joint in normal adults. *Acta Radiol. Diagn.*, **17**, 249–56

Gartland, J.J. and Werley, C.W. (1951) Evaluation of healed Colles' fractures. *J. Bone Joint Surg.*, **33A**, 895–907

Gibbon, J.H. (1926) Lucas-Championnière and mobilisation in the treatment of fractures. *Surg. Gynecol. Obstet.*, **63**, 271–8

Gibson, A.G.F. and Bannister, G.C. (1983) Bracing or plaster for Colles' fractures? A randomised prospective controlled trial. *J. Bone Joint Surg.*, **66B**, 221

Goyrand, G. (1832) Mémoire sur les fractures de l'extrémité inférieure du radius qui simulent les luxations du poignet. *Gaz. Med.*, **3**, 664–7

Gupta, A. (1991) The treatment of Colles' fracture. Immobilisation with the wrist dorsiflexed. *J. Bone Joint Surg.*, **73B**, 312–5

Hirasawa, Y., Katsumi, Y., Akiyoshi, T. *et al.* (1990) Clinical and microangiographic studies on rupture of the EPL tendon after distal radial fractures. *J. Hand Surg.*, **15B**, 51–7

Hove, L.M. (1994) Delayed rupture of the thumb extensor tendon. A 5 year study of 18 consecutive cases. *Acta Orthop. Scand.*, **65**, 199–203

Hove, L.M. and Mölster, A.O. (1994) Surgery for post-traumatic wrist deformity. *Acta Orthop Scand.*, **65**, 434–8

Hove, L.M., Neilson, P.T., Furnes, O. *et al.* (1997) Open reduction and internal fixation of displaced intra-articular fractures of the distal radius. *Acta Orthop. Scand.*, **68**, 59–63

Howard, P.N., Stewart, H.D., Hind, R.E. and Burke, F.D. (1989) External fixation or plaster for severely displaced comminuted Colles' fractures? *J. Bone Joint Surg.*, **71B**, 68–73

Husband, J.B., Cassidy, C., Leinberry, C.F. *et al.* (1997) Multicenter clinical trial of Norian SRS versus conventional therapy in the treatment of unstable distal radius fractures: preliminary results. *Proc. AAOS*, 64th Annual Meeting, 212

Jenkins, N.H. (1989) The unstable Colles' fracture. *J. Hand Surg.*, **14B**, 149–54

Jenkins, N.H., Jones, D.G., Johnson, S.R. and Mintowt-Czyt, W.J. (1987) External fixation of Colles' fractures. An anatomical study. *J. Bone Joint Surg.*, **69B**, 207–11

Jonsson, U. (1983) External fixation for redislocated Colles' fractures. *Acta Orthop Scand.*, **54**, 878–83

Jupiter, J.B., Fernandez, D.L., Toh, T.-L. *et al.* (1996) Operative treatment of volar intra-articular fractures of the distal end of the radius. *J. Bone Joint Surg.*, **78A**, 1817–28

Jupiter, J.B. and Lipton, H. (1993) The operative treatment of intraarticular fractures of the distal radius. *Clin. Orthop.*, **292**, 48–61

Jupiter, J.B. and Ring, D. (1996) A comparison of early and late reconstruction of malunited fractures of the distal end of the radius. *J. Bone Joint Surg.*, **78A**, 739–48

Jupiter, J.B., Seiler, J.G. and Sienowicz, R. (1994) Sympathetic maintained pain (causalgia) associated with a demonstrable peripheral nerve lesion. Operative treatment. *J. Bone Joint Surg.*, **76A**, 1376–84

Kapandji, A. (1976) Ostéosynthèse par double embrochage intra-focal. Traitement fonctionnel des fractures non articulaires de l'extrémité inférieure du radius. *Ann. Chir.*, **30**; 903–8

Keating, J.F., Court-Brown, C.M. and McQueen, M.M. (1994) Internal fixation of volar displaced distal radial fractures. *J. Bone Joint Surg.*, **76B**, 401–5

Knirk, J.L. and Jupiter, J.B. (1986) Intra-articular fractures of the distal end of the radius in young adults. *J. Bone Joint Surg.*, **68A**, 647–9

Kongsholm, J. and Olerud, C. (1986) Carpal tunnel pressure in the acute phase after Colles' fracture. *Arch Orthop. Trauma Surg.*, **105**, 183–6

Kongsholm, J. and Olerud, C. (1989) Plaster cast versus external fixation for unstable intra-articular Colles' fractures. *Clin. Orthop.*, **241**, 57–65

Kopylov, P., Jonsson, K., Thorngren, K.G. and Aspenberg, P. (1996) Injectable calcium phosphate in the treatment of distal radial fractures. *J. Hand Surg.*, **21B**, 768–71

Kyoshige, Y. (1995) Bone cementing of distal radial fractures in the elderly. In *Fractures of the Distal Radius* (O. Saffar and W.P. Cooney, eds) pp. 84–8, London: Martin Dunitz

Lafontaine, M., Hardy, D. and Delince, P. (1989) Stability assessment of distal radial fractures. *Injury*, **20**, 208–10

Lenoble, E., Dumontier, C., Youtallier, D. and Apoil, A. (1995) Fracture of the distal radius. A prospective comparison between trans-styloid and Kapandji fixations. *J. Bone Joint surg.*, **77B**, 562–67

Lugger, L.J. and Pechlaner, S. (1984) Schrennrupturen als Komplikation nach Osteosynthese am distalen Radius. *Unfallchirurgie*, **10**, 266–70

Mah, E.T. and Atkinson, R.N. (1992) Percutaenous Kirschner wire stabilisation following closed reduction of Colles' fractures. *J. Hand Surg.*, **17B**, 55–62

Mannerfelt, L., Oltker, R., Ostlund, B. and Elbert, B. (1990) Rupture of the extensor pollicis longus tendon after Colles' fracture and by rheumatoid arthritis. *J. Hand Surg.*, **15B**, 49–50

Marx, R.G. and Axelrod, T.S. (1996) Intraarticular osteotomy of distal radial malunions. *Clin. Orthop.*, **327**, 152–7

McAuliffe, T.B., Hilliar, K.M., Coates, C.J. and Grange, W.J. (1987) Early mobilisation of Colles' fractures. A prospective trial. *J. Bone Joint surg.*, **69B**, 727–9

McBirnie, J., Court-Brown, C.M. and McQueen, M.M. (1995) Early open reduction and bone grafting for unstable fractures of the distal radius. *J. Bone Joint Surg.*, **77B**, 571–5

McCarroll, H.R. (1984) Nerve injuries associated with wrist trauma. *Orthop. Clin. North*, **15**, 217–36

McQueen, M.M. (1998) Redisplaced unstable fractures of the distal radius. A randomised prospective study of bridging versus non-bridging external fixation. *J. Bone Joint Surg.*, **80B**, 665–9

McQueen, M.M. and Caspers, J. (1988) Colles' fracture: does the anatomical result affect final function? *J. Bone Joint Surg.*, **70B**, 649–51

McQueen, M.M., Hajducka, C. and Court-Brown, C.M. (1996) Redisplaced unstable fractures of the distal radius: a prospective randomised comparison of four methods of treatment. *J. Bone Joint Surg.*, **78B**, 404–9

McQueen, M.M., MacLaren, A. and Chalmers, J. (1986) The value of remanipulating Colles' fractures. *J. Bone Joint Surg.*, **68B**, 232–3

McQueen, M.M., MacLennan, W. and Latta, L. (1994) Fractures of the distal radius in elderly people. In: *Skeletal Trauma in Old Age* (Rowley, D.I. and Clift, B. eds) pp. 109–23. London: Chapman & Hall

McQueen, M.M., Michie, M. and Court-Brown, C.M. (1992) Hand and wrist function after external fixation of unstable distal radial fractures. *Clin. Orthop.*, **285**, 200–4

Melendez, E.M., Dehne, D.K. and Posner, M.A. (1989) Treatment of unstable Colles' fractures with a new radius mini-fixator. *J. Hand Surg.*, **14A**, 807–1

Melone, C.P. (1984) Articular fractures of the distal radius. *Orthop. Clin. North Am.*, **15**, 217–36

Mjoberg, B., Pettersson, H., Rosenquist, R. and Rydholm, A. (1984) Bone cement, thermal injury and the radiolucent zone. *Acta Orthop. Scand.*, **55**, 597–600

Moir, J.S., Murali, S.R., Ashcroft, G.P. *et al.* (1995) A new functional brace for the treatment of Colles' fractures. *Injury*, **26**, 587–93

Older, T.M., Stabler, E.V. and Cassebaum, W.H. (1965) Colles' fracture: evaluation and selection of therapy. *J. Trauma*, **5**, 469–76

Pool, C. (1973) Colles' fracture: a prospective study of treatment. *J. Bone Joint Surg.*, **55B**, 544

Porter, M. and Stockley, I. (1987) Fractures of the distal radius. Intermediate and end results in relation to radiologic parameters. *Clin. Orthop.*, **220**, 241–52

Porter, M.L. and Tillman, R.M. (1992) Pilon fractures of the wrist. *J. Hand Surg.*, **17B**, 63–8

Pouteau, C. (1783) *Oeuvres posthumes de M. Pouteau. Mémoire, contenant quelques réflexions sur quelques fractures de l'avant-bras sur les luxations incomplètes du poignet et sur le diastasis.* Paris: PhD. Pierres

Prince, H. and Worlock, P. (1988) The small AO external fixator in the treatment of unstable distal forearm fractures. *J. Hand Surg.*, **13B**, 294–7

Rikli, D.A. and Regazzoni, P. (1996) Fractures of the distal end of the radius treated by internal fixation and early function. *J. Bone Joint Surg.*, **78B**, 588–92

Roumen, R.M.H, Hesp, W.L.R.M and Bruggink, E.D.M. (1991) Unstable Colles' fractures in elderly patients. A randomised trial of external fixation for redisplacement. *J. Bone Joint Surg.*, **73B**, 307–11

Sarmiento, A. (1965) The brachioradialis as a deforming force in Colles' fractures. *Clin. Orthop.*, **38**, 86–92

Sarmiento, A., Pratt, G.W., Berry, N.C. and Sinclair, W.F. (1975) Colles' fracture: functional bracing in supination. *J. Bone Joint Surg.*, **57A**, 311–7

Schmalholz, A. (1990) External skeletal fixation versus cement fixation in the treatment of redislocated Colles' fractures. *Clin. Orthop.*, **254**, 236–41

Seitz, W.H., Froimson, A.J., Leb, R. and Shapiro, J.D. Augmented external fixation of unstable distal radius fractures. *J. Hand Surg.*, **16A**, 1010–16

Simpson, N.S., Wilkinson, R., Barbenel, J.C. and Kinnimonth, A.W.G. (1994) External fixation of the distal radius. A biomechanical study. *J. Hand Surg.*, **19B**, 188–92

Smith, R.W. (1847) *A treatise on fractures in the vicinity of joints and on certain forms of accidental and congenital dislocations.* Dublin: Hodges & Smith

Solgaard, S. (1988) Function after distal radius fracture. *Acta Orthop. Scand.*, **59**, 39–42

Sommerkamp, T.G., Seeman, M., Silliman, J. *et al.* (1994) Dynamic external fixation of unstable fractures of the distal part of the radius. A prospective randomised comparison with static external fixation. *J. Bone Joint Surg.*, **76A**, 149–61

Stein, A.H. and Katz, S.F. (1975) Stabilisation of comminuted fractures of the distal inch of the radius: percutaneous pinning. *Clin. Orthop.*, **108**, 174–81

Stewart, H.D., Innes, A.R. and Burke, F.D. (1984) Functional cast bracing for Colles' fractures. A comparison between cast bracing and conventional plaster casts. *J. Bone Joint Surg.*, **66B**, 749–53

Stewart, H.D., Innes, A.R. and Burke, F.D. (1985a) Factors affecting the outcome of Colles' fracture: an anatomical and functional study. *Injury*, **16**, 289–95

Stewart, H.D., Innes, A.R. and Burke, F.D. (1985b) The hand complications of Colles' fractures. *J. Hand Surg.*, **10B**, 103–6

Stoffelen, D. (1997) *Fractures of the distal radius: an experimental and clinical approach*. Thesis, University of Leuven

Taleisnik, J. and Watson, H.K. (1984) Midcarpal instability caused by malunited fractures of the distal radius. *J. Hand Surg.*, **9A**, 350–7

Van der Linden, W. and Ericson, R. (1981) Colles' fracture: how should its displacement be measured and how should it be immobilized? *J. Bone Joint Surg.*, **63A**, 1285–8

Veldman, P.H.J.M., Reynen, H.M., Arntz, I.E. and Goris, R.J.A. (1993) Signs and symptoms of reflex sympathetic dystrophy: prospective study of 829 patients. *Lancet*, **342**, 1012–6

Vidal, J., Buscayret, C., Fischbach, C. *et al.* (1977) Une méthode originale dans le traitement des fractures comminutives de l'extrémité inférieure du radius: le taxis ligamentaire. *Acta Orthop. Belg.*, **43**, 781–9

Villar, R.N., Marsh, D., Rushton, N. and Greatorex, R.A. (1987) Three years after Colles' fracture. A prospective review. *J. Bone Joint Surg.*, **69B**, 635–8

Wagner, W.F., Tencer, A.F., Kiser, P. and Trimble, T.E. (1996) Effects of intra-articular distal radius depression on wrist joint contact characteristics. *J. Hand Surg.*, **21A**, 554–60

Warwick, D., Prothero, D., Field, J. and Bannister, G. (1993) Radiological measurement of radial shortening in Colles' fracture. *J. Hand Surg.*, **18B**, 50–2

Weber, S.C. and Szabo, P.M. Severely comminuted distal radial fracture as an unsolved problem: complications associated with external fixation and pins and plaster technique. *J. Hand Surg.*, **11A**, 157–65

5

Forearm fractures

T. Rüedi

Introduction

Conservative treatment of forearm shaft fractures usually results in a poor functional outcome, with the exception of the rare case of an undisplaced and stable fracture (Hughston, 1957; Charnley, 1961), and up to 27% of malunion and non-union or the formation of bridging callus or synostosis

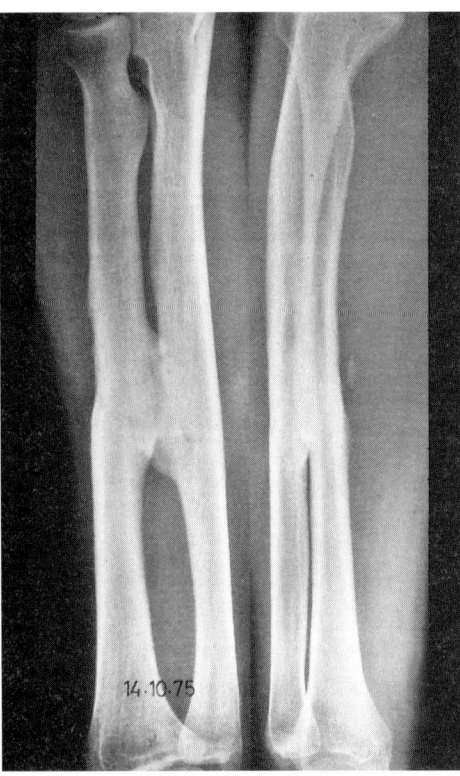

Figure 5.1 Bridging callus or synostosis after conservative treatment of both bone forearm fracture

(Figure 5.1). Charnley (1961) and similarily McLaughlin (1965) therefore strongly recommended the operative approach with plate fixation as early as the 1960s. Sarmiento's (1975) recommendations of functional bracing apply mostly to non-displaced fractures, especially of the ulna, while for unstable and displaced both bone fractures he also advocates open reduction and internal fixation (ORIF).

The reasons for the high rate of non-union and malunion as well as poor functional outcome are that the entire forearm has a quite complex anatomical structure with a sophisticated coordination between muscles, tendons, bones and joints, which is ultimately responsible for the multifold functions of the arm and hand including pronation and supination, where the radius rotates around the ulna. These movements rely on five articulations which are the ulnohumeral, the radiohumeral and the proximal radioulnar joints at the elbow as well as the distal radioulnar and the radiocarpal articulation at the wrist. Forearm fractures have therefore also been classified among the group of 'articular fractures' always requiring surgery.

According to Schneidermann (1993), the interosseous membrane as well as the triangular fibrocartilage complex of the distal radioulnar joint appear to have important functions as force transmitters between the two forearm bones, while the exact role of the strong annular ligament has – to my knowledge – not yet been fully investigated. Nevertheless, a major fracture displacement at any level of one or both forearm bones results in a considerable imbalance of the fine-tuned movements of the arm and hand. Furthermore, if fracture healing results in an angulation or shortening of the ulna, malrotation or straightening of the normal dorsoradial bow of the radius, then func-

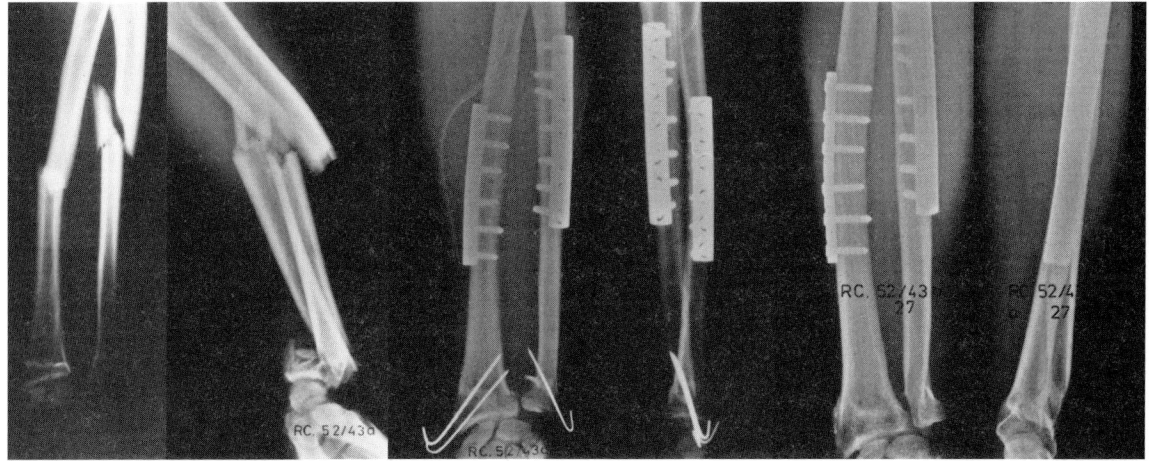

Figure 5.2 Complex forearm and wrist fracture in a young male with multiple injuries. Anatomical reconstruction can only be achieved by ORIF allowing for early movement of arm and hand

tional impairment, especially of pronation and supination, will be the consequence.

Indications for surgery

It is generally accepted today that anatomical reconstruction of both radius and ulna, including the adjacent elbow and wrist joints, is absolutely essential in order to obtain full functional restitution not only in adults but also in adolescents (over the age of 12 years) (Figure 5.2).

To guarantee the exact anatomical reconstruction of displaced fractures and to allow for early motion, rigid internal fixation must be provided, which is probably best obtained by plates and screws according to AO principles (Figure 5.3). If surgery is performed, it must eliminate the need for external immobilization, casts, splints or braces, but must allow for functional aftercare. ORIF, as described in the following pages, appears to achieve a high percentage of union in about 96–98% of cases (Anderson *et al.*, 1975; Hadden *et al.*, 1983) and restitution of function in up to 90% of cases (Tile and Petrie, 1969). Tscherne *et al.* (1978) reported on a large series of 143 fractures including 43% of open cases with an infection rate of 4.9% and good and excellent results in 85%. Oestern and Tscherne (1983) collected details of 664 forearm shaft fractures from German AO clinics, with 2.7% of infections and 3.7% of non-unions. The functional results were good and excellent in 74%, satisfactory in 13% and poor in 12% of the cases.

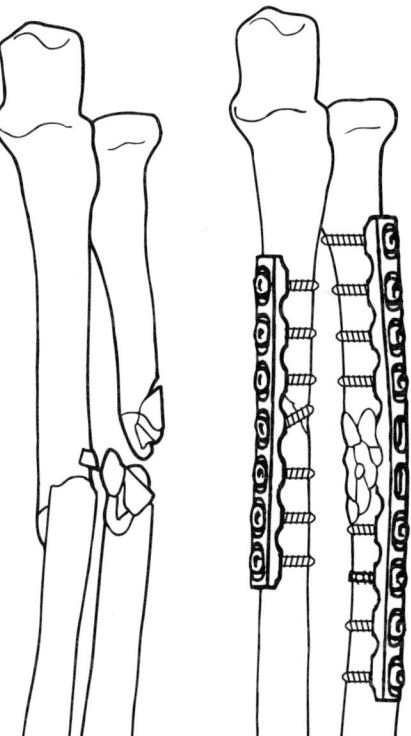

Figure 5.3 Complex fracture of radius with simple ulnar fracture. The ulnar fracture is approached first and reduced anatomically and stabilized with a 3.5 mm LC-DCP including an interfragmentary lag screw. Fixation of the radius is done with a long (10 hole) 3.5 mm LC-DCP as a bridging plate. The cancellous autograft should not be placed on to the interosseous membrane (From *AO Manual*, 3rd edn., Springer-Verlag, Heidelberg, Berlin, New York)

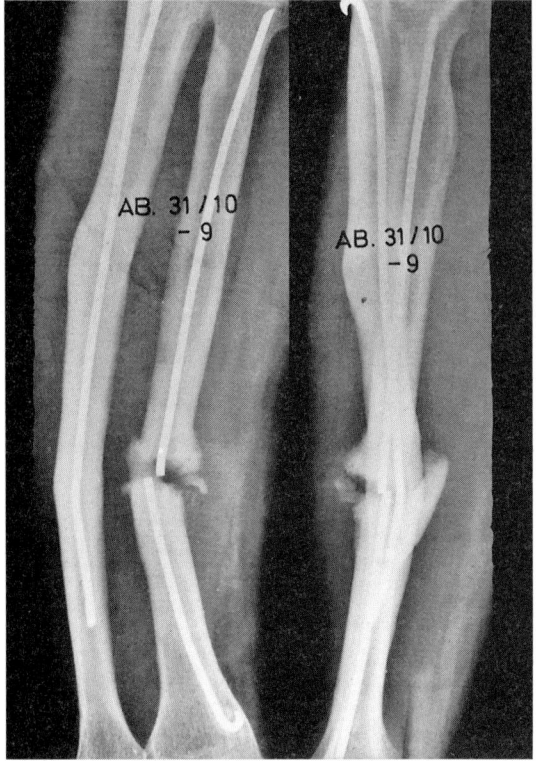

Figure 5.4   Malunion of the ulna and non-union of the radius after internal splinting with an intramedullary device

The intramedullary devices currently available are not yet able to fulfil these requirements as consistently and reliably as plates. Malunions and non-unions (in up to 50% after nailing) are still a problem, as the different forces acting on the bones are difficult to neutralize by an intramedullary nail (Figure 5.4). Jones *et al.* (1995) have shown in a biomechanical comparison on cadaver specimens that from a mechanical point of view, plates still appear to provide significantly greater stiffness in torsion and distraction/compression than the IM rods that they tested.

Clinical assessment and classification

A fracture of one or both forearm bones can usually be diagnosed clinically, but only an X-ray examination will give more detailed information. As well as looking for neurovascular deficiencies, it is essential to assess the soft tissues carefully, especially for swelling and compartment syndrome and to decide the timing of surgery.

In order to assess an injury in its full extent and to discuss and plan the treatment modalities as well as to give some prognosis to the patient, it is important to classify these fractures in detail. A prerequisite for a correct classification is adequate X-rays in two different planes which must include the elbow and wrist joints. The most appropriate classification presently available is the one developed by Müller and co-workers of the AO/ASIF (Association for the Study of Internal Fixation) group (1990) (see Chapter 2).

Planning of surgery

Once the clinical picture has been evaluated and the fracture has been classified, the operative procedure must be planned carefully in order to anticipate any surprises during surgery. This planning must include proper timing, choice of the best approach and positioning of the patient, choice of reduction technique and selection of the implant.

While an open fracture, an imminent compartment syndrome or a fracture dislocation at either joint is considered as an absolute indication for emergency surgery, all other forearm fractures may be approached electively. An open fracture of a forearm bone is handled by the same principles as are applied to other open long bone injuries: careful debridement of all damaged and dead tissue; copious irrigation; stabilization of the bone; and open wound management. Antibiotics are administered as a single dose, and if possible the tourniquet should not be inflated. For closed fractures we also advocate early surgery within 24–48 hours after injury. The advantages of early ORIF are easier and gentler reduction manoeuvres, more precise fragment adaptation and better evacuation of haematoma; factors which possibly all contribute to a reduction of the incidence of bridging callus. Furthermore, the immediate return of active motion considerably improves functional recovery (Oestern and Tscherne, 1983).

Surgical approaches

The patient is preferably placed in a supine position, the arm lying on a radiolucent arm board. While the prone position may be more convenient for the surgeon, it puts the patient at an unnecessary risk of having to be turned around under

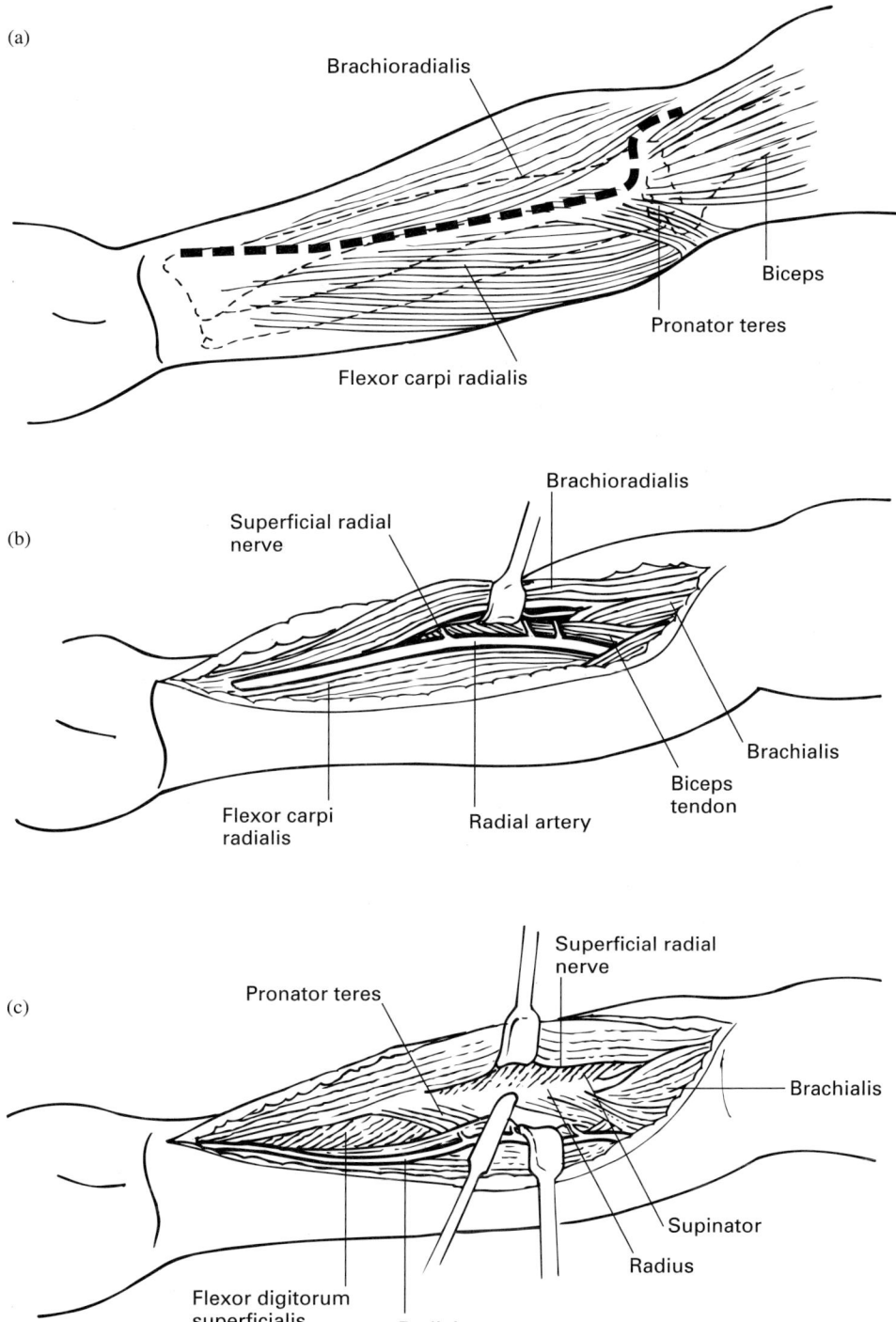

Figure 5.5   Volar approach to the radius as described by Henry (1970). The arm lies in supination. (a) Skin incision medial to the mobile wad (M. brachioradialis, M. extensor carpi radialis longus and brevis). Retract M. brachioradialis with the superficial branch of the radial nerve, which exposes the radial artery. (b) Close to the biceps tendon the recurrent branch of the artery has to be ligated, allowing exposure of the underlying supinator muscle, which 'contains' the deep branch of the radial nerve. (c) The supinator is now carefully detached from the radius at its origin and reflected. By pronating the arm, the entire radius is now exposed (From Müller *et al.*, (1991). *AO-Manual*, 3rd edn, Springer-Verlag, Heidelberg, Berlin, New York)

anaesthesia. A sterile tourniquet is always applied after draping to the humerus; it should, however, only be inflated if, after careful haemostasis, continuous bleeding obscures adequate reduction.

Depending on the site and level of fracture – especially of the radius – different approaches have to be considered. Every approach should provide a good exposure of the fracture area without interfering with important neurovascular structures, and should be extensile as well. Excessive stripping of surrounding tissue and periosteum is harmful to bone vascularity and may result in delayed healing.

The approach to the ulna is relatively simple due to its subcutaneous position. We choose a straight incision, parallel to but not directly over the palpable ulnar crest on either the volar or the dorsal side, elevating the flexor carpi or extensor carpi ulnaris muscles, but not both. The position of the plate on the bone, and accordingly the approach, should be chosen in relation to the fracture pattern (e.g. possibility of interfragmentary compression, lag screw placement, etc.).

For the radius, we have different options of approaches and in cases of both bone fractures we generally need two separate incisions for ulna and radius. Care has to be taken that a wide enough soft tissue bridge is preserved between the two. Henry (1970) described a volar or anterior approach to the radius which gives a perfect exposure of the entire length of the bone from elbow to wrist (Figure 5.5). It requires, however, careful study of the normal anatomy, some practice, and familiarity with a few tricks as well as a gentle dissection technique.

The advantages of Henry's technique are that, especially in the proximal third of the bone, the deep branch of the radial nerve remains well protected by the supinator muscle. The dorsal or posterior approach as described by Thompson (1918) is used for middle and proximal third fractures (Figure 5.6a). The plate is easily contoured to the bow of the radius and ideally lies on the tension side. In the case of distal extension of a fracture, the outcropping tendons to the thumb (abductor policis longus and extensor policis brevis) may be tunnelled for plate placement without disturbing later function. The same incision may be extended proximally. Care must be taken to identify the space between the mobile wad (extensor carpi radialis longus and brevis and brachioradialis muscles) anteriorly and extensor digitorum communis posteriorly. Between the two muscle groups, the supinator muscle is exposed as it wraps around the radius (Figure 5.6b). Three finger breadths from the radial head, the deep branch of the radial nerve can be palpated within the supinator muscle as it circles around the bone at right angles to the muscle fibres. Having identified the position of the nerve, the supinator can be cut or elevated and a plate placed beneath it.

Both bone fractures

Reduction technique and choice of implant

The most important, and at the same time most difficult, part in fracture surgery is the atraumatic and yet anatomical reduction of the fracture. The exposure of the bone and the fracture must be limited to the area where the plate is going to be placed. While in the original teaching the plate had always to lie on the tension side of the bone, we now consider the fracture pattern and the choice of the approach to be rather more important than the tension side. However, the natural bow of the radius must always be respected by contouring the plate accordingly.

Transverse or short oblique fractures of the radius or ulna are mostly easy to reduce and should always be approached first. Reduction can usually be obtained by manipulating either fragment with fine-pointed reduction forceps. Another way is to loosely fix the plate (minimum 7 hole 3.5 mm limited contact–dynamic compression plate (LC–DCP)) with one screw to the main fragment and to subsequently reduce the opposite fragment to the plate. Attention must be given to any rotational malalignment. We therefore fix simpler fractures initially with a plate and two screws and then approach the other bone with the more complex fracture pattern. Once both bones have been stabilized, pronation and supination are controlled. As soon as there are one or more intermediate fragments, the indirect reduction technique as described by Mast *et al.* (1989) should be applied. Again a usually quite long 8–10 hole plate is fixed to one main fragment with one screw only (Figure 5.7.1). Close to the opposite end of the plate, a 3.5 mm cortical screw is introduced into the other main fragment. With the help of a small lamina spreader, which is placed between that screw head and the free end of the plate, distraction of the fracture can be obtained, thus allowing the fragments to fall into place or to be gently manipulated without stripping their soft tissue attachments. The plate can then be fixed to

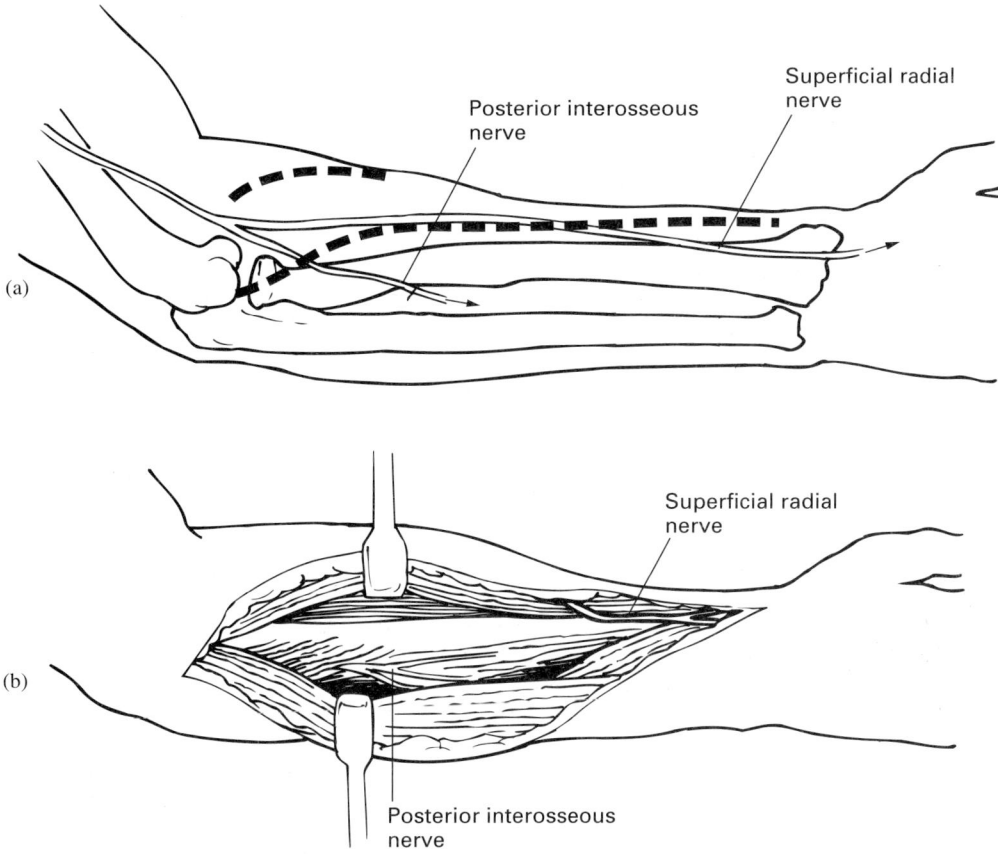

Figure 5.6   (a) Skin incision for dorsal or posterior approach to the middle and proximal third of the radius. The landmark is the mobile wad (M. brachioradialis and M. extensor carpi radialis longus and brevis). (b) The supinator muscle is exposed through the interval between the mobile wad and the extensor digitorum communis muscle. Within the supinator we can palpate the deep branch of the radial nerve, as it curves at right angles to the muscle fibres around the bone

the bone as a bridge plate not interfering with the comminuted area at all. If, on the other hand, some larger fragments can be fitted back anatomically, they should be fixed by small interfragmentary lag screws, while axial compression may be added by eccentric screw placement. Today the implant of choice for both forearm bones is the 3.5 mm LC-DCP or DCP (Figure 5.7.2) in combination with 3.5 mm cortical screws; 4.5 mm implants are only exceptionally applied in very heavy bones, while smaller and tubular plates do not provide sufficiently rigid fixation.

Autologous bone grafts are only advocated if there is a substantial bony defect or if the vitality of the fracture zone appears questionable (damage or exposure too extensive). Bone grafts should not, however, be placed close to the interosseous membrane.

Fractures of one forearm bone

Even more so than in both bone fractures, isolated injuries to either ulna or radius appear to have a poor prognosis if treated non-operatively. Any major displacement at the fracture site is almost always associated with some kind of injury and dislocation at either the proximal (Monteggia) or the distal (Galeazzi) radioulnar joint (Figure 5.8). These fracture dislocations, if not reduced early and anatomically, will always result in considerable and permanent functional impairment, especially of pronation and supination. On the other hand, if such complex injuries are approached immediately and the fractured ulna or radius is reconstructed anatomically, the dislocated radial head or distal ulna will usually reduce itself 'automatically' without any need for a local exposure

(a)

(b)

(c)

(d)

(e)

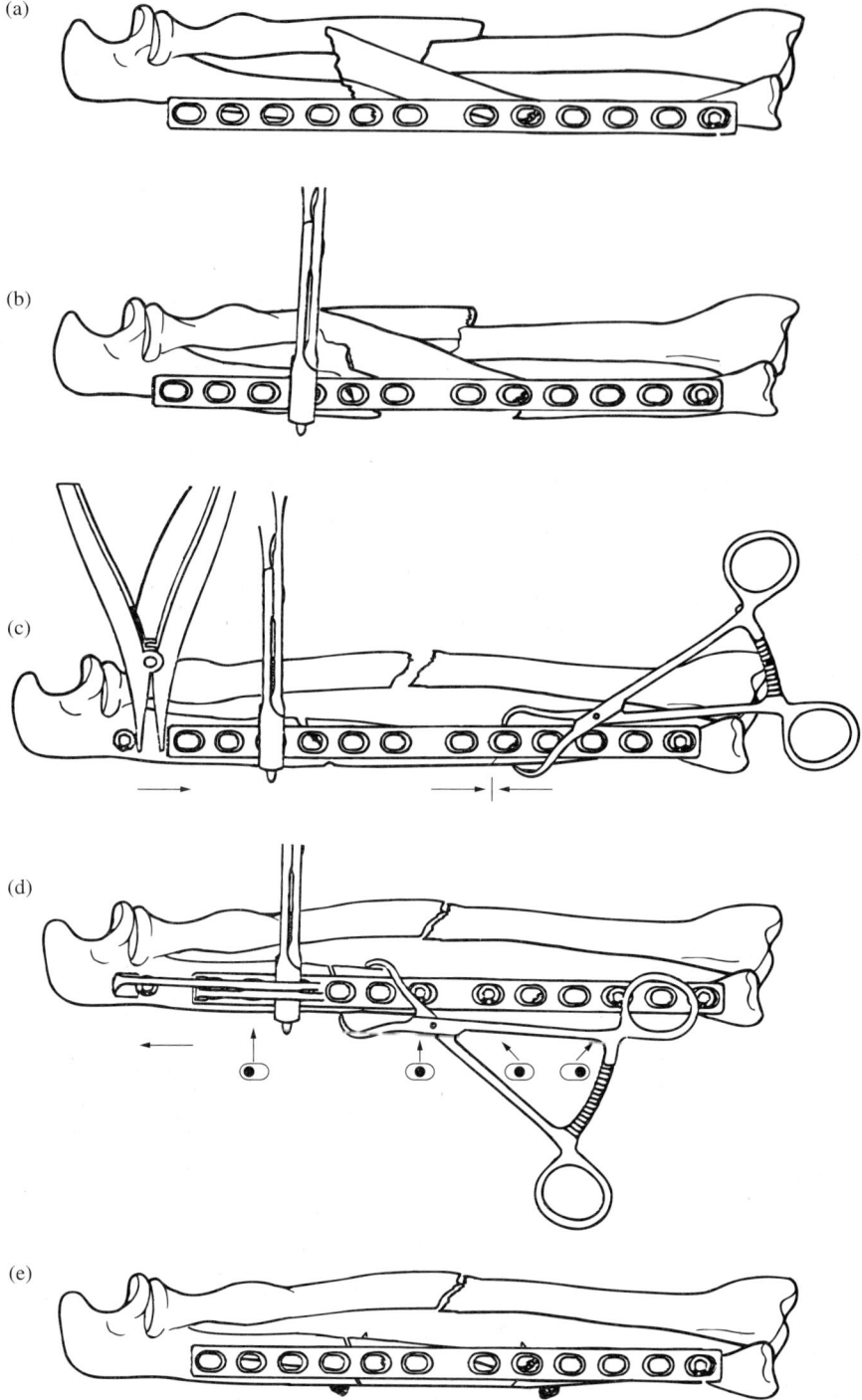

Figure 5.7.1   Indirect reduction technique of segmental ulnar fracture as described by Mast (1989). (a) A 10 hole 3.5 mm DCP or LC-DCP is loosely fixed to the distal main fragment with one screw. (b) A Verbrugge clamp grabs the proximal main fragment and the plate. (c) A plate-independent screw has been placed close to and in the same axis as the plate. With lamina spreader put between this screw head and the plate, end distraction is exerted, which allows the intermediate fragment to slip into place. Final alignment, without exposing more than the fracture gap, can be made with small pointed reduction clamps. (d) By reverting the distraction mechanism into compression and additional plate-independent lag screws a very stable fixation can be obtained, as shown in (e)

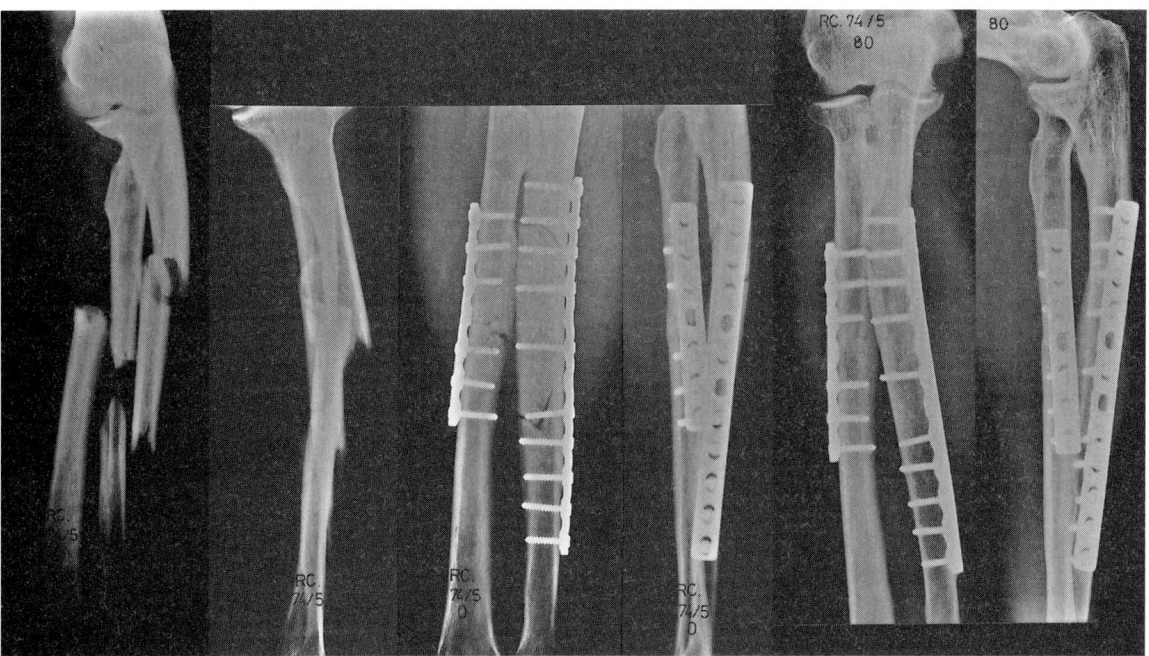

Figure 5.7.2 Clinical example of a complex, segmental fracture of the ulna and a simple transverse fracture of the radius. Both bones were fixed with 3.5 mm LC-DCP plates, giving uneventful, callus-free bony union and an excellent functional result

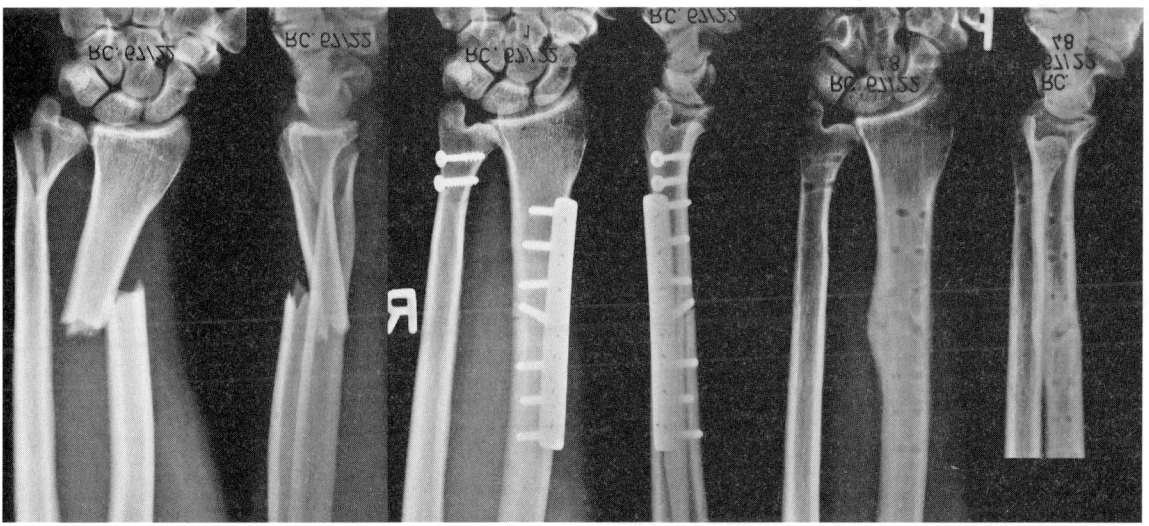

Figure 5.8 Galeazzi-type fracture of distal radius and fracture dislocation of the ulna. The radius was fixed with a 3.5 mm 7 hole DCP. With the anatomical reconstruction of the radius, the distal ulna is also reduced. In order to be able to treat the patient without a cast by early motion, two small tag screws were used to fix the ulna head process. The final outcome shows a good functional result

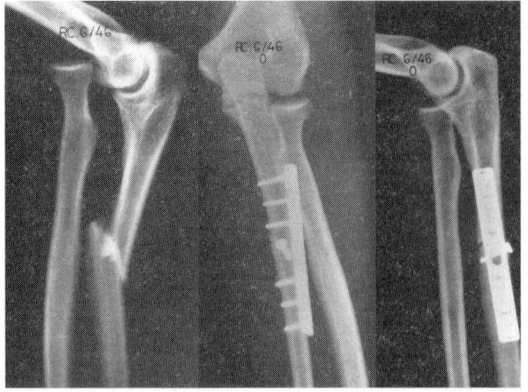

Figure 5.9   Monteggia fracture of the proximal ulna with dislocation of the radial head. After emergency ORIF of the ulna, the radial head usually reduces itself automatically without additional measures. The correct position and stability of the radial head must be checked under fluoroscopy

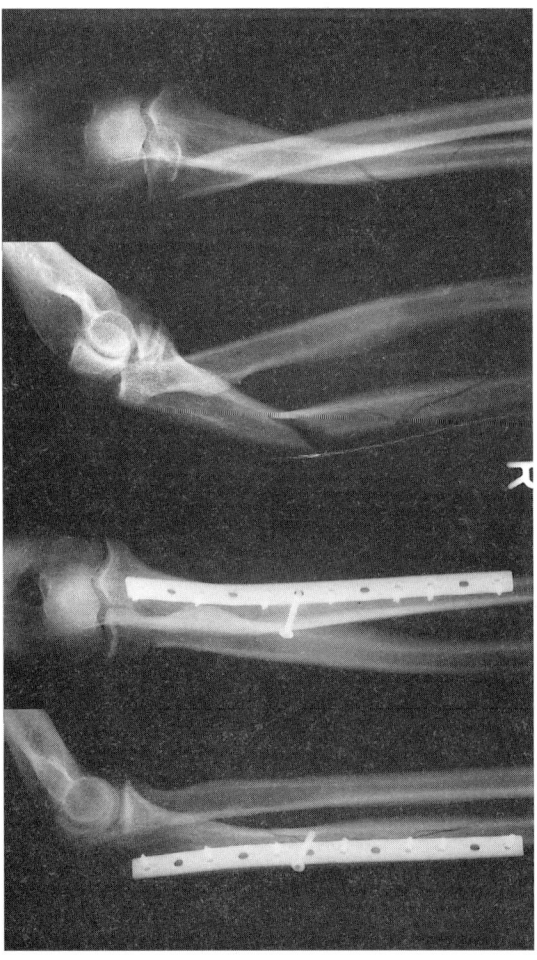

or ligament repair (Figure 5.9). The 'stability' of the reduced position must, however, be checked under fluoroscopy and under pronation and supination movements, while the patient is still anaesthetized.

If the 'associated' injury at the radial head or distal ulna appears to be irreducible, the fracture alignment (length and rotation) has to be re-assessed and eventually corrected, or a local exposure will be required. After open reduction of the radial head, the torn annular ligament must be sutured. At the distal radioulnar junction there is

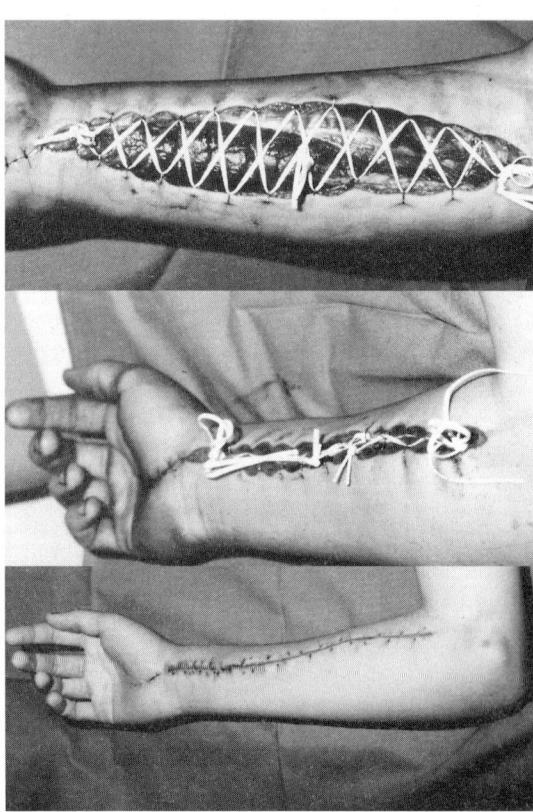

Figure 5.11   Forearm fracture with severe swelling or impending compartment syndrome. After ORIF the long volar incision is left open. The skin edges are held together loosely by fine rubber 'shoelaces'. Within 5 days the skin is slowly closed and finally a secondary suture is possible

◀ **Figure 5.10**   An unusual Monteggia-type injury in a polytrauma patient with a radial head fracture dislocation. The complex ulnar fracture was fixed with the new unicortical PC-Fix plate (PC = point contact). As the radial head and the fracture both slipped back into anatomical position, no further fixation of the radial head was attempted. The patient started immediately with pain-free pronation and supination

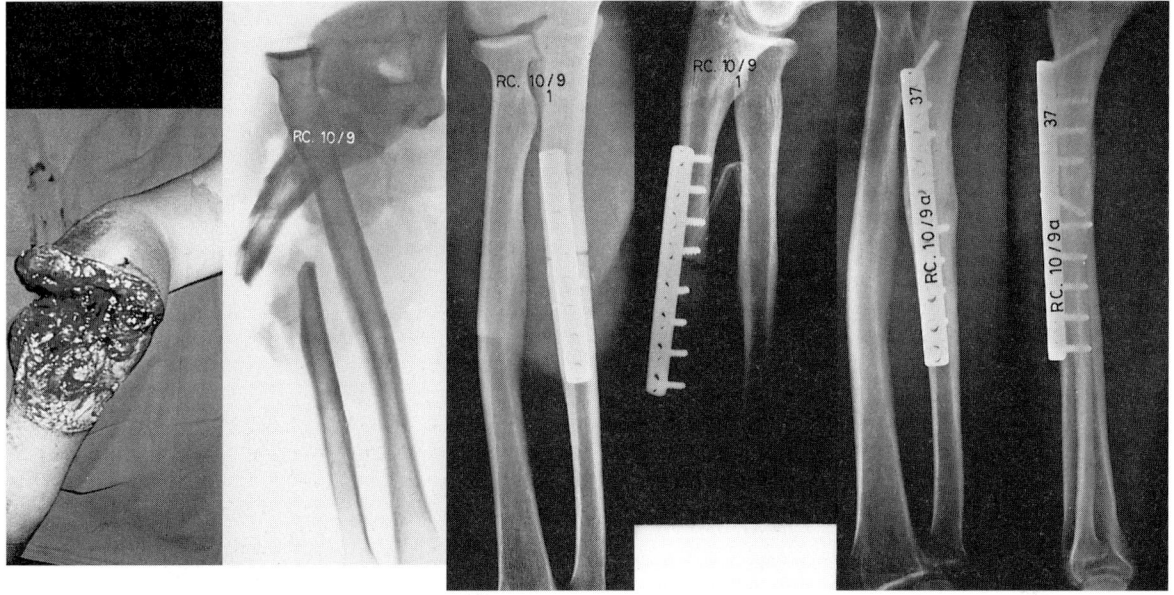

(a)

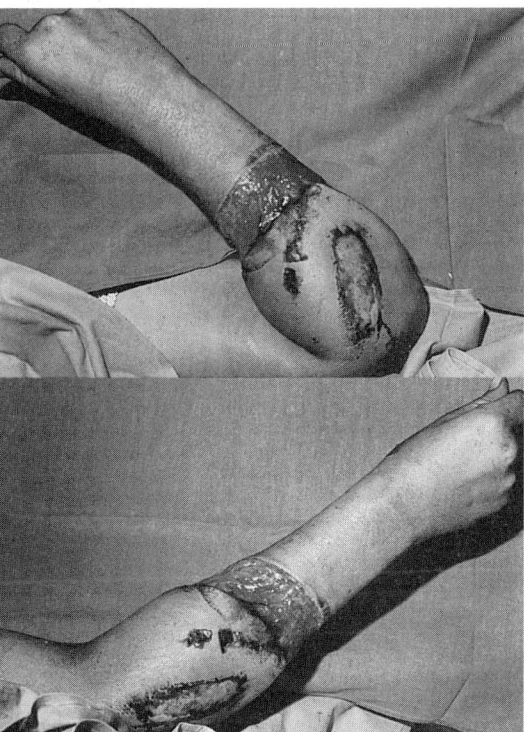

(b)

Figure 5.12   (a) Type IIIa open Monteggia fracture after a road traffic accident. Emergency plating of the ulna, debridement and washout were carried out, and the wound was left open. The patient started with immediate active movements of elbow, wrist and pronation and supination. A secondary split skin graft was done. After 20 weeks the 8 hole 3.5 mm DCP was exchanged for a 10 hole plate because of delayed healing. The patient made a full functional recovery and returned to his former occupation as a manual worker. (b) The same patient as in (a): extent of active elbow motion 10 days after the accident, and ORIF

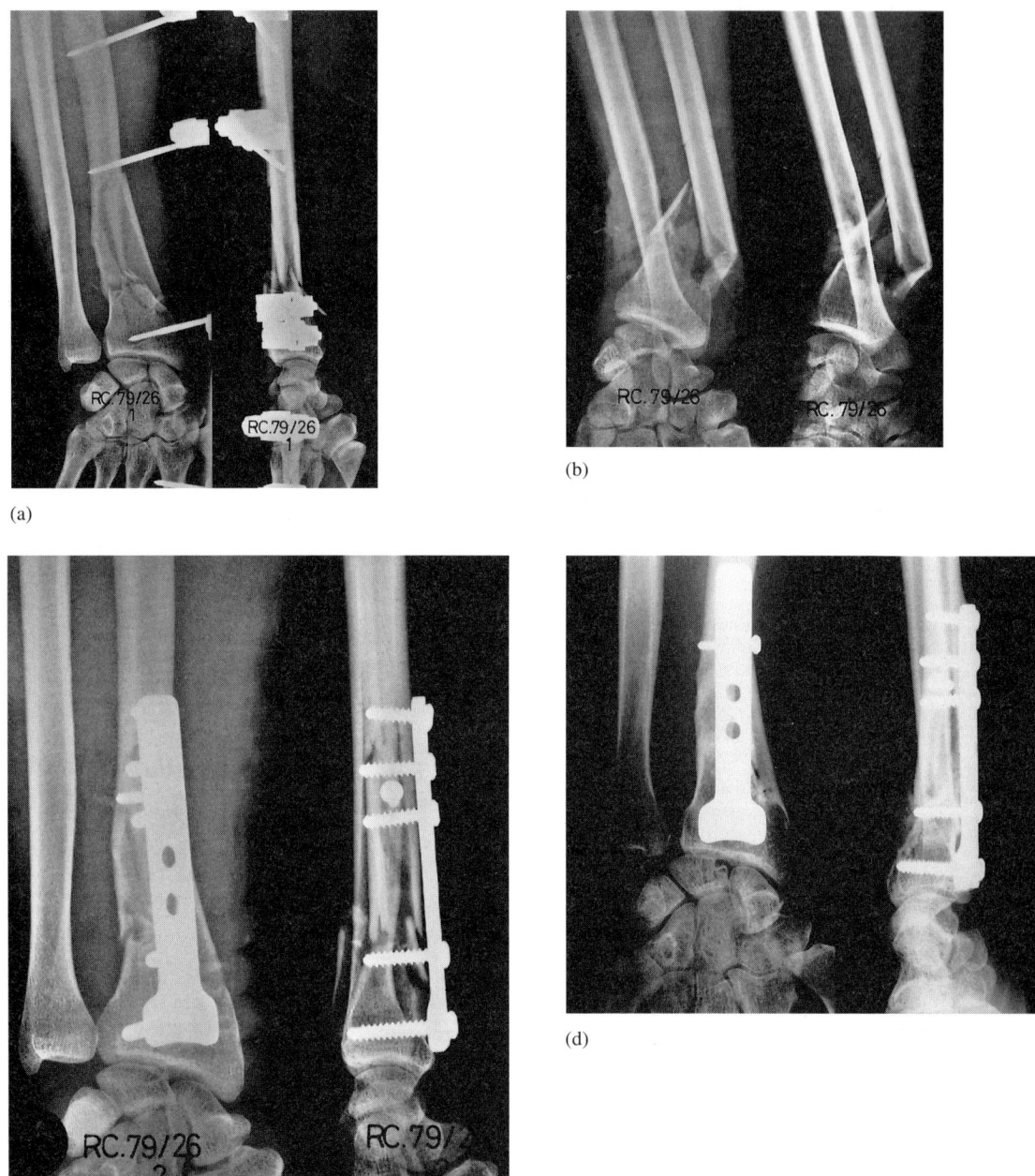

(a)

(b)

(c)

(d)

Figure 5.13 Second-degree open distal radius fractures with ulnar dislocation (Galeazzi type) in a 60-year-old woman. Immediate reduction and temporary joint bridging external fixation were performed. After 2 weeks, internal fixation with a small T plate and release of the median nerve in the carpal tunnel were done. A removable splint was used. Uneventful healing but severe algodystrophy occurred

usually not much ligamentous repair required. In cases of instability we may temporarily transfix the ulna to the radius with a Kirschner wire for 3–6 weeks and add a cast. In the case of a fracture dislocation of the radial head, the reduced radial head fracture may be quite stable (Figure 5.10). If not, an ORIF should be performed. Only exceptionally in severely comminuted fractures do we use a prosthetic replacement of the radial head.

Wound closure and aftercare

Skin closure, as well as all soft tissue handling, must be done with utmost care and atraumatic technique. Any tension on the skin edges is to be avoided. The bone and implant material should preferably be covered by soft tissue, especially over the ulnar crest. There is, however, no obligation to approximate the skin primarily over the radius, which is usually well covered by muscles. If there is severe swelling, moist dressings can be applied to the open wound while waiting for secondary skin closure. Rarely is there an indication for a primary split skin graft or for a more substantial free tissue transfer (Figure 5.11).

Postoperatively the arm is placed in an elevated position, and early active movement (as of the first postoperative day) is encouraged with the help of a physiotherapist or even better by the surgeon him- or herself. A protective splint is only applied if the patient appears unreliable and uncooperative or if the fixation is not judged stable enough. Normal daily activities may be resumed after 2–3 weeks, while any manual labour or sports involving the arms must be delayed until bony union has occurred clinically and radiologically, usually after 8–12 weeks.

Open forearm fractures

Open fractures of the radius and ulna are approached along the same principles as in other open long bone injuries (Moed *et al.*, 1986). While external fixation is widely used in the lower limb, we only exceptionally and at most temporarily apply a standard or small external fixator in diaphyseal fractures of the forearm (Knopp *et al.*, 1988). Plating is almost always possible, if not primarily (Figure 5.12) then as a secondary procedure after external fixation (Figure 5.13). Care must be taken to cover the plates with muscle,

while the skin may be left open for secondary closure or split skin grafting.

Implant removal

In the early years of ORIF, all metallic implants were removed routinely. Today we only remove plates that seem to interfere with normal activity, irritate or are prominent underneath delicate skin. Hidaka and Gustilo (1984) report on 7 refractures after removal of 32 plates from forearm bones, in spite of protection with a cast for 6 weeks. In our own as well as the German experience (Oestern and Tscherne, 1983), however, the incidence of refractures after implant removal does not exceed 1–2% and it is seen mostly after new injury.

Recent investigations have shown that the main reason for the poor reconstruction of the cortex observed beneath the plate is not the so-called mechanical stress protection but rather a biological one. Minimal stripping of soft tissue and periosteum during surgery and preservation of optimal blood supply to the bone are probably the most essential factors. The newly designed LC-DCP, which reduces the plate–bone contact area, appears to be another major contribution to the optimization of cortical blood supply and thereby the enhancement of sound fracture healing (Perren, 1991).

References

Anderson, L.D., Sish, T.D., Tooms, R.E. and Park, W.I. (1975). Compression plate fixation in acute diaphyseal fractures of radius and ulna. *J. Bone Joint Surg.*, **57A**, 287–97

Charnley, J. (1961). *Closed Treatment of Common Fractures*, 3rd edn. Edinburgh: Churchill Livingstone

Hadden, W.A., Reschauer, R. and Seggl, W. (1983) Results of AO-plate fixation of forearm shaft fractures in adults. *Injury*, **15**, 44–52

Henry, A.K. (1970) *Extensile Exposure*, 3rd edn. Edinburgh: E. & S. Livingstone

Hidaka, S. and Gustilo, R.B. (1984) Refracture of bones of the forearm after plate removal. *J. Bone Joint Surg.*, **66A**, 1241–2

Hughston, J.D. (1957) Fractures of the distal radial shaft, mistakes in management. *J. Bone Joint Surg.*, **39A**, 249–64

Jones, D.J., Henley, M.B., Schemitsch, E.H. and Trencer, A.F. (1995) A biomechanical comparison of two methods of fixation of fractures of the forearm. *J. Orthop. Trauma*, **9**, 198–206

Knopp, W., Neumann, K. and Muhr, G. (1988) Management des komplizierten Unterarmbruchs. Externe Fixation und frühzeitiger Verfahrenswechsel. *Unfallchirurg*, **91**, 539–44

Mast, J., Jakob, R. and Ganz, R. (1989) *Planning and Reduction Techniques in Fracture Surgery*. Berlin: Springer-Verlag

McLaughlin, H.L. (1965) Prevention and repair of non-union of fractures in the adult forearm. *Clin. Orthop.* 55–63

Moed, B.R., Kellam, J.F., Foster, R.J. *et al.* (1986) Immediate internal fixation of open fractures of the diaphysis of the forearm. *J. Bone Joint Surg.*, **68A**, 1008–17

Müller, M.E., Nazarian, S. and Koch, P. (1990) *The Comprehensive Classification of Fractures of Long Bones.* Berlin: Springer Verlag

Müller, M.E., Allgöwer, M., Schneider, R. and Willenegger, H. (1991) *Manual of Internal Fixation*, 3rd edn. Berlin: Springer Verlag

Oestern, H-J. and Tscherne, H. (1983) Ergebnisse der AO-Sammelstudie über Unterarmschaftfrakturen. *Unfallheilkunde*, **86**, 136–42

Perren, S.M. (1991) The concept of biological plating using the limited contact–dynamic compression plate (LC–DCP). Scientific background, design and application. *Injury*, **22**, Suppl 1, 1–41

Rüedi, T., von Hochstetter, A.H. and Schlumpf, R. (1984) *Surgical Approaches for Internal Fixation.* Heidelberg: Springer Verlag

Sarmiento, A., Cooper, J.S. and Sinclair, W.F. (1975) Forearm fractures. *J. Bone Joint Surg.*, **57A**, 297–304

Schneidermann, G., Meldrum, R.D., Bloebaum, R.D. *et al.* (1993) The interosseous membrane of the forearm: structure and its role in Galeazzi fractures. *J. Trauma*, **35**, 879–85

Thompson, J.E. (1918) Anatomical methods of approach in operations on the long bones of the extremities. *Ann. Surg.*, **68**, 309

Tile, M. and Petrie, D. (1969) Fractures of the radius and ulna. *J. Bone Joint Surg.*, **51B**, 193

Tscherne, H., Oestern, H.-J. and Sander, U. (1978) Technik und Ergebnisse der Plattenosteosynthese am Unterarmschaft. *Unfallheilkunde*, **81**, 332–43

6

Isolated fractures of the bones of the forearm

C.S. Mudgal and J.B. Jupiter

To ensure satisfactory function following fracture of the forearm diaphysis, it is necessary that the management be precise and meticulous. The recovery of function is dependent not only upon the maintenance of rotation of the forearm, but also on a functional range of motion at the wrist and elbow and the recovery of grip strength.

Isolated fractures of the radius

We have chosen to divide isolated fractures of the radius on the basis of the anatomical location of the fracture. This will include those fractures involving the proximal third, mid-diaphysis and distal third. Lesions of the distal third will be subdivided into the Galeazzi fracture as well as the complex albeit uncommon short oblique fracture of the distal radius occurring at the junction of the metaphysis and diaphysis.

Specific lesions

Galeazzi fracture

This uncommon injury involves a fracture of the shaft of the radius at the junction of the middle and distal thirds in association with dislocation at the distal radioulnar joint (Figure 6.1). A description of such a lesion was first made by Sir Astley Cooper in 1822 (Reckling and Peltier, 1965; Reckling and Cordell, 1968). However, following the publication of a series of 18 cases in 1934 by Galeazzi, it is his name that has become commonly associated with this injury (Galeazzi, 1934; Reckling and Peltier, 1965; Reckling and Cordell, 1968). Other names eponymously associated with

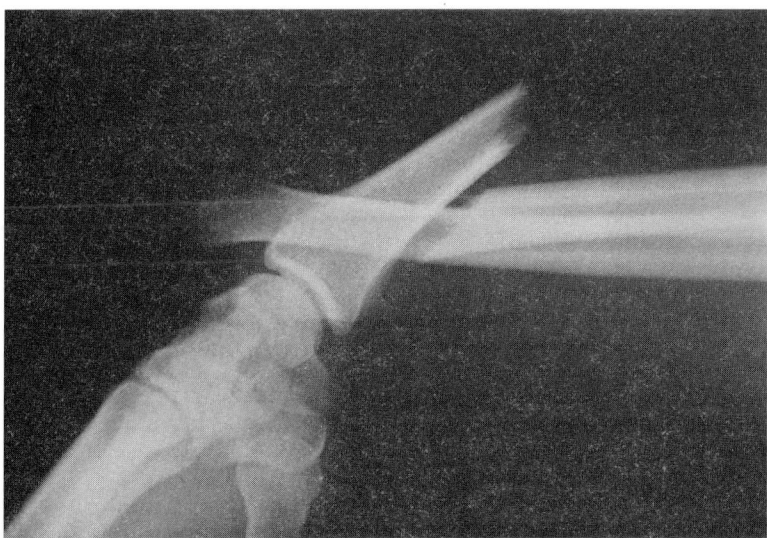

Figure 6.1 The Galeazzi fracture seen on lateral radiograph reveals a short, sharp fracture line of the distal third of the radius and dorsal dislocation of the distal radioulnar joint

this fracture include the Darrach–Hughston–Milch fracture (Reckling and Peltier, 1965), Piedmont fracture (Hughston, 1957) or reverse Monteggia fracture (Vince and Miller, 1987). Mikic (1975) has recommended that all fractures of the shaft of the radius associated with dislocation of the distal radioulnar joint be thought of as Galeazzi lesions. The reported incidence of the Galeazzi fracture varies from 3% to 7% of forearm fractures (Galeazzi, 1934; Wong, 1967; Sarmiento and Cooper, 1975; Moore *et al.*, 1985a). It is found more often in males (Dodge and Cady, 1972; Mikic, 1975; Moore *et al.*, 1985a) and is most common between the third to fifth decades of life (Mikic, 1975; Moore *et al.*, 1985a). An unusual variety that has been described includes a radial fracture at the junction of the middle and distal thirds, along with a fracture of the distal end of the ulna. This has been called the Galeazzi-equivalent lesion (Reckling, 1982).

Mechanics   The classic location of the Galeazzi fracture is at the junction of the middle and distal thirds of the radius situated between the insertions of the pronator teres and the pronator quadratus (Mikic, 1975; Reckling, 1982). Hughston (1957), in a study concentrating on problems identified in the management of this fracture, felt the most usual cause of injury to be a direct blow to the dorsoradial aspect of the forearm. Currently, most would favour a mechanism which includes axial loading of a hyperpronated forearm (Wong, 1967; Mikic, 1975). As displacement continues, the force may be transmitted via the interosseous membrane to the ulna causing dislocation of the ulnar head and tearing of the triangular fibrocartilage complex, with resultant loss of its stabilizing influence on the distal radioulnar joint (Mikic, 1975). Rotational stresses on the forearm would seem essential for dislocation of the distal radioulnar joint. It has been shown clinically and experimentally that incompetence of the triangular fibrocartilage complex is the first step in distal radioulnar joint dislocation and occurs at extremes of wrist extension and forearm pronation (Coleman, 1960; Reckling and Peltier, 1965; Reckling and Cordell, 1968). Most often the distal fragment is displaced ulnarwards (Mikic, 1975).

Hughston further suggested that there are four major deforming forces that can cause deformity and loss of reduction. These include:

1. The insertion of the pronator quadratus on the volar surface of the distal fragment rotates it ulnarwards and pulls it proximally and volarly.

2. The brachioradialis, which inserts onto the radial styloid, causes the distal fragment to pivot on the distal radioulnar joint, leading to its rotation and proximal displacement.

3. The abductors and extensors of the thumb cause shortening and relaxation of the radial collateral ligament, resulting in an inability to keep the radial soft tissues at length.

4. Gravity acting through the weight of the hand (even if in a cast) tends to cause subluxation of the distal radioulnar joint and dorsal angulation at the fracture.

In addition to these features, the tearing of the triangular fibrocartilage complex renders the entire distal radioulnar joint complex unstable (Valande, 1929; Veseley, 1967; Guistra *et al.*, 1974; Vince and Miller, 1987). The clinical relevance of these observations is important.

It has been shown that radial shortening of less than 5 mm does not require disruption of the distal radioulnar joint (McDougall and White, 1957; Morrey *et al.*, 1979). However, for shortening greater than 10 mm, there has to be injury to the interosseous membrane in addition to tearing of the triangular fibrocartilage complex (Moore *et al.*, 1985b).

Tearing of the triangular fibrocartilage complex is suggested radiologically by an avulsion of the ulnar styloid which is volarly displaced (Mikic, 1975; Bruckner *et al.*, 1992). Despite extensive knowledge of these mechanics, a Galeazzi fracture has not been produced experimentally (Moore *et al.*, 1985b).

Clinical features   The clinical presentation will vary with severity and mechanism of injury and degree of fracture displacement. Swelling and tenderness in the vicinity of the fracture, contusions or lacerations in cases of a direct blow to the forearm and shortening of the radius with dorsal angulation will at present with visible deformity. The ulnar head may or may not be dorsally prominent, which would suggest injury to the distal radioulnar joint.

Radiographic findings   These include:

1. A short oblique fracture of the radius with dorsal angulation as seen on the lateral radiograph (Figure 6.2)
2. Shortening of the radius in relationship to the distal ulna on the anteroposterior radiograph
3. Fracture of the ulnar styloid at its base

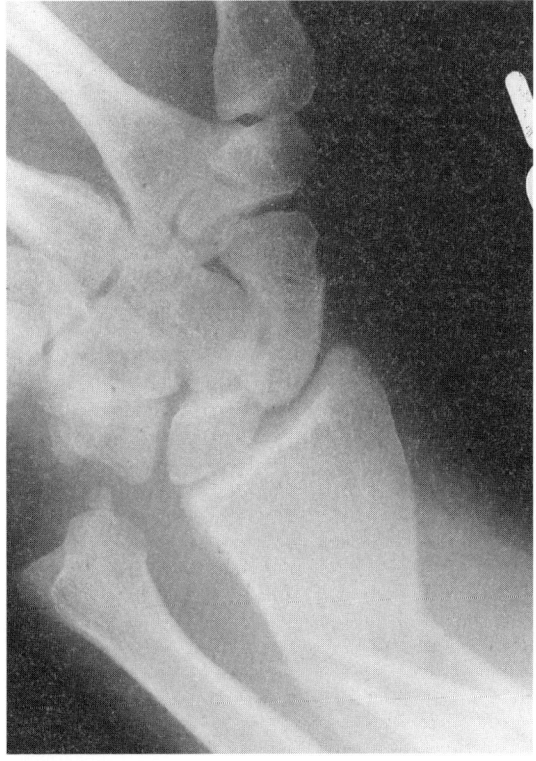

Figure 6.2 An oblique view of a Galeazzi fracture dislocation of the distal radius. Note the oblique fracture of the radius, disruption of the distal radioulnar joint, and ulnar styloid fracture

4. Widening of the distal radioulnar joint space on the anteroposterior radiograph
5. Dorsal displacement of the ulna relative to the radius, seen best on a true lateral radiograph.

The last three findings, if associated with radial shortening of more than 5 mm, are suggestive of traumatic disruption of the distal radioulnar joint (Moore *et al.*, 1985a,b).

Methods of treatment   Historically, numerous methods have been described for the management of this fracture, many with limited success and an unacceptable rate of complications. They include non-operative treatment (Hughston, 1957; Sarmiento *et al.*, 1975; Samiento and Latta, 1981; Shang *et al.*, 1987), intramedullary pin fixation (Smith and Sage, 1957; Sage, 1959; Marek, 1961; Mikic, 1975), Kirschner wires with onlay grafts (Hughston, 1957), interfragmentary screws, wire loops, Kuntscher nails (Hughston, 1957; Mikic, 1975) and silver pins (Schone, 1988), among other

methods. Conservative treatment has met with very poor results. Hughston (1957) analysed a series of 62 Galeazzi fractures. He identified that the vast majority of fractures resulted in an unfavourable outcome. Others have also noted similarly discouraging results (Wong, 1967; Reckling and Cordell, 1968). While reports from China regarding integration of traditional Chinese manipulative techniques with Western medicine are very encouraging regarding non-operative treatment of forearm fractures, with nearly 85% good to excellent results (Shang *et al.*, 1987), these reports may not include unstable Galeazzi fractures.

Intramedullary nailing of the radius has also met with limited success. The medullary canal at the level of the fracture is wide and flares distally. The curve of the radius is such that the concavity faces ulnarwards. Hence, any intramedullary fixation must maintain this bend in the radius (the radial bow). In addition, a round device is unsatisfactory; in view of the fact that the medullary canal of the radius is circular in cross-section, a device with a similar cross-section would offer very poor rotational control. Finally, the biggest disadvantage of all methods mentioned thus far is their inability to prevent shortening at the fracture site. Consequently, the potential for redislocation of the distal radioulnar joint is very high (Mikic, 1975).

Authors' preferred method of treatment   Our policy for the treatment of Galeazzi fractures has been open reduction and stable internal fixation using AO/ASIF compression plates (Müller *et al.*, 1991). Campbell has been quoted to have called this 'a fracture of necessity' (Stewart, 1957), by which he meant that operative treatment with stable internal fixation was necessary to achieve an acceptable outcome. We prefer to use the anterior approach of Henry (Henry, 1966; Hoppenfeld and de Boer, 1984; Crenshaw, 1991). Once the fracture has been anatomically reduced, stable fixation is achieved using a dynamic compression plate (DCP). All attempts are made to engage a minimum of six cortices on either side of the fracture. A longer plate is recommended in the presence of a comminuted fracture (Figure 6.3).

The plate is secured provisionally with two screws, and intraoperative radiographs are obtained in two planes to check accuracy of fracture reduction and, just as critically, the accuracy of reduction of the distal radioulnar joint. If the

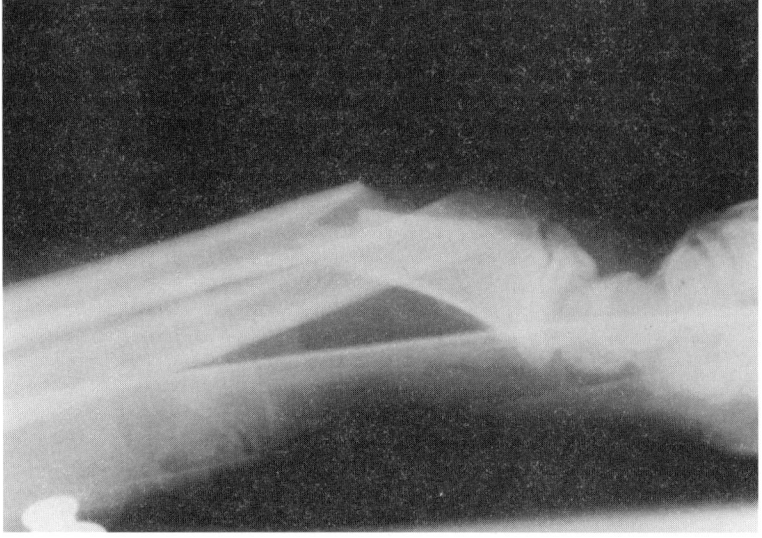

(a)

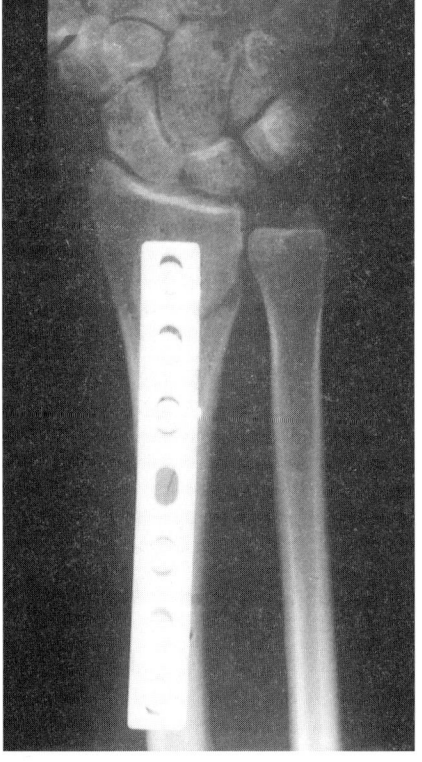

(b)

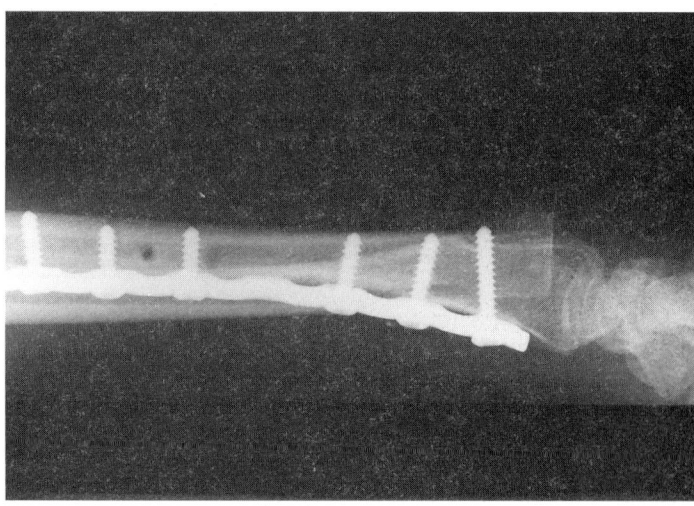

(c)

Figure 6.3   A Galeazzi fracture dislocation of the distal radius in a 28-year-old nurse. (a) The lateral radiograph reveals the fracture dislocation. (b, c) Anteroposterior and lateral radiographs following internal fixation with an LC–DCP plate. Excellent functional motion was the outcome

reduction of the distal radioulnar joint is satisfactory in both radiographic views, then stability of the distal radioulnar joint is tested clinically by passively rotating the forearm.

If the joint is stable through the entire range of rotation, no postoperative immobilization is necessary and early functional rehabilitation may be started.

Inability to reduce the distal radioulnar joint or its redislocation was originally thought to be an infrequent occurrence. However, in recent years there is a growing number of reports of such a clinical situation (Stewart, 1957; Mikic, 1975; Alexander and Lichtman, 1981; de Carvalho *et al.*, 1987; Jenkins *et al.*, 1987; Biyani and Bhan, 1989; Bruckner *et al.*, 1992). Bruckner *et al.* (1992) have designated such cases as 'complex' dislocations of the distal radioulnar joint. These include all cases in which the distal radioulnar joint is obviously irreducible or extremely unstable after reduction or the joint reduction has an ill-defined end-point.

If the distal radioulnar joint is reducible but unstable with forearm rotation, there exist several treatment alternatives. In those cases associated with a fracture of the ulnar styloid at its base, we recommend open reduction and internal fixation of the styloid fracture, using either Kirschner wires in conjunction with a tension band or a small screw (Mikic, 1975; Bruckner *et al.*, 1992) (Figure 6.4). Should there be no associated ulnar styloid fracture, the distal ulna may be transfixed to the radius using Kirschner wires with the forearm in

40° of supination. In this case, an above-elbow cast is recommended for 6 weeks. The Kirschner wires are removed after 6 weeks (Kellam and Jupiter, 1991).

Almost all irreducible distal radioulnar joint dislocations reported to date have been caused by tendon entrapment. The tendons implicated include the extensor carpi ulnaris (Stewart, 1957; Alexander and Lichtman, 1981), extensor digiti minimi, or both (Biyani and Bhan, 1989). In these cases, operative exposure of the distal radioulnar joint is recommended. Through a dorsal approach, the entrapped tendon is elevated, the joint reduced, and if possible the triangular fibrocartilage repaired (Figure 6.6*b*). The forearm should be immobilized in an above-elbow cast for 6 weeks in 40° of supination (Veseley, 1967; Dameron, 1972; Mikic, 1975; de Carvalho *et al.*, 1984; Jenkins *et al.*, 1987).

Complications Complications associated with Galeazzi fractures include not only delayed union, non-union, malunion and sepsis of the radius fracture, but also instability or dysfunction of the distal radioulnar joint (Anderson and Meyer, 1991). In fact, the latter may prove the most difficult of all complications.

Moore *et al.* (1985a) reviewed a series of 36 Galeazzi fractures and noted a complication rate of 39%. Included in these were seven injuries to the radial nerve, with six involving injury to the sensory branch and one being a posterior interosseous nerve palsy.

Non-union and malunion have been associated most commonly with plaster cast treatment or with inadequate or inappropriate internal fixation (Hughston, 1957; Grace and Eversmann, 1980; Stern and Drury, 1983; Moore *et al.*, 1985a, Schatzker and Tile, 1987). Most cases of persistent distal radioulnar joint subluxation are a result of failure to accurately assess the reduction or stability of the distal radioulnar joint, rather than a failure of adequate treatment of the radius fracture.

Malunion results most frequently from a combination of the following: (1) inability to restore the radial bow; and (2) alteration of location of the maximum radial bow. It has been shown that in patients with a malunion of the radius, there was a change in the magnitude of the radial bow and in its location (Matthews *et al.*, 1982; Tarr *et al.*, 1984a; Schemitsch and Richards, 1992). Schemitsch and Richards (1992) showed that this usually resulted in forearm rotation of less than 80% of normal with the difference being statisti-

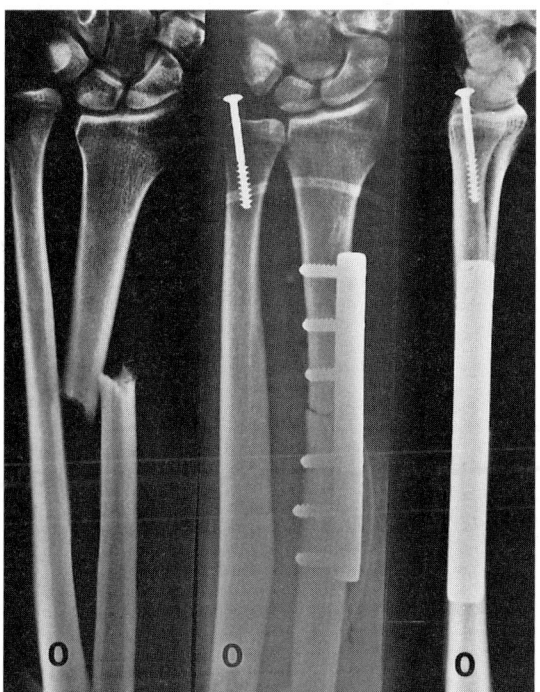

Figure 6.4 Following plate fixation of the radial shaft, an unstable distal radioulnar joint was treated by screw fixation of the ulnar styloid

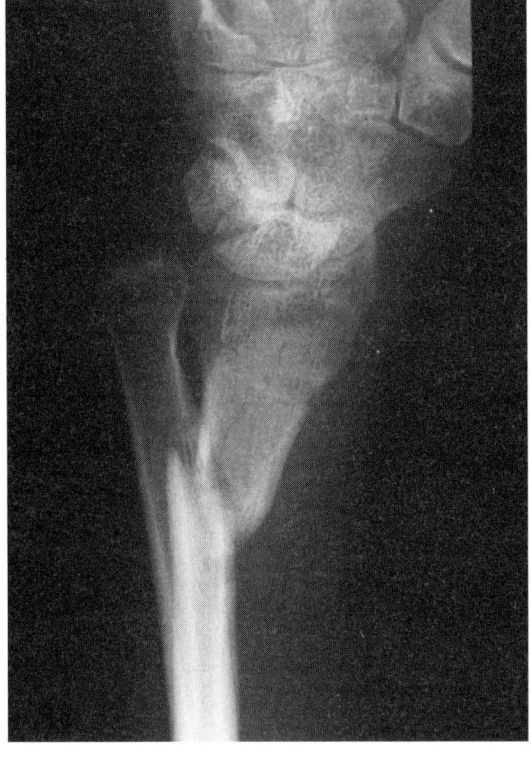

(a)

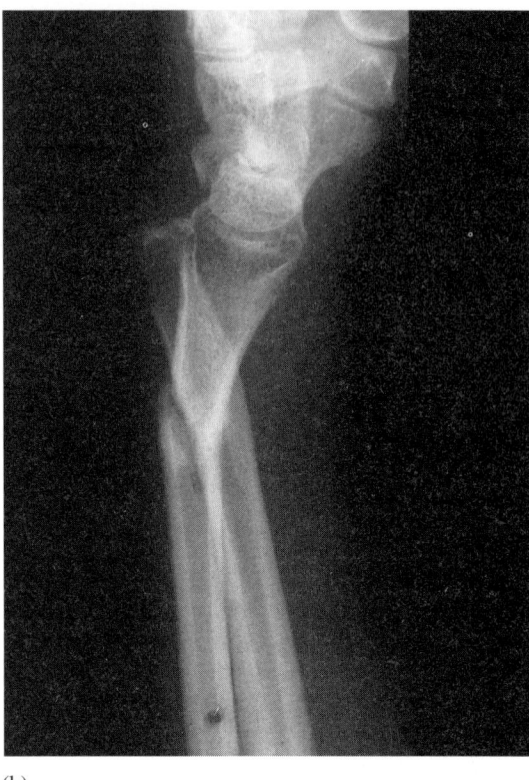

(b)

Figure 6.5   (a) Anteroposterior and (b) lateral radiographs of a malunited Galeazzi fracture in a 26-year-old woman. The patient was treated with an external fixation device. Marked limitation of forearm rotation resulted

cally significant. In their study, malunion also resulted in a statistically significant deterioration of function, grip strength and pinch strength. These authors showed that the relationship between the radial bow and both rotation of the forearm and grip strength was not linear. This suggested that with mild deformity a good functional result could be expected. However, once a certain point of deformity was reached, there was a rapid decline in the outcome (Figure 6.5).

Other complications described include cross-union and refracture after removal of plates (Hidaka and Gustilo, 1984; De Luca *et al.*, 1988; Langkamer and Ackroyd, 1991; Rosson *et al.*, 1991; Rosson and Shearer, 1991; Schemitsch and Richards, 1992).

Fractures of the distal third

These fractures of the radius are found in the 'watershed' area at the junction of the distal metaphysis with the diaphysis. These will prove difficult as they may not initially appear complex, and may

be confused with a Colles' type fracture. These fractures are most common in elderly osteoporotic patients and can also be seen after a direct blow to the distal metaphysis or in patients involved in high-velocity motor vehicle accidents. They do not tend to follow the age distribution and mechanism of injury of the Galeazzi fracture (Mikic, 1975).

The most difficult fractures to treat are the short oblique fractures. These prove to be unstable, not amenable to non-operative treatment, intramedullary or extramedullary fixation, and offer a limited area distal to the fracture to which a plate may be secured. Frequently, disruption of the distal radioulnar joint will coexist (Schatzker and Tile, 1987) (Figure 6.6).

We recommend open reduction and internal fixation with either a 3.5 mm DCP or a buttress T plate with a long stem placed on the anterior surface. Whenever possible, an interfragmentary screw should be placed across the oblique fracture line. The management of the associated disruption of the distal radioulnar joint is similar to that described in the section on Galeazzi fractures.

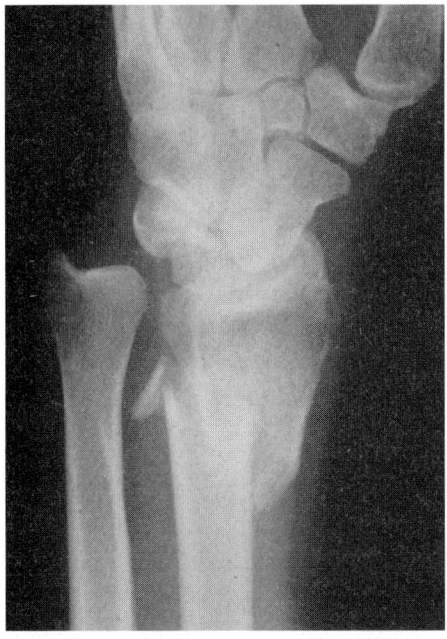

(a)

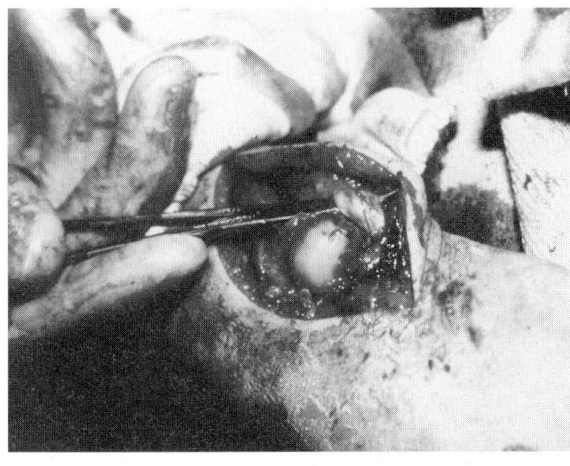

(b)

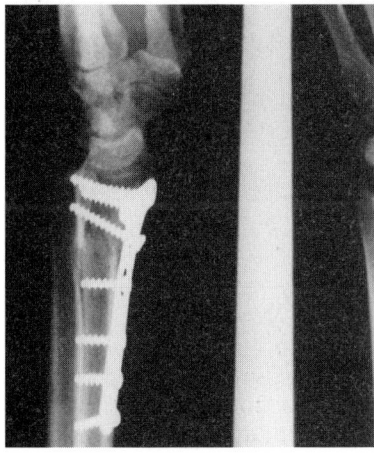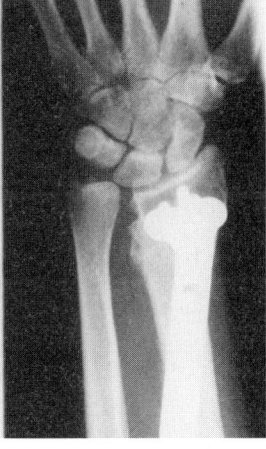

(c)

Figure 6.6   A short oblique fracture of the distal third of the radius at the junction of the diaphysis and metaphysis in a 42-year-old woman. (a) Anteroposterior radiograph 3 weeks after injury reveals the persistent displacement and disruption of the distal radioulnar joint. (b) The extensor carpi ulnaris was found displaced into the distal radioulnar joint, blocking reduction of the radius. (c) Anteroposterior and lateral radiographs following open reduction, fracture realignment and stable plate fixation

Fractures of the proximal radius

This is a type of injury which has been considered by some authors to be not unlike its counterpart, the Galeazzi lesion (Mikic, 1975). Unlike the Galeazzi fracture, however, these injuries are less frequently associated with disruption of the distal radioulnar joint. It must be stressed that the forearm is a two-bone structure, and a displaced diaphyseal fracture in any one bone is usually associated with injury to one or the other radioulnar joint. If these fractures are associated with disruption of the distal radioulnar joint, the management is the same as for a Galeazzi fracture (Figure 6.7).

Most radial fractures in the proximal portion are usually associated with an ulnar shaft fracture at the same level. The management of these injuries has been covered elsewhere in the text (see Chapter 5). It would suffice to say that the treatment recommended is open reduction and internal fixation of both bones using compression plates (Müller *et al.*, 1991).

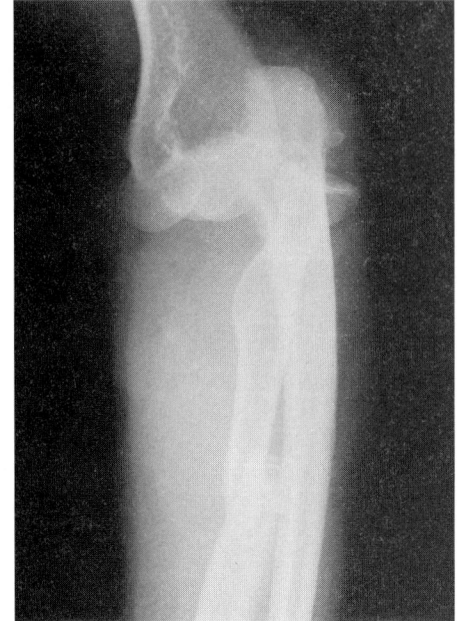

(a)

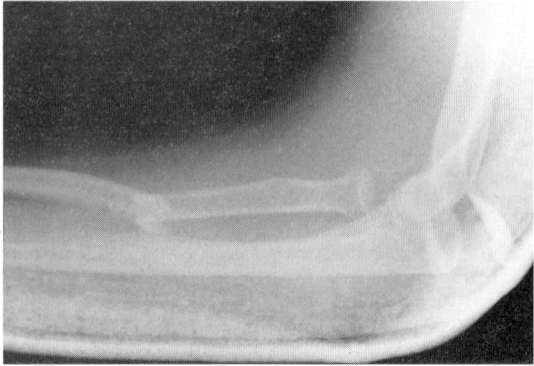

(b)

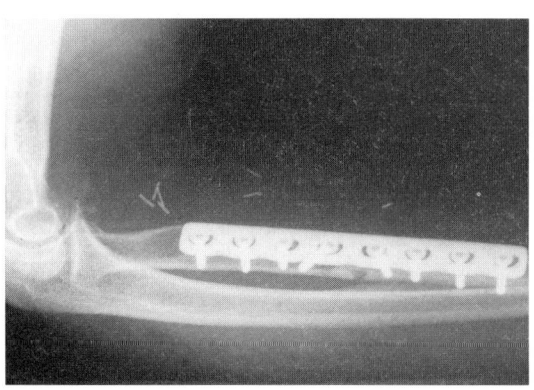

(c)

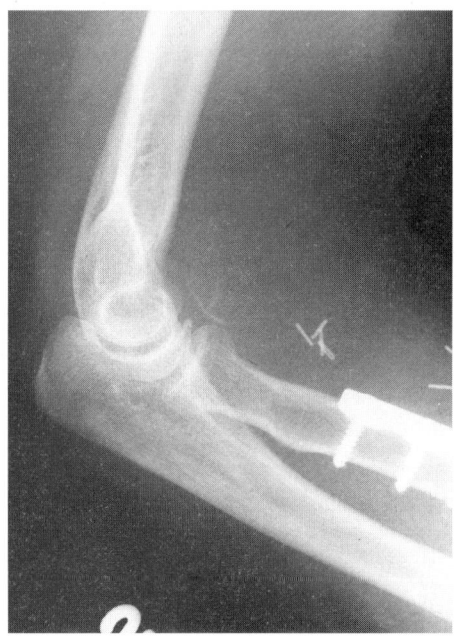

(d)

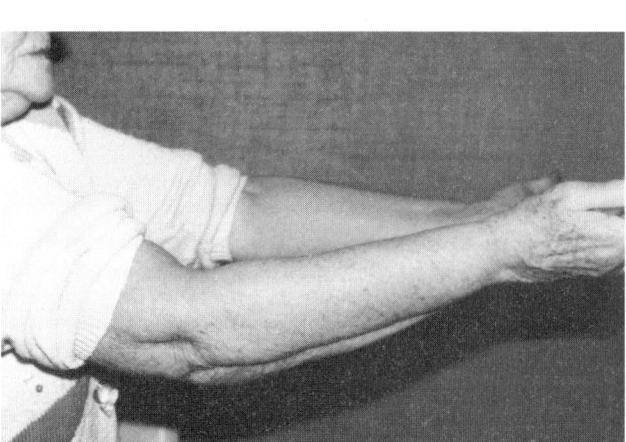

(e)

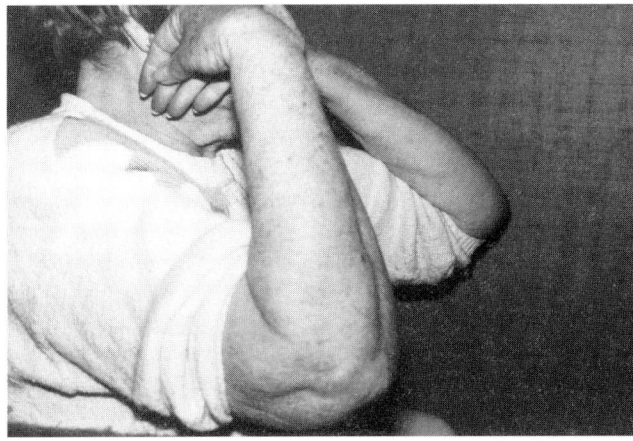

(f)

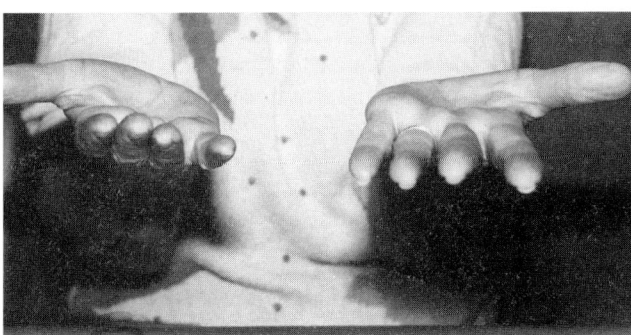

(g)

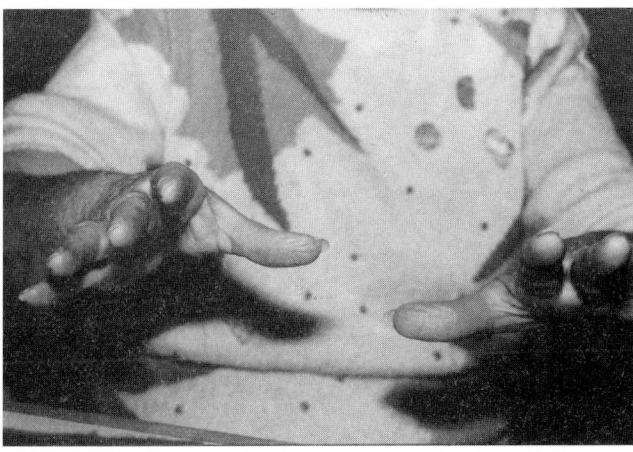

(h)

Figure 6.7 A proximal third radius fracture
associated with an elbow dislocation in a 63-year-old
woman. (a, b) Anteroposterior and lateral radiographs
of the complex fracture dislocation. (c, d) Through an
anterior approach, the radius fracture was stabilized
with an LC–DCP. The lateral elbow soft tissue
complex was repaired to the distal humerus. (e–h) At
4 months post-injury, the patient had regained a
functional arc of forearm and elbow mobility

Closed methods of treatment in proximal fractures of the adult radius are prone to failure. The reasons for this include the muscular envelope around the radius and the muscular forces acting on the radius. The muscular envelope may pre-clude satisfactory moulding of a cast or functional brace and causes a loss of reduction. Muscle forces acting on the radius include the pronator teres, the supinator and the biceps. Depending on the location of the radial fracture, each of these muscles is capable of producing a rotary movement on the fracture fragments, causing a loss of reduction or malrotation, or both.

We prefer, as do others (Chapman *et al.*, 1989), the anterior approach to most proximal radial fractures. The individual choice of surgical approach, however, depends on the surgeon's familiarity with the regional anatomy and experience. As far as possible, all fractures are stably fixed with a DCP.

The Essex-Lopresti lesion

Fractures of the radial head are the most common type of fracture around the elbow. On occasion, radial head fractures are associated with injury to the interosseous membrane and disruption of the distal radioulnar joint. These injuries can be difficult to diagnose accurately and treat effectively. A combination of radial head fracture with acute distal radioulnar joint disruption was first described by Curr and Coe in 1946. The classic description of this injury, as we understand it today, was given by Peter Essex-Lopresti to the British Orthopaedic Association in 1950 (Essex-Lopresti, 1951). Since

then, this injury has been eponymously associated with his name (Figure 6.8).

It must be remembered that the diagnosis of distal radioulnar joint disruption at the time of injury is the *sine qua non* of the Essex-Lopresti lesion. In fact, in a number of reports in the literature which deal with the results of the radial head excision for fracture, a high incidence of distal radioulnar joint pain and proximal radial migration has been reported (Brockman, 1930; Essex-Lopresti, 1951; McDougall and White, 1957; Mikic and Vukadinovic, 1983; Goldberg *et al.*, 1986).

Failure to diagnose the distal injury may be further compounded by an ill-advised radial head excision, which would lead to chronic wrist symptoms and an extremely difficult (to treat) clinical situation. Essex-Lopresti (1951) described two patients. The first had a comminuted fracture treated by radial head excision. Limitation of wrist motion and forearm rotation resulted due to proximal migration of the radius. The second case was successfully managed by open reduction and internal fixation of the radial head.

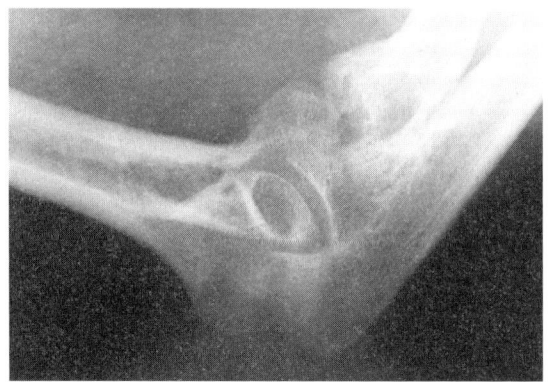

(a)

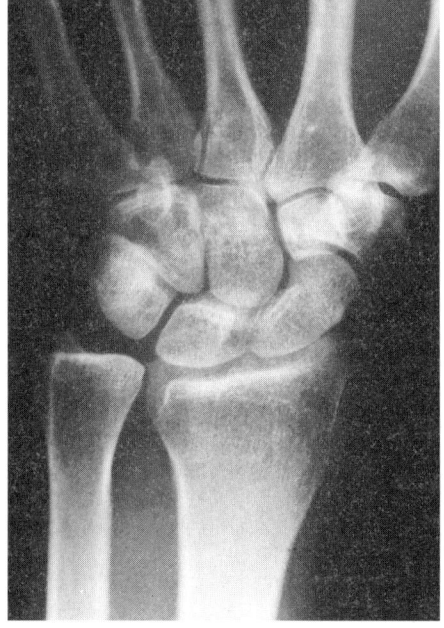

(b)

Figure 6.8 A 22-year-old woman fell from a height, landing on her outstretched hand with her elbow semi-flexed. She presented 3 months after injury, complaining of wrist pain and limited forearm rotation. (a) The oblique radiograph of the elbow demonstrated a distorted and collapsed radial head. (b) Anteroposterior radiographs of the wrist revealed proximal migration of the radius and disruption of the distal radioulnar joint

Mechanism of injury Most of these injuries are caused by a fall onto the outstretched hand with the forearm pronated. In this position, it is possible to disrupt the distal radioulnar joint as well as the interosseous membrane. In addition, the degree of radiocapitellar contact is high in pronation (Morrey *et al.*, 1988). The axial force is therefore transmitted to the radial head which abuts against the capitellum leading to fracture of the radial head (Essex-Lopresti, 1951; De Lee, 1984; Hotchkiss and Green, 1991). With sufficient force, the fracture fragments are displaced, and due to the concomitant disruption of the distal radioulnar joint and interosseous membrane, the radius will have a tendency to migrate in a proximal direction. This observation has been verified experimentally (Hotchkiss *et al.*, 1989).

Associated injuries include articular fractures of the capitellum, fracture of the radial shaft, dislocation of the elbow and fracture of the ulnar shaft (Morrey *et al.*, 1981; Edwards and Jupiter, 1988; Khurana *et al.*, 1988; Bock *et al.*, 1992). Khurana *et al.* (1988) have reported the only documented case wherein disruption of the interosseous membrane was visualized intraoperatively.

Clinical presentation Symptoms and signs at the elbow are similar to those of radial head fractures. A careful examination of the forearm and especially the wrist is mandatory for all radial head fractures. Any evidence of instability of the ulnar head or asymmetric prominence of the ulnar head should be suggestive of acute disruption of the distal radioulnar joint. In addition, alteration of the normal relationship between the radial and ulnar styloid processes (as compared to the normal side), restriction of ulnar deviation at the wrist or swelling of the forearm are corroborative of distal radioulnar joint injury and suggestive of an unstable forearm with the potential for proximal radial migration (Edwards and Jupiter, 1988; Hotchkiss and Green, 1991).

Radiographic examination It is imperative that all cases of comminuted radial head fractures should have the wrist examined radiographically. Recommended views include a posteroanterior view and a lateral view centred on the wrist. A posteroanterior projection and a zero-rotation lateral projection as described by Epner *et al.* (1982) are extremely helpful in accurate determination of any proximal radial migration. Corroborative evidence may be gained by similar views of the contralateral wrist, which will provide information regarding the patient's normal ulnar variance. The upper limit of proximal migration of the radius during normal rotation of the forearm has been shown by Morrey *et al.* (1979) to be 2 mm.

Classification McDougall and White (1957) were among the earliest authors to put forth a classification of distal radioulnar joint displacement associated with damage to the radial head. However, they did not restrict themselves to acute distal radioulnar joint injuries, but included the entire spectrum of radial head injuries associated with distal radioulnar joint disruptions, as well as delayed proximal migration of the radius after excision of the radial head.

Edwards and Jupiter (1988) have proposed a working classification of the Essex-Lopresti lesion, wherein all groups involved have distal radioulnar joint displacement *at the time of injury*. Different types of displaced radial head fractures were recognized, based on the degree of comminution and size of fracture fragments. This classification was developed to help determine an optimal method of treatment:

Type I: radial head fractures with large displaced fragments amenable to internal fixation (comminution absent or minimal).
Type II: comminuted radial head fractures, not amenable to internal fixation, which require excision of the radial head.
Type III: chronic cases with old proximal migration of the radius, which is irreducible.

In Edwards and Jupiter's (1988) review of seven patients, there were three type I, two type II and two type III fractures.

Method of treatment The goals of treatment include restoration of radial length and stabilization of the distal radioulnar joint. This will depend to a large degree on the type of radial head fracture.

Type I fractures These are fractures with large fragments which are frequently amenable to internal fixation (Shmueli and Herold, 1981; Heim and Pfeiffer, 1982; Bunker and Newman, 1985). Stable internal fixation will assure the restoration of normal anatomy of the radiocapitellar joint, as well as radial length (Heim and Pfeiffer, 1982). Most studies which recommend internal fixation of radial head fractures have shown a high percentage of good to excellent results (Shmueli and Herold, 1981; Heim and Pfeiffer, 1982; Bunker and Newman, 1985; Edwards and Jupiter, 1988; Geel and Palmer, 1992).

The choice of implant to be used depends on its availability and the experience of the surgeon. Use of the small AO screws has been recommended (Shmueli and Herold, 1981; Heim and Pfeiffer, 1982). The problem with these screws tends to be the occasional inability to countersink the screw head completely. The screw may therefore act as a mechanical block to pronation and supination. The Herbert screw has been reported to be quite effective due to its ability to be buried beneath the chondral surface. Bunker and Newman (1985) reported excellent results after using it to fix non-comminuted as well as comminuted fractures. The advantages of the Herbert screw include the lack of a head, which allows it to be buried completely, and its differential pitch, which allows some compression across the fracture (Herbert and Fisher, 1984).

Stability of the distal radioulnar joint must be tested. If stable, then external immobilization in supination should be extended for 4 weeks. A Munster splint for the forearm is recommended, which will permit flexion and extension of the elbow. If the ulnar head is not stable after reduction, it may be pinned to the radius with the forearm in 40° of supination for a similar period of 4 weeks, after which the pin is removed and forearm rotation is commenced.

Type II fractures A radial head which is fractured into more than two main fragments may prove difficult to fix securely with internal fixation. Excision may prove to be the only available treatment. Considerable debate exists regarding the management of the unstable radius following radial head excision. Some authors recommend insertion of a silicone prosthesis to prevent proximal migration at a later date (Levin, 1973; Mackay *et al.*, 1979; Swanson *et al.*, 1981). Others have noted a high incidence of failure of the prosthesis and do not recommend its use (Morrey *et al.*, 1981), while a few others believe radial head excision to be an entirely adequate procedure (Morrey *et al.*, 1979; Broberg and Morrey, 1986; Goldberg *et al.*, 1986).

To help reduce compressive loading on the silicone prosthesis and maintain radial length in the early postoperative period, the radius and ulna are temporarily transfixed with pins after making certain that the distal radioulnar joint is aligned. The pins are maintained for 4 weeks. Further management is similar to that for type I fractures.

The silicone radial head prosthesis is very flexible compared to bone, and its ability to transfer dynamic forces across the radiocapitellar joint has been questioned (Carn *et al.*, 1986). Several studies have shown that a silicone implant is incapable of adequate load transfer across the elbow, which results in rarefaction of the capitellum, a feature reported by other authors as well (Mackay *et al.*, 1979; Carn *et al.*, 1986). In addition, under dynamic loading conditions, this implant does not completely prevent proximal radial migration (Mackay *et al.*, 1979; Morrey *et al.*, 1981; Swanson *et al.*, 1981; Carn *et al.*, 1986). It would seem therefore that a radial head prosthesis is needed which has a modulus of elasticity similar to that of bone (Carn *et al.*, 1986). However, in the acute situation, despite all its drawbacks, the silicone implant is the best implant available and most authors concur on its use (Mackay *et al.*, 1979; Harrington and Tountas, 1981; Morrey *et al.*, 1981; Swanson *et al.*, 1981; Edwards and Jupiter, 1988; Carn *et al.*, 1986). Its role in chronic injury is unpredictable and does not meet with universal acceptance.

Complications Complications associated with the Essex-Lopresti lesion can be divided into those resulting from the severe radial head fracture alone and those resulting from its excision and instability of the forearm articulation (Jupiter, 1991). Among the most troublesome complications are proximal radial migration, restriction of forearm rotation and wrist pain associated with considerable loss of grip strength (Curr and Coe, 1946; Essex-Lopresti, 1951; McDougall and White, 1957; Taylor and O'Connor, 1964; Swanson *et al.*, 1981; Edwards and Jupiter, 1988). In addition, excision of the radial head may be associated with valgus instability of the elbow (Hotchkiss and Weiland, 1986) and heterotopic bone formation (Thompson and Garcia, 1967). Failure to achieve accurate reduction and fixation of the radial head or an untreated fracture of the radial head can lead to both post-traumatic arthritis and stiffness of the elbow (Mikic and Vukadinovic, 1983; Morrey, 1985).

Isolated fractures of the ulna

Fractures of the ulnar shaft can be divided into two distinct groups:

1. Those in which the ulna is fractured, but the radial head is *in situ* ('nightstick' fracture).

2. Ulnar fractures associated with dislocation and/or fracture of the radial head (Monteggia fracture).

Nightstick fractures

These fractures are the result of a direct blow on the ulna, as can happen when defending against an attacker (wielding a hard object) by raising one's forearm. The resulting fracture can be placed at any site along the ulnar length, but is more commonly seen in the distal half of the bone (Pollock *et al.*, 1983; Zych *et al.*, 1987). These fractures are often displaced, and the usual method to quantify displacement of the fracture has been in fractions of the width of the bone. Using this method, these fractures are usually divided into two groups: those displaced less than 50% of the width of the shaft and those displaced more than 50% (Pollock *et al.*, 1983; Dymond, 1984; Zych *et al.*, 1987) (Figure 6.9).

Fractures that are displaced less than 50% should be considered stable and effectively treated by non-operative means. Immobilization recom-

mendations have varied from an above-elbow plaster cast (Pollock *et al.*, 1983; Dymond, 1984) or fracture bracing (Sarmiento *et al.*, 1976; Sarmiento and Latta, 1981; Zych *et al.*, 1987) to minimal or no immobilization (Pollock *et al.*, 1983).

Stable fractures will have a high rate of union with predictable functional results. Fractures of the proximal half are more likely to be displaced due to the bending strain imposed on the ulna by the movement of muscles which form motors for the extrinsics to the hand (Lanyon *et al.*, 1975). In addition, these proximal fractures tend to be more difficult to hold in a reduced position with external immobilization (Anderson and Meyer, 1991). Hence, proximal fractures which are treated non-operatively need to be followed up carefully with frequent radiographs. A loss of position will necessitate operative intervention.

Reports of isolated stable ulnar fractures treated by external immobilization in a cast or functional brace have suggested union to occur between 6 and 12 weeks post-injury (Sarmiento *et al.*, 1976; Brakenbury *et al.*, 1981; Pollock *et al.*, 1983; Dymond, 1984; Zych *et al.*, 1987). Failure to unite has been reported to occur in 0.6% to 14% of cases (Grace and Witner, 1980; Corea *et al.*, 1981; Holst-Nielsen and Jensen, 1984; Tarr *et al.*, 1984b). Pollock *et al.* (1983) in a study emphasizing little or no immobilization, identified a non-union rate of 5%.

Fractures with displacement greater than 50% of the diameter of the ulna and/or with angulation of greater than 10° in any plane should be considered unstable.

Based on a clinical and cadaveric study, Dymond (1984) suggested that these fractures are unstable due to the disruption of the interosseous membrane and ulnar periosteum, which otherwise act as stabilizers of the fracture fragments. The instability of these fractures supports the role of operative treatment. Fracture stability can be predictably achieved with either a 2.7 mm or 3.5 mm DCP.

Complications associated with nightstick fractures include non-union, delayed union, malunion and synostosis (Brakenbury *et al.*, 1981; Matthews *et al.*, 1982; Pollock *et al.*, 1983; Dymond, 1984; Zych *et al.*, 1987).

The Monteggia lesion

In 1814, Giovanni Battista Monteggia described an injury of the forearm in which the proximal ulna

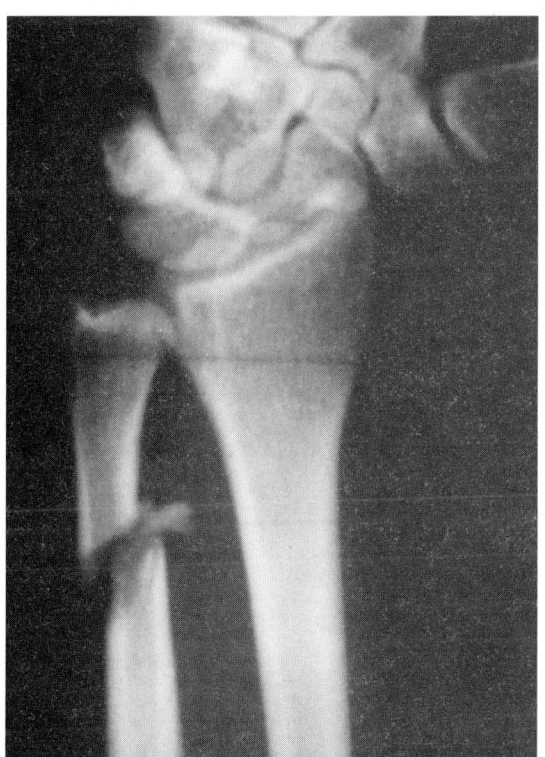

Figure 6.9 A distal third fracture of the ulna due to a 'nightstick' injury

fractured and the radial head was dislocated. Since then, this injury has been eponymously associated with his name. Subsequently, Bado (1967) coined the term 'Monteggia lesion', to encompass a number of traumatic lesions having in common disruption of the radiohumeroulnar joint in conjunction with a diaphyseal fracture of the ulna. The Monteggia lesion is an uncommon injury with incidences quoted in literature varying from 1% to 7% of all elbow injuries (Edwards, 1952; Caden, 1961; Burwell and Charnley, 1964; Reckling and Cordell, 1968; Sarmiento and Latta, 1981).

Classification

The internationally accepted classification of the Monteggia lesion was devised by Jose Luis Bado

(1967). He classified this lesion into four distinct types: percentages mentioned are those from his series (Figure 6.10).

Type 1:   anterior dislocation of the radial head. Fracture of the ulnar diaphysis at any level with anterior angulation (60%).

Type 2:   posterior or posterolateral dislocation of the radial head. Fracture of the ulnar diaphysis with posterior angulation (15%).

Type 3:   lateral or anterolateral dislocation of the radial head. Fracture of the ulnar metaphysis (20%).

Type 4:   anterior dislocation of the radial head. Fracture of the proximal third of the radius. Fracture of the ulna at the same level (5%).

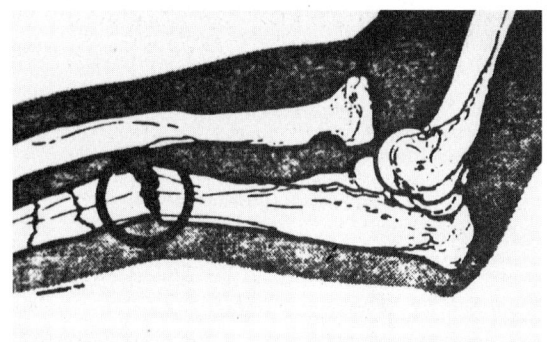

(a)

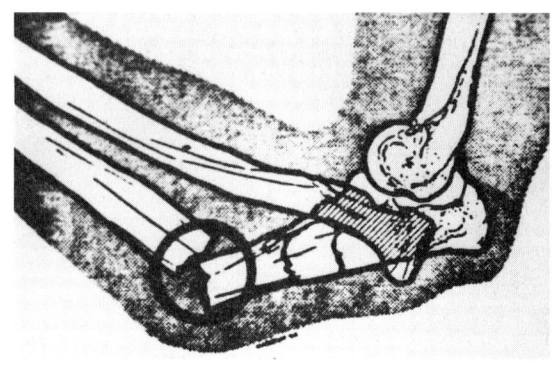

(b)

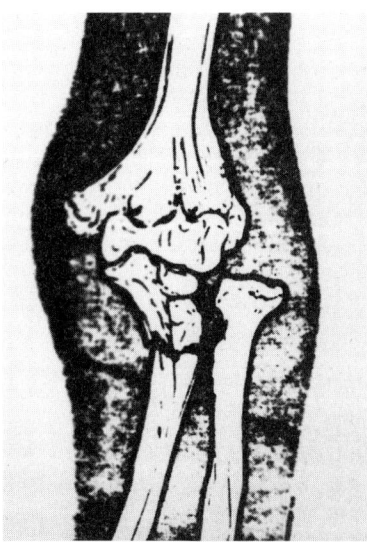

(c)

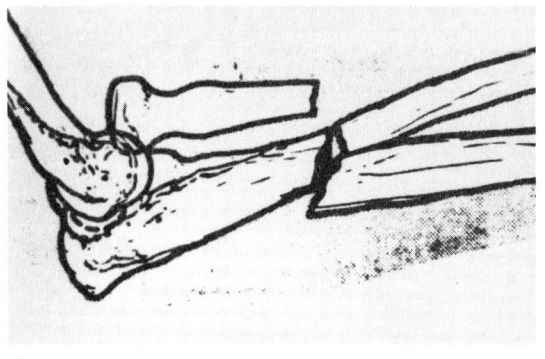

(d)

Figure 6.10   Bado classified the Monteggia lesion into four distinct types: (a) Type 1: anterior dislocation of the radial head with ulnar fracture with anterior angulation. (b) Type 2: posterior or posterolateral dislocation of the radial head with ulnar fracture with posterior angulation. (c) Type 3: lateral or anterolateral dislocation of the radial head with ulnar metaphyseal fracture. (d) Type 4: anterior dislocation of the radial head with fracture of the proximal third of the radius

Mechanism of injury

The different types of Monteggia lesions reflect different mechanisms of injury. There is some debate in the literature surrounding the mechanism of the type 1 lesion. The other types, fortunately, do not suffer from the same fate!

Historically, the type 1 lesion was thought to result from a direct blow to the ulnar border of the forearm, resulting in a fracture of the ulna with anterior angulation in conjunction with an anterior dislocation of the radial head (Speed and Boyd, 1940; Smith, 1947). Evans (1949), however, did not agree, suggesting instead that a type 1 lesion was the outcome of a hyperpronation injury to the forearm. After demonstrating this mechanism of injury on cadavers, he further postulated that this injury was caused by a fall on the outstretched and maximally pronated forearm.

On contact with the ground, the hand remains relatively fixed, while the trunk continues to rotate. This subjects the pronated forearm to even greater pronation, ending with an ulnar fracture. At the same time, the hyperpronated radius lies across the fractured ulna, and becomes anteriorly displaced by the force of the ulnar fracture. As additional evidence, Evans offered two more points in favour of this hypothesis, including (1) the infrequent, if not rare, occurrence of either a contusion or haematoma over the fractured ulnar border, and (2) the posterior position of the bicipital tuberosity on initial radiographs, suggesting that this position of the bicipital tuberosity is possible only in extreme pronation. We believe that both mechanisms are capable of causing this injury.

In contrast to the percentages suggested by Bado, Pavel *et al.* (1965) and Penrose (1951) found the type 2 lesion to occur more frequently. Penrose felt that this injury was more likely to be seen in middle-aged women, unlike the type 1 lesion, which occurs more frequently in children. He described three distinct components of this injury, which include: (1) a comminuted fracture of the proximal ulna near the coronoid, frequently including a triangular or quadrangular fragment; (2) a posterior dislocation of the radial head; and (3) a triangular fracture of the anterior aspect of the radial head, which results when the radial head shears against the capitellum (Figure 6.11). Penrose additionally suggested that this injury could be considered a variant of the posterior dislocation of the elbow. Based on cadaveric studies, he further suggested that this lesion was most

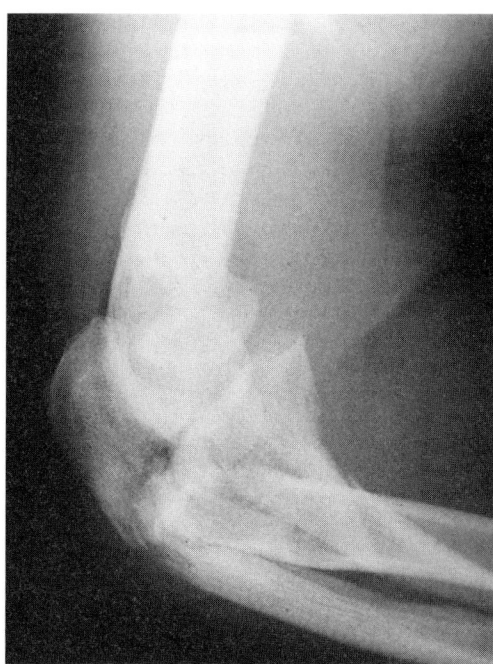

Figure 6.11 The morphology of the posterior Monteggia lesion commonly includes: a comminuted fracture of the ulna near or including the coronoid process; a posterior dislocation of the radial head; and a triangular fracture of the anterior aspect of the radial head

likely to be seen in a middle-aged adult sustaining a fall on the outstretched hand, with the elbow flexed approximately 60° and the forearm pronated around 30°. The high incidence of the type 2 lesion has been verified independently by the senior author. Among 18 Monteggia lesions treated operatively at the Massachusetts General Hospital over a period of 8 years, 13 were type 2 lesions (Jupiter *et al.*, 1991).

A type 3 lesion can occur in adults as well as in children (Penrose, 1951; Mullick, 1977). It has been suggested that this injury may result either from an adduction injury (Mullick, 1977) or from a combination of angulation and rotation (Pollen, 1973). No mechanism of injury is as yet forthcoming for the type 4 lesion.

Injuries associated with Monteggia lesions include nerve palsies (Spinner *et al.*, 1968; Stein *et al.*, 1971; Spar, 1977; Nunley and Urbaniak, 1980; Engber and Keene, 1983), scaphoid and distal radial fractures (Mullan *et al.*, 1980), elbow dislocation, and fractures of the radial shaft with disruption of the distal radioulnar joint (Odena, 1952; Jupiter *et al.*, 1994). The latter lesion is in effect a combination of the Monteggia and

Galeazzi lesions, and has been labelled as the 'bipolar fracture-dislocation of the forearm' by Odena (1952). In the largest reported series concerning this lesion, Jupiter *et al.* (1994) found its hallmark to be an unstable or 'floating' radial diaphysis. The associated severe soft tissue disruption, commonplace in this injury, proved to be the primary cause of unsatisfactory results in 30% of cases.

Clinical features

The feature common to all types of Monteggia injuries is the degree of pain about the elbow and a mechanical block to forearm rotation. In addition, elbow movements are also very painful. The clinical presentation will depend in part on the type of the Monteggia lesion. In patients with a type 1 lesion, the radial head may be more prominent over the anterior aspect of the elbow. There will most likely be a contusion or haematoma seen over the anteriorly angulated ulna. In type 2 lesions, the radial head can be palpated posteriorly and inferior to the lateral epicondyle, and there is posterior angulation of the ulna.

Assessment of the patient's neurological status is of paramount importance. The posterior interosseous nerve is the nerve most frequently reported to be involved (Spinner *et al.*, 1968; Boyd and Boats, 1969; Stein *et al.*, 1971; Jessing, 1975; Spar, 1977; Anderson and Meyer, 1991), with an incidence reported in one series to be as high as 17% (Bruce *et al.*, 1994). The aetiology for this, as suggested by Spar (1977), is either from direct trauma or stretching of the nerve by a dislocated radial head or entrapment between the radius and ulna. The last mechanism was also seen in the case described by Morris (1974).

It is extremely important to make a note of any neurological deficit and also make the patient aware of its existence prior to any procedure performed under anaesthetic. It is also important to warn the patient of the possibility of a postoperative neurological deficit. Incidences as high as 14% have been reported for the same (Bruce *et al.*, 1974). Other nerve injuries have also been reported in association with Monteggia lesions and include injuries to the anterior interosseous nerve (Engber and Keene, 1983), median nerve (Stein *et al.*, 1971; Bruce *et al.*, 1974; Nunley and Urbaniak, 1980) and ulnar nerve (Stein *et al.*, 1971; Bruce *et al.*, 1974), although much less commonly than injuries to the posterior interosseous nerve.

Radiological features

The diagnosis on standard anteroposterior and lateral radiographs is clear only as far as the ulnar fracture is concerned. Dislocation of the radial head may not be as apparent (Guistra *et al.*, 1974) (Figure 6.12). In fact, in one study, 52% of these lesions were not accurately assessed on initial radiographs (Speed and Boyd, 1940). Two points should be emphasized when analysing any forearm fracture. These include:

1. The forearm represents a two-bone unit and therefore a diaphyseal fracture of one bone may well be associated with an injury to either of the radioulnar joints. Good-quality films centred on the wrist and elbow will be imperative to evaluate the entire forearm.
2. A line through the radial shaft and head should intersect the capitellum for any position of the elbow (McLaughlin, 1959).

In addition, marked swelling in the vicinity of the proximal forearm or spontaneous reduction of the radial head dislocation (Anderson and Meyer, 1991), either immediately after injury or during temporary splintage (Schatzker and Tile, 1987), have all been described as causes contributing towards failure to accurately diagnose a Monteggia lesion.

Methods of treatment

While it is accepted that paediatric lesions seldom require little else than manipulative reduction and a cast, it is also agreed that adult lesions are an entirely different entity and the results of manipulative reduction prove unsatisfactory (Monteggia, 1813–1815; Speed and Boyd, 1940). As with other forearm fractures, anatomical repositioning is mandatory to assure recovery of function. Compression plate fixation would appear to offer the most favourable method of maintaining the realignment of the ulnar fracture (Boyd and Boals, 1969; Reckling, 1982; Jupiter *et al.*, 1991, 1994).

Authors' method of treatment

In the treatment of an adult Monteggia lesion, we believe that open reduction and stable internal fixation are mandatory. The ulnar fracture is to be reduced subsequent to the relocation of the radial head. The latter must be done gently with the patient under anaesthetic. With a fully relaxed patient, gentle traction on the forearm may be all

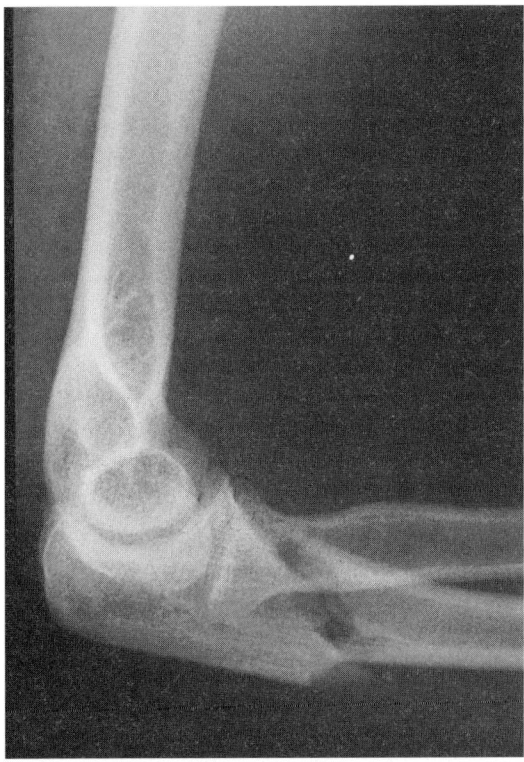

(a)

Figure 6.12   Diagnosis of the posterior Monteggia lesion is not always clear. (a) The lateral radiograph upon presentation in the emergency room may not be suggestive of radiocapitellar instability. (b) the flexion deformity of the ulnar fracture and posterior radiocapitellar subluxation was evident only after the limb was immobilized in a cast

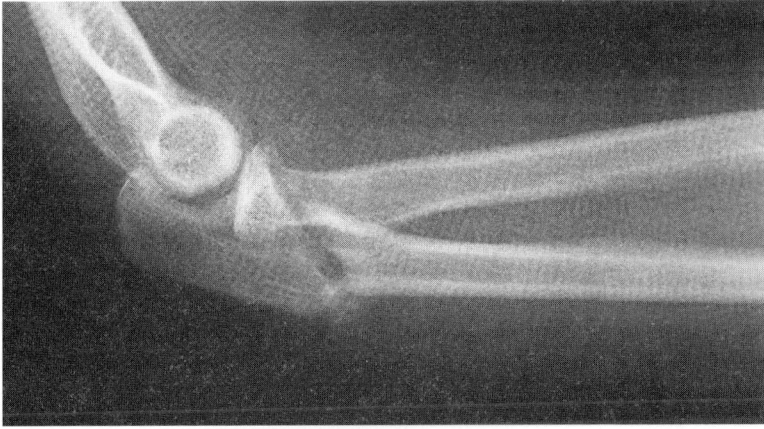

(b)

that is required to reduce the radial head. Gentle supination of the forearm may also help in relocation of the radial head. Only in rare circumstances will the radial head be irreducible due to interposition of the capsule, annular ligament or the posterior interosseous nerve (Morris, 1974; Spar, 1977; Reckling, 1982; Kellam and Jupiter, 1991). In such a situation, forcible attempts to reduce the radial head are ill-advised and likely to complicate

matters. Hence, it will be prudent to proceed to open reduction of the radial head.

The ulna can be approached through an extensile incision over the posterior border. Should the radial head be irreducible, this same incision can be extended as a Boyd–Thompson approach, reflecting the anconeus, extensor carpi ulnaris and supinator muscles from the ulna to expose the radiohumeral joint.

We recommend at least a 6 hole, 3.5 mm DCP to stabilize the ulna, with six cortices to be engaged on either side of the fracture. Following fixation of the ulna, the stability of the radial head is verified. For this it is useful to use the image intensifier, as it helps not only to visualize the stability of the radial head with rotation of the forearm but also to ascertain the accuracy of reduction and adequacy of fixation. Any instability of the radial head or inability to achieve a congruous reduction suggests the possibility of a malreduction of the ulnar fracture (Schatzker and Tile, 1987).

Failure to achieve an anatomical reduction of the proximal ulna is a particular problem with type 2 lesions. In these, the anterior triangular or

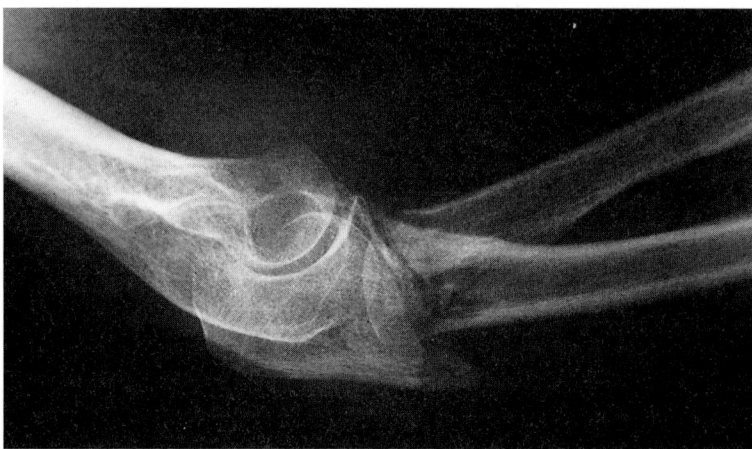

(a)

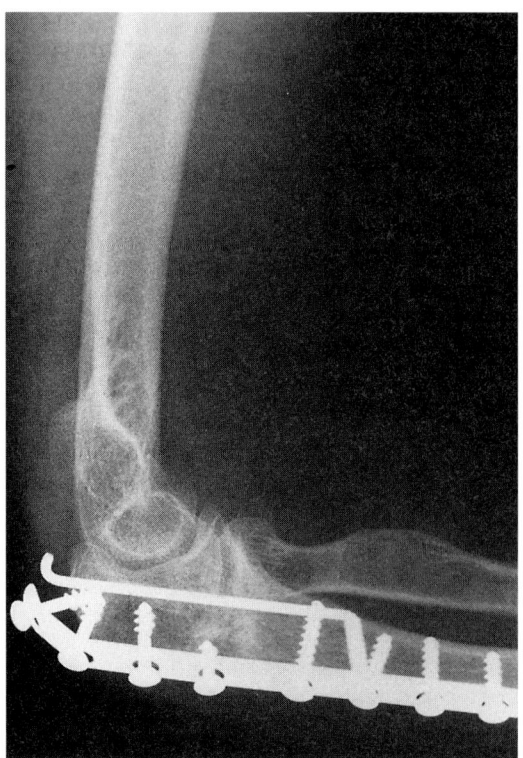

(b)

Figure 6.13   A complex unstable posterior Monteggia lesion in an active 70-year-old opera singer. (a) The lateral radiograph reveals the complex fracture pattern. (b) Operative stabilization consisted of a dorsally applied plate and screws with an adjuvant Kirschner wire

quadrangular fracture fragment represents a defect in the anterior cortex. This defect encourages posterior angulation of the ulna, which in turn leads to posterior radial head displacement. In such a situation, where the anterior buttress may be deficient or incompletely restored, any fixation with an intramedullary device is bound to fail. Our preference is to stabilize these fractures with a dorsally applied, contoured 3.5 mm DCP. In this location, the plate can behave to some extent as a tension band plate, provided the anterior cortex is restored by engaging the anterior fragment with a screw through the plate or by means of a wire cerclage around the plate (Figure 6.13).

Should the annular ligament be ruptured, it should be repaired. Fascial reconstruction of the annular ligament, as described by Boyd and Boals (1969), will not be necessary (Reckling, 1982), and furthermore is fraught with the possibility of residual scarring and loss of forearm rotation. It is extremely uncommon to find isolated instability of the radial head if the ulnar fracture has been anatomically reduced.

Associated fractures of the radial head are managed based upon the morphology of the injury. A single large fragment is fixed internally. In case of a comminuted fracture, if it is possible to restore the radial head, we still recommend internal fixation. If restoration of the radial head is not possible, it is excised and a silicone implant of the appropriate size is inserted.

Following stable internal fixation of a Monteggia lesion in an adult, little in the way of postoperative immobilization will be needed.

Complications

Numerous complications have been reported in association with the Monteggia lesion. These include infection, non-union, malunion, restriction of forearm rotation, redislocation of the radial head, nerve palsies (Boyd and Boals, 1969; Bruce *et al.*, 1974) and synostosis (Bruce *et al.*, 1974). A postoperative palsy of the posterior interosseous nerve is the most commonly reported nerve palsy (Spinner *et al.*, 1968; Boyd and Boals, 1969; Stein *et al.*, 1971; Jessing, 1974; Spar, 1977; Anderson and Meyer, 1991). In the experience of most authors, there is evidence of nerve recovery within 6–8 weeks and complete recovery may be expected by 12–16 weeks. At the end of this period, if there is no evidence of recovery then exploration is advisable.

Tardy palsies of the posterior interosseous nerve following old unreduced Monteggia fractures have also been reported (Austin, 1975; Lichter and Jacobsen, 1975; Holst-Nielsen and Jensen, 1984). Results of treating these with exploration and excision of the radial head have been satisfactory.

Malunion may result from inappropriate reduction and/or fixation of the ulnar fracture, and failure to diagnose dislocation of the radial head. Long-standing dislocations of the radial head are almost always impossible to reduce and require excision of the radial head. The ulnar malunion may be resolved by doing an appropriate osteotomy followed by stable internal fixation.

Non-union is usually the end result of problems related to the operative technique and/or inadequate internal fixation (Langkamer and Ackroyd, 1991). In the absence of infection, it is possible to achieve union in most cases by open reduction, stable internal fixation and supplemental bone graft, as necessary.

Results

Bruce *et al.* (1974) reported only 25% good results in their series with no excellent results. In Jupiter *et al.*'s (1991) series, there were 54% good to excellent results among patients with type 2 lesions. Boyd and Boals (1969) achieved 77% good to excellent results in their series, which included all types of Monteggia lesions. It is imperative that diagnosis and treatment of these injuries is prompt. An experienced surgeon, meticulous attention to detail and technique, and adherence to principles of internal fixation are more than likely to provide a satisfactory result (Anderson *et al.*, 1975; Moore *et al.*, 1985a; Chapman *et al.*, 1989; Langkamer and Ackroyd, 1991; Müller *et al.*, 1991).

References

Alexander, A.H. and Lichtman, D.M. (1981) Irreducible distal radio-ulna joint occurring in a Galeazzi fracture – case report. *J. Hand Surg.*, **6**, 258

Anderson, L.D. and Meyer, F.N. (1991) Fractures of the shafts of the radius and ulna. In *Fractures*, vol. 1, 3rd edn. (C.A. Rockwood, D.P. Green and R.W. Bucholz, eds) pp. 679–737, Philadelphia: Lippincott

Anderson, L.D., Sisk, T.D., Tooms, R.E. and Park, W.I. III. (1975) Compression-plate fixation in acute diaphyseal fractures of the radius and ulna. *J. Bone Joint Surg.*, **57A**, 287–97

Austin, R. (1975) Tardy palsy of the radial nerve from a Monteggia fracture. *Injury*, **7**, 202

Bado, J.L. (1967) The Monteggia lesion. *Clin. Orthop.*, **50**, 71

Biyani, A. and Bhan, S. (1989) Dual extensor tendor entrapment in Galeazzi fracture-dislocation. A case report. *J. Trauma*, **29**, 1295

Bock, G.W., Cohen, M.S. and Resnick, D. (1992) Fracture-dislocation of the elbow with inferior radioulnar dislocation. A variant of the Essex-Lopresti injury. *Skel. Radiol.*, **21**, 315–17

Boyd, H.B. and Boals, J.C. (1969) The Monteggia lesion. A review of 159 cases. *Clin. Orthop.*, **66**, 94

Brakenbury, P.H., Corea, J.R. and Blakemore, M.E. (1981) Non-union of the isolated fracture of the ulnar shaft in adults. *Injury*, **12**, 371

Broberg, M.A. and Morrey, B.F. (1986) Results of delayed excision of the radial head after fracture. *J. Bone Joint Surg.*, **68A**, 669–74

Brockman, E.P. (1930) Two cases of disability at the wrist joint following excision of the head of the radius. *Proc. R. Soc. Med.*, **24**, 904

Bruce, H.E. Harvey, J.P. and Wilson, J.C. (1974) Monteggia fractures. *J. Bone Joint Surg.*, **56A**, 1563

Bruckner, J.D. Lichtman, D.M. and Alexander, A.H. (1992) Complex dislocations of the distal radioulnar joint. Recognition and management. *Clin. Orthop.*, **275**, 90–103

Bunker, T.D. and Newman, J.H. (1985) The Herbert differential pitch bone screw in displaced radial head fractures. *Injury*, **16**, 621–4

Burwell, H.N. and Charnley, A.D. (1964) Treatment of forearm fractures in adults with particular reference to plate fixation. *J. Bone Joint Surg.*, **46B**, 404–25

Caden, J.G. (1961) Internal fixation of fractures of the forearm. *J. Bone Joint Surg.*, **43A**, 1115–21

Carn, R.M., Medige, J., Curtain, D. and Koenig, A. (1986) Silicone rubber replacement of the severely fractured radial head. *Clin. Orthop.*, **209**, 259–69

de Carvalho, A., Moller, J.T. and Vestergard-Andersen, T. (1984) Radiologic aspects of the Galeazzi lesion. *Eur. J. Radiol.*, 4, 169

Chapman, M.W., Gordon, J.E. and Zissimos, A.G. (1989) Compression-plate fixation of acute fractures of the diaphyses of the radius and ulna. *J. Bone Joint Surg.*, **71A**, 159–69

Coleman, H.M. (1960) Injuries of the articular disc at the wrist. *J. Bone Joint Surg.*, **42B**, 522

Corea, J.R., Brakenbury, P.H. and Blakemore, M.E. (1981) The treatment of isolated fractures of the ulnar shaft in adults. *Injury*, **12**, 365

Crenshaw, A.H. Jr. (1991) Surgical approaches. In *Campbell's Operative Orthopaedics*, vol. 1, 8th edn. (A.H. Crenshaw, ed.) pp. 102–113, St. Louis: Mosby-Yearbook

Curr, J.F. and Coe, W.A. (1946) Dislocation of the inferior radio-ulna joint. *Br. J. Surg.*, **34**, 74–7

Dameron, T.B. (1972) Traumatic dislocation of the distal radioulnar joint. *Clin. Orthop.*, **83**, 55

De Lee, J.C., Green, D.P. and Wilkins, K.E. (1984) Fractures and dislocations of the elbow. In (eds.). *Fractures*, vol. 1, 2nd edn. (C.A. Rockwood and D.P. Green, eds) pp. 636–8, Philadelphia: Lippincott

Dodge, H.S. and Cady, G.W. (1972) Treatment of fractures of the radius and ulna with compression plates. A retrospective study of one hundred and nineteen fractures in seventy eight patients. *J. Bone Joint Surg.*, **54A**, 1167–76

Dymond, I.W.D. (1984) The treatment of isolated fractures of the distal ulna. *J. Bone Joint Surg.*, **66B**, 408

Edwards, E.G. (1952) The posterior Monteggia fracture. *Am. Surg.*, **18**, 323–7

Edwards, G.S. and Jupiter, J.B. (1988) Radial head fracture with acute radioulnar dislocation. Essex-Lopresti revisited. *Clin. Orthop.*, **234**, 61–9

Engber, W.D. and Keene, J.S. (1983) Anterior interosseous nerve palsy associated with a Monteggia fracture. A case report. *Clin. Orthop.*, **174**, 133–7

Epner, R.A., Bowers, W.H. and Guildford, W.B. (1982) Ulnar variance – the effect of wrist positioning and roentgen filming technique. *J. Hand Surg.*, **7A**, 298–305

Essex-Lopresti, P. (1951) Fractures of the radial head with distal radio-ulnar dislocation. Report of two cases. *J. Bone Joint Surg.*, **33B**, 244–7

Evans, E.M. (1949) Pronation injuries of the forearm with special reference to the anterior Monteggia fracture. *J. Bone Joint Surg.*, **31B**, 578

Galeazzi, R. (1934) Ueber ein besonderes syndrom bei verletzungen im Bereich der unterarmknocken. *Arch. Orthop. Unfallchir.*, **35**, 557–62

Geel, C.W. and Palmer, A.K. (1992) Radial head fractures and their effect on the distal radioulnar joint. A rationale for treatment. *Clin. Orthop.*, **275**, 79–84

Giustra, P.E., Killoran, P.J., Furman, R.S. and Root, J.A. (1974) The missed Monteggia fracture. *Radiology*, **10**, 45

Goldberg, I., Peylan, J. and Yosipovitch, Z. (1986) Late results of excision of the radial head for an isolated closed fracture. *J. Bone Joint Surg.*, **68A**, 675–9

Grace, T.G. and Eversmann, W.W. Jr. (1980) Forearm fractures. Treatment by rigid fixation with early motion. *J. Bone Joint Surg.*, **62A**, 433

Grace, T.G. and Witmer, B.J. (1980) Isolated fractures of the ulnar shaft. *Orthop. Trans.*, **4**, 299

Harrington, I.J. and Tountas, A.A. (1981) Replacement of the radial head in the treatment of unstable elbow fractures. *Injury*, **12**, 405

Heim, U. and Pfeiffer, K.M. (1982) *Small Fragment Set Manual*. p. 85, New York: Springer-Verlag

Henry, A.K. (1966) *Extensile Exposure*, 2nd edn. Edinburgh: E&S Livingstone

Herbert, T.G. and Fisher, W.E. (1984) Management of the fractured scaphoid using a new bone screw. *J. Bone Joint Surg.*, **66B**, 114–23

Hidaka, S. and Gustilo, R.B. (1984) Refracture of bones of the forearm after plate removal. *J. Bone Joint Surg.*, **66A**, 1241–3

Holst-Nielsen, F. and Jensen, V. (1984) Tardy posterior interosseous nerve palsy as a result of an unreduced radial head dislocation in Monteggia fractures. A report of two cases. *J. Hand Surg.*, **9A**, 572–5

Hoppenfeld, S. and de Boer, P. (1984) *Surgical Exposures in Orthopaedics: The Anatomic Approach*. Philadephia: Lippincott

Hotchkiss, R.N., An, K.-N., Sowa, D.T. *et al.* (1989) An anatomic and mechanical study of the interosseous membrane of the forearm. Pathomechanics of proximal migration of the radius. *J. Hand Surg.*, **14A**, 256–61

Hotchkiss, R.N. and Green, D.P. (1991) Fractures and dislocations of the elbow. In *Fractures*, vol. 1, 3rd edn. (C.A. Rockwood, D.P. Green and R.W. Bucholz, eds) pp. 805–24, Philadelphia: Lippincott

Hotchkiss, R.N. and Weiland, A.J. (1986) Valgus stability of the elbow. *Orthop. Trans.*, **10**, 224

Hughston, J.C. (1957) Fracture of the distal radius shaft. Mistakes in management. *J. Bone Joint Surg.*, **39A**, 249–64

Jenkins, N.H., Mintowt-Czyz, W.J. and Fairclough, W.A. (1987) Irreducible dislocation of the distal radioulnar joint. *Injury*, **18**, 40

Jessing, P. (1975) Monteggia lesions and their complicating nerve damage. *Acta Orthop. Scand.*, **46**, 601

Jupiter, J.B. (1991) Trauma to the adult elbow and fractures of the distal humerus. In *Skeletal Trauma*, vol. 2, 1st edn (B.D. Browner, J.B. Jupiter, A.M. Levine and P.G. Trafton, eds) pp. 1126–34, Philadephia: Saunders

Jupiter, J.B., Kour, A.K., Richards, R. and Nathan, J. (1994) The floating radius in bipolar fracture-dislocation of the forearm. *J. Orthop. Trauma*, **8**, 99–106

Jupiter, J.B., Leibovic, S.J., Ribbans, W. and Wilk, R. (1991) The posterior Monteggia lesion. *J. Orthop. Trauma*, **5**, 395

Kellam, J.F. and Jupiter, J.B. (1991) Diaphyseal fractures of the forearm. In *Skeletal Trauma*, vol. 2, 1st edn. (B.D. Browner, J.B. Jupiter, A.M. Levine and P.G. Trafton, eds) pp. 1095–124, Philadelphia: Saunders

Khurana, J.S., Kattapuram, S.V., Becker, S. and Mayo-Smith, W. (1988) Galeazzi injury with an associated fracture of the radial head. *Clin. Orthop.*, **234**, 70–1

Langkamer, V.G. and Ackroyd, C.E. (1991) Internal fixation of forearm fractures in the 1980's; lessons to be learnt. *Injury*, **22**, 97–102

Lanyon. L.E., Hampson, W.G.J., Goodship, A.E. and Shah, J.S. (1975) Bone deformation recorded *in vivo* from strain gauges attached to the human tibial shaft. *Acta Orthop. Scand.*, **46**, 256

Levin, P.D. (1973) Fracture of the radial head with dislocation of the distal radio-ulnar joint. Case report. Treatment by prosthetic replacement of the radial head. *J. Bone Joint Surg.*, **55A**, 837–40

Lichter, R. and Jacobsen, T. (1975) Tardy palsy of the posterior interosseous nerve with a Monteggia fracture. *J. Bone Joint Surg.*, **57A**, 124

Mackay, I., Fitzgerald, B. and Miller J.H. (1979) Silastic replacement of the radial head in trauma. *J. Bone Joint Surg.*, **61B**, 494–7

Marek, F.M. (1961) Axial fixation of forearm fractures. *J. Bone Joint Surg.*, **43A**, 1099–114

Matthews, L.S., Kaufer, H., Garver, D.F. and Sontesgard, D.A. (1982) The effect on supination–pronation of angular malalignment of fractures of both bones of the forearm. An experimental study. *J. Bone Joint Surg.*, **64A**, 14–17

McDougall, A. and White, J. (1957) Subluxation of the inferior radio-ulna joint complicating fracture of the radial head. *J. Bone Joint Surg.*, **39B**, 278–87

McLaughlin, H.L. (1959) *Trauma*. p. 225, Philadelphia: Saunders

Mikic, Z.D. (1975) Galeazzi fracture-dislocations. *J. Bone Joint Surg.*, **57A**, 1071–80

Mikic, Z.D. and Vukadinovic, S.M. (1983) Late results in fractures of the radial head treated by excision. *Clin. Orthop,*, **181**, 220–8

Monteggia, G.B. (1813–1815) *Instituzione Chirugiche*, 2nd edn. Milan: G. Maspero

Moore, T.M., Klein, J.P., Patzakis, M.J. and Harvey, J.P. Jr. (1985a) Results of compression plating of closed Galeazzi fractures. *J. Bone Joint Surg.*, **67A**, 1015–21

Moore, T.M., Lester, D.K. and Sarmiento, A. (1985b) The stabilising effect of soft tissue constraints in artificial Galeazzi fractures. *Clin. Orthop.*, **194**, 189–94

Morrey, B.F., An, K.-N. and Stormont, T.J. (1988) Force transmission through the radial head. *J. Bone Joint Surg.*, **70A**, 250–6

Morrey, B.F., Askew, L. and Chao, E.Y. (1981) Silastic prosthetic replacement for the radial head. *J. Bone Joint Surg.*, **63A**, 454–8

Morrey, B.F., Chao, E.Y. and Hui, F.C. (1979) Biomechanical study of the elbow following excision of the radial head. *J. Bone Joint Surg.*, **61A**, 63–8

Morris, A. (1974) Irreducible Monteggia lesion with radial nerve entrapment. *J. Bone Joint Surg.*, **56A**, 1744

Mullan, G.B., Franklin, A. and Thomas, N.P. (1980) Adult Monteggia lesion with ipsilateral wrist injuries. *Injury*, **12**, 413–16

Müller, M.E., Allgower, M., Schneider, R. and Willeneger, H. (1991) *Manual of Internal Fixation*, 3rd edn. pp. 466–74, Berlin: Springer-Verlag

Mullick, S. (1977) The lateral Monteggia fracture. *J. Bone Joint Surg.*, **33A**, 543

Nunley, J.A. and Urbaniak, J.R. (1980) Partial bony entrapment of the medan nerve in a greenstick fracture of the ulna. *J. Hand Surg.*, **5A**, 557

Odena, I.C. (1952) Bipolar fracture-dislocation of the forearm. *J. Bone Joint Surg.*, **34A**, 968

Palmer, A.K. and Werner, F.W. (1984) Biomechanics of the distal radioulnar joint. *Clin. Orthop.*, **187**, 26–35

Pavel, A., Pitman, J.M., Lance, E.M. and Wade, P.A. (1965) The posterior Monteggia fracture. A clinical study. *J. Trauma*, **5**, 185

Penrose, J.H. (1951) The Monteggia fracture with posterior dislocation of the radial head. *J. Bone Joint Surg.*, **33B**, 65

Pollen, A.G. (1973) *Fractures and Dislocations in Children*. pp. 61–8, Edinburgh: Churchill Livingstone

Pollock, F.H., Pankovich, A.M., Prieto, J.J. and Lorenz, M. (1983) The isolated fracture of the ulnar shaft. Treatment without immobilisation. *J. Bone Joint Surg.*, **65A**, 339

Reckling, F.W. (1982) Unstable fracture-dislocations of the forearm (Monteggia and Galeazzi lesions). *J. Bone Joint Surg.*, **64A**, 857

Reckling, F.W. and Cordell, L.D. (1968) Unstable fracture-dislocations of the forearm. The Monteggia and Galeazzi lesions. *Arch. Surg.*, **96**, 999–1007

Reckling, F.W. and Peltier, L.F. (1965) Ricardo Galeazzi and Galeazzi's fracture. *Surgery*, **58**, 453

Rosson, J.W., Petley, G.W. and Shearer, J.R. (1991) Bone structure after removal of internal fixation plates. *J. Bone Joint Surg.*, **73B**, 65–7

Rosson, J.W. and Shearer, J.R. (1991) Refracture after the removal of plates from the forearm. An avoidable complication. *J. Bone Joint Surg.*, **73B**, 415–17

Sage, F.P. (1959) Medullary fixation of fractures of the forearm. A study of the medullary canal of the radius and a report of fifty fractures of the radius treated with a prebent triangular nail. *J. Bone Joint Surg.*, **41A**, 1489–516

Sarmiento, A , Cooper, J.S. and Sinclair, W.F. (1975) Forearm fractures. Early functional bracing – a preliminary report. *J. Bone Joint Surg.*, **57A**, 297–304

Sarmiento, A., Kinman, P.B., Murphy, R.B. and Phillips, J.G. (1976) Treatment of ulnar fractures by functional bracing. *J. Bone Joint Surg.*, **58A**, 1104

Sarmiento, A. and Latta, L. (1981) *Closed Functional Treatment of Fractures*. pp. 460–88, Berlin: Springer-Verlag

Schatzker, J. and Tile, M. (1987) *The Rationale of Operative Fracture Care*. pp. 103–29, Berlin: Springer-Verlag

Schemitsch, E.H. and Richards, R.R. (1992) The effect of malunion on functional outcome after plate fixation of fractures of both bones of the forearm in adults. *J. Bone Joint Surg.*, **74A**, 1068–78

Schone, G. (1988) The treatment of forearm fractures with pins. *Clin. Orthop.*, **234**, 2

Shang, T.-Y., Gu, Y.-W. and Dong, F.-H. (1987) Treatment of forearm bone fractures by an integrated method of traditional Chinese and western medicine. *Clin. Orthop.*, **215**, 56–64

Shmueli, G. and Herold, H.Z. (1981) Compression screwing of displaced fractures of the head of the radius. *J. Bone Joint Surg.*, **63B**, 535

Smith, F.M. (1947) Monteggia fractures. An analysis of 25 consecutive fresh injuries. *Surg. Gynecol. Obstet.*, **85**, 630–640

Smith, H. and Sage, F.P. (1957) Medullary fixation of forearm fractures. *J. Bone Joint Surg.*, **39A**, 91–8

Spar, I. (1977) A neurologic complication following Monteggia fracture. *Clin. Orthop.*, **122**, 207

Speed, J.S. and Boyd, H.B. (1940) Treatment of fracture of the ulna with dislocation of the head of the radius (Monteggia fracture). *J.A.M.A.*, **115**, 1699

Spinner, M. (1978) *Injuries to the Major Branches of Peripheral Nerves of the Forearm*, 2nd edn. pp. 117–21, Philadelphia: Saunders

Spinner, M., Freundlich, B.D. and Teicher, J. (1968) Posterior interosseous nerve palsy as a complication of Monteggia fractures in children. *Clin. Orthop.*, **58**, 141

Stein, F., Grabias, S.L. and Deffer, P.A. (1971) Nerve injuries complicating Monteggia lesions. *J. Bone Joint Surg.*, **53A**, 1432

Stern, P.J. and Drury, W.J. (1983) Complications of plate fixation of forearm fractures. *Clin. Orthop.*, **175**, 25–9

Stewart, M. (1957) In discussion of paper, 'Fracture of the distal radius shaft. Mistakes in management'. *J. Bone Joint Surg.*, **39A**, 264

Strachan, J.C.H. and Ellis, B.W. (1971) Vulnerability of the posterior interosseous nerve during radial head resection. *J. Bone Joint Surg.*, **53B**, 320

Swanson, A.B., Jaeger, S.H. and La Rochelle, D. (1981) Comminuted fractures of the radial head. *J. Bone Joint Surg.*, **63A**, 1039–49

Tarr, R.R., Garfinkel, A.I. and Sarmiento, A. (1984a) The effects of angular and rotational deformities of both bones of the forearm. An *in vitro* study. *J. Bone Joint Surg.*, **66A**, 65–70

Tarr, R.R., Sew Hoy, A.L., Racette, W.L. and Sarmiento, A. (1984b) Evolution and current status of functional fracture bracing. *Orthop. Rev.*, **13**, 25

Taylor, T.K.F. and O'Connor, B.T. (1964) The effect upon the inferior radio-ulnar joint of excision of the head of the radius in adults. *J. Bone Joint Surg.*, **46B**, 83–8

Thompson, H.C. III and Garcia, A. (1967) Myositis ossificans. Aftermath of elbow injuries. *Clin. Orthop.*, **50**, 129

Valande, M. (1929) Luxation en arriere du cubitus avec fracture de la diaphyse radiale. *Bull. Membres Soc. Nat. Chir.*, **55**, 435–7

Veseley, D.G. (1967) The distal radioulnar joint. *Clin Orthop.*, **51**, 75

Vince, K.G. and Miller, J.E. (1987) Cross-union complicating fracture of the forearm. Part I: Adults. *J. Bone Joint Surg.*, **69A**, 640–53

Wong, P.C.N. (1967) Galeazzi fracture-dislocation in Singapore 1960–1964. Incidence and results of treatment. *Singapore Med. J.*, **8**, 186–93

Zych, G.A., Latta, L. and Zagorski, J.B. (1987) Treatment of isolated ulna shaft fractures with prefabricated functional fracture braces. *Clin. Orthop.*, **219**, 194

7

Complications of forearm fractures in adults

D. Ring and J.B. Jupiter

The rotary movements (pronation/supination) provided by the unique two-bone, dual intra-articulation structure of the forearm greatly expand the variety of ways in which objects can be positioned and manipulated by the hand. Loss of this motion as a result of malunion, prolonged immobilization and/or proximal or distal radioulnar joint incongruity following trauma to the adult forearm can be extremely disabling. The gradual improvement in functional outcomes and decrease in the rate of complications associated with the management of forearm fractures during this century parallels the history of the development of sound, stable techniques of internal skeletal fixation which permit mobility while assuring maintenance of skeletal alignment during fracture union. Treatment-related complications are fortunately not common but include iatrogenic injury to nerves or blood vessels, radioulnar synostosis, and refracture following plate removal. Infection is unusual, even in the case of an open fracture, due in part to the relative ease of wound debridement as well as the well-perfused forearm musculature.

Forearm fractures are often the sequelae of high-energy injury and a relatively large percentage are open fractures. Injury-related complications, to which these high-energy injuries are prone, remain a common source of difficulty in the management of forearm fractures. These include compartment syndrome, neurovascular injury, soft tissue loss, bone loss and post-traumatic radioulnar synostosis (more common in high-energy injuries, especially those associated with head injury).

Treatment

For decades, forearm fracture treatments were judged by their ability to achieve union. Results were inconsistent so that any mode of treatment with a demonstrated non-union rate of 10% or under gained advocacy, including closed reduction and cast immobilization (Whipple and St. John, 1917; Magnuson, 1922; Buxton, 1925, 1939; Bagley, 1926; Eliason et al., 1937; Carrell, 1938; Evans, 1945; Compere, 1948; Bolton and Quinlan, 1952), plate and screw fixation (Knight and Purvis, 1949; Lowie, 1956; Jinkins et al., 1960; Caden, 1961; Hicks, 1961), intramedullary fixation (Marek, 1961; Sage, 1982; Street, 1986) and even external fixation (Schuind et al., 1991). The functional results of forearm fracture treatment remained inconsistent at best when the patient's arm was immobilized in a long-arm cast (regardless of whether or not the fractures had been internally fixed). In addition to loss of forearm rotation, wrist and elbow motion were in jeopardy.

Despite numerous descriptions of the relevant anatomy and proper methods of reducing forearm fractures which appeared early in this century (Whipple and St. John, 1917; Magnuson, 1922; Buxton, 1925, 1939; Bagley, 1926; Eliason et al., 1937; Carrell, 1938; Evans, 1945; Compere, 1948; Bolton and Quinlan, 1952), these fractures remained 'problem' fractures and attracted a variety of early attempts at operative treatment. The necessity of anatomically reducing and stably fixing these fractures internally was first recognized in injuries associated with either proximal or distal radioulnar joint dislocation (Galeazzi and Monteggia fractures), where malunion was common and residual subluxation of a radioulnar joint led to predictable, severe loss of pronation and supination (Speed and Boyd, 1940; Hughston, 1957). However, similar acceptance of routine operative treatment of both bone forearm fractures was not forthcoming because early attempts at internal fixation, although sufficient to hold open reductions, did not preclude the need for external immobilization, resulting in comparably poor functional outcomes in fractures treated by open or closed methods (Compere, 1948; Knight and Purvis, 1949).

The development of larger, corrosion-resistant compression plates led to a dramatic decrease in the rate of fracture non-union while simultaneously providing sufficient stability for confident early mobilization of the forearm (Anderson *et al.*, 1975; Grace and Eversmann, 1980; Chapman *et al.*, 1989). The substantial improvements in functional outcome have led to the wide application of plate fixation in forearm fracture treatment.

On the other hand, treatment of forearm fractures by other methods has remained problematic. Sarmiento *et al.* (1975) have demonstrated success with early (6 weeks post-injury) fracture-brace mobilization of forearm fractures which can be stably reduced in a long-arm cast as initial treatment. However, the difficulty of obtaining and maintaining adequate, stable reduction of a two-bone unit with complex muscular attachments and rotational instability, as well as the inability of others to reproduce the successful closed treatment reported by Sarmiento's group, have hindered the acceptance of these methods.

The weaknesses of the intramedullary approach have included difficulty maintaining rotational alignment, failure to restore the anatomical radial bow, the need for supplemental external fixation, a high non-union rate and the technical difficulty of device insertion with numerous related complications such as splitting of the cortex and protrusion of the nail through the cortex or into a joint (Smith and Sage, 1957; Caden, 1961; Sage, 1982; Street, 1986). The results of intramedullary nailing of forearm fractures have improved modestly following the introduction of nails of square and triangular cross-section intended to better control rotation, improved nail design and insertion techniques intended to restore the anatomical radial bow, and closed nailing under fluoroscopic guidance (Marek, 1961; Sage, 1982; Street, 1986). However, despite these improvements, intramedullary nailing continues to lack the predictability and stability of modern plate and screw fixation which have essentially solved the 'problem' of forearm fractures and made plate fixation the treatment of choice.

Compartment syndrome

Reconstructive procedures can only minimally address the disabling combination of fibrosis and contracture of the extrinsic hand musculature and irreversible neurological deficits which result from tissue ischaemia as a consequence of elevated intracompartmental pressure (Parks, 1951; Page, 1956; Tsuge, 1975). The emphasis therefore falls upon prompt recognition and treatment of compartment syndrome.

Early descriptions indicate that so-called ischaemic paralysis was initially recognized far more commonly in association with upper than lower extremity injury, and was primarily associated with excessively tight bandaging (Volkmann, 1881). While compartment syndrome is now encountered far more frequently in the lower extremity, it still represents a potential complication of forearm fracture, especially a fracture associated with crush injury, gunshot wound and other high-energy mechanisms (Eaton and Green, 1975; Gelberman *et al.*, 1981; Moore *et al.*, 1985a; Brostrom *et al.*, 1990).

Prompt diagnosis is dependent upon clinical suspicion and frequent, careful examination focusing on pain out of proportion to injury, pain with passive stretch of the fingers and compartment palpation. Of note is the fact that a change in two-point discrimination was found by Gelberman *et al.* (1981) to be associated with compartment syndrome. Weakness is usually a late finding.

A reliable examination is dependent upon patient understanding and cooperation. In a patient with altered mental status (e.g. due to associated traumatic injury, intoxication or narcotic medication), in the presence of associated neurological deficit, or in any case in which the physician is not confident of his or her examination, serial measurement of intracompartmental forearm pressures is useful (Mubarak *et al.*, 1978). Elevated or rising pressures indicate the need for fasciotomy. The absolute pressure at which fasciotomy is indicated is debated. Pressures above 30 mmHg are of concern and at pressures above 45 mmHg one should strongly consider forearm fascial release.

Gunshot fractures of the forearm are particularly prone to compartment syndrome. Moed and Fakouri (1991) recorded a 15% overall incidence among 60 gunshot fractures of the forearm. Comminuted and severely displaced fractures were more commonly associated with compart-

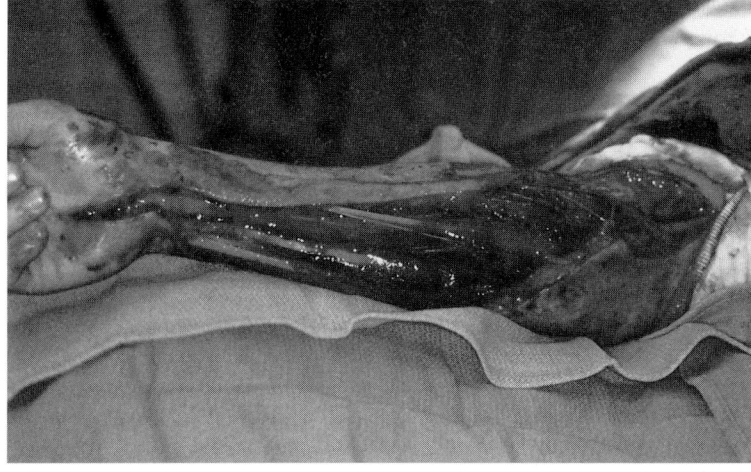
(a)

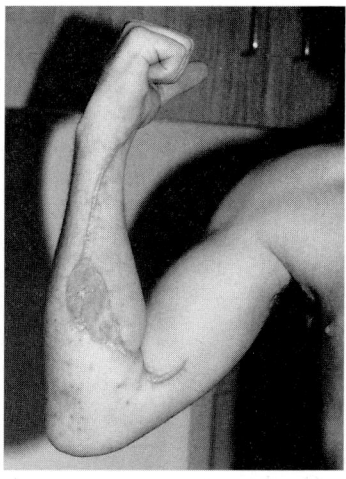
(b)

Figure 7.1 A 24-year-old male was buried in a mine accident. A closed fracture of the ulna was accompanied by elevated interstitial pressures in the forearm compartments. (a) The forearm compartment was released through an ulnar-based McConnell incision. (b) excellent function resulted

ment syndrome. They also found gunshot fracture of the proximal third of the forearm to be associated with a substantial risk of compartment syndrome. Eight of the nine compartment syndromes they documented were associated with gunshot fracture to the proximal third of the forearm.

In cadaver studies, the volar and dorsal forearm compartments, as well as the mobile wad region, were found to lack firm fascial divisions (Gelberman *et al.*, 1978a). As a result, the pressures in these compartments can be considered interdependent, as reflected in the fact that release of the volar compartment is usually sufficient to decrease pressures throughout the forearm.

Release of the volar compartment is achieved via an incision which begins proximal to the humeral epicondyles, crosses the antecubital fossa obliquely releasing the lacertus fibrosis, and then takes either a straight course down the ulnar aspect of the forearm (e.g. McConnell's combined exposure of the ulnar and median nerves as described by Henry (1973) (Figure 7.1)), or a curvilinear course over the mobile wad (Gelberman *et al.*, 1978a). With either approach, the incision then returns to the midline and crosses the wrist crease ending in the mid-palm and allowing release of the carpal tunnel and Guyon's canal (Henry, 1973; Gelberman *et al.*, 1978a). This fasciotomy can be accomplished as a part of Henry's anterior expo-

sure of the radius during open reduction and internal fixation (Henry, 1973) (Figure 7.2). If the compartment syndrome has progressed to include median nerve weakness, Gelberman *et al.* (1978a, 1981) suggest exploring and releasing the nerve where it dives beneath the pronator teres and flexor superficialis muscles. Following volar compartment release, dorsal pressures are measured and if they are still elevated, the dorsal compartment is released via a midline longitudinal incision.

Neurovascular injury

The forearm and hand are perfused by two major arteries (radial and ulnar) which terminate in the hand as the superficial and deep palmar arches. As a result, in the vast majority of patients, a single patent artery is sufficient to adequately perfuse the hand. Most major vascular injuries associated with forearm fracture involve a single arterial laceration without threatening the viability of the hand. In fact, some studies have suggested that repair of the damaged vessel in this situation may not prove worthwhile as thrombosis may develop due to backflow from the patent artery (Gelberman *et al.*, 1983; Trumble *et al.*, 1987).

Median, ulnar and radial nerve palsies have been reported in association with forearm fractures

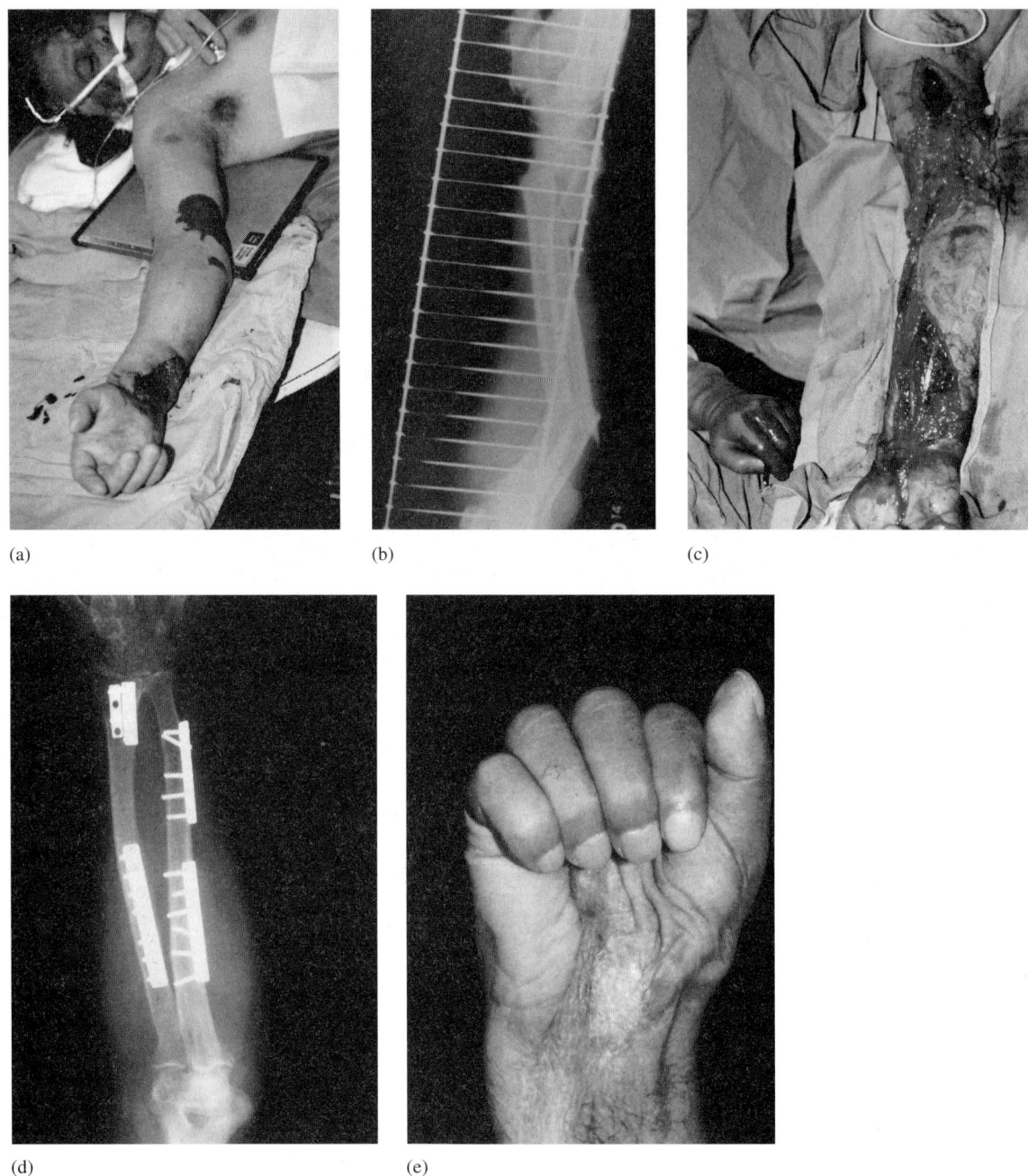

(a)   (b)   (c)

(d)   (e)

Figure 7.2   A 60-year-old assembly line worker had his dominant limb caught in a machine rolling hot plastic sheets. Median nerve function was absent on presentation. (a) Massive forearm swelling was observed. Note the open wound at the distal forearm. (b) Radiographs reveal segmental fractures of both ulna and radius. (c) Through an extensile operative approach, the forearm compartments were decompressed. (d) The fractures were stabilized with plates and screws. (e) Satisfactory hand function resulted despite some loss of median nerve function

(Spinner *et al.*, 1968; Stein *et al.*, 1971; Morris, 1974; Wolfe and Eyring, 1974; Jessing, 1975; Nunley and Urbaniak, 1980; Prosser and Hooper, 1986; Geissler *et al.*, 1990). One of the most common nerve injuries is a posterior interosseous nerve palsy in association with a Monteggia frac-

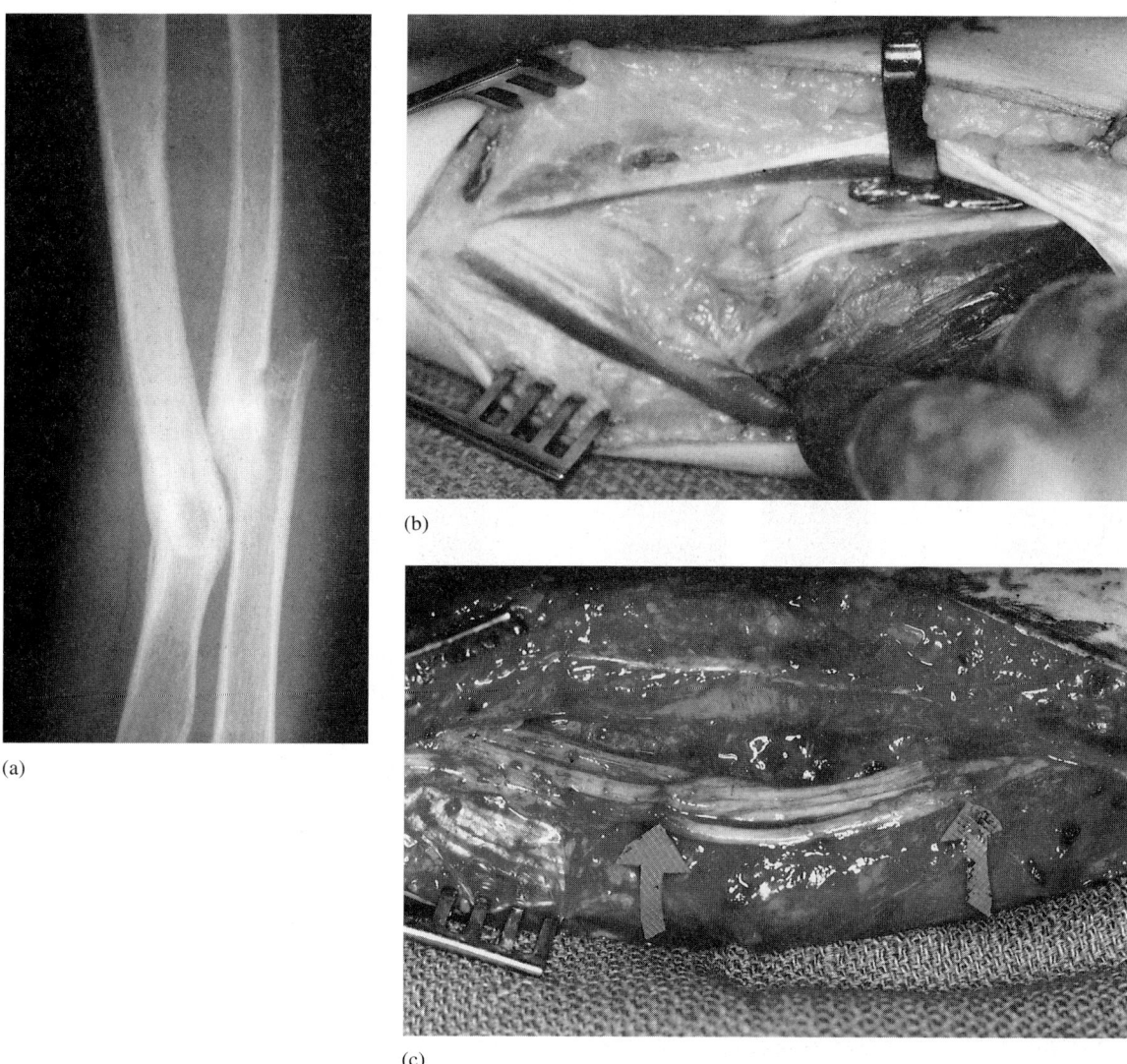

(a)

(b)

(c)

Figure 7.3 An 11-year-old male sustained a closed ulnar shaft fracture. Over the course of 6 months a complete low ulnar nerve palsy was noted. (a) Radiograph of the healed fracture. (b) At surgery the ulnar nerve was found impacted and nearly completely severed at the fracture site. (c) Interfascicular nerve grafts were required to bridge the neural defect

ture dislocation (Spinner *et al.*, 1968; Stein *et al.*, 1971; Jessing, 1975). Nerve palsy associated with forearm fracture usually represents a neurapraxic injury due to nerve contusion or traction which will recover spontaneously. Rarely, a nerve will be trapped in the fracture site or even transected by a sharp fracture fragment (Gurdjian and Smathers, 1945; Merris, 1974; Wolfe and Eyring, 1974; Nunley and Urbaniak, 1980; Geissler *et al.*, 1990) (Figure 7.3). In general, early nerve exploration in the management of fracture-associated nerve

deficit is not indicated unless there is a change in neurological exam following reduction, there is an associated vascular injury or open wound, or if the fracture is irreducible (Seigel and Gelberman, 1991). Most forearm fractures will be treated operatively, and a nerve exploration can usually be performed without considerable additional surgical trauma.

Even when associated with a high-velocity bullet wound, the majority of neural injuries recover spontaneously (Omer, 1974, 1982; Sunderland,

1978). Nerve exploration can be carried out during the initial irrigation and debridement. However, even with early exploration the extent of nerve damage can be difficult to determine. In the case of complete nerve transection, the nerve ends can be sutured to adjacent soft tissue to prevent retraction and delayed repair is performed when the wound is clean and adequately debrided (Omer, 1974, 1982).

Decision making in the management of a neuroma in continuity at the site of nerve damage can be difficult. Determination of the electrical transmission across this nerve segment can help determine whether to attempt neurolysis or to proceed with resection and grafting (Kline and De Jong, 1968). Recovery following neurorrhaphy can be disappointing. In separate series, Omer (1974, 1982) and Brown (1970) each recorded a 40–45% rate of progressive functional return following nerve suture of nerves injured in gunshot wounds.

Complex combined neurovascular injury is but one of the adverse sequelae of severe mangling upper extremity trauma. The approach to vascular injuries is modified somewhat by the presence of an associated nerve lesion, as there is some evidence to suggest that recovery of nerve function is improved by optimization of blood flow (Gelbermann *et al.*, 1978b; Engkvist *et al.*, 1985; Freelander, 1986; Nylander *et al.*, 1987). Hand viability may be maintained even when both radial and ulnar arteries have been damaged, because, in addition to the dual major arterial supply to the forearm, there are a number of longitudinally oriented collateral vessels in the upper extremity (Strandness, 1969; Jupiter and Kleinert, 1988). In order to optimize nerve recovery, recommendations state that, in the presence of associated nerve injury, if both major arteries are damaged, both should be repaired, regardless of the apparent vascular status of the hand (Gelbermann *et al.*, 1982). If one major supplier remains intact and provides adequate perfusion to the entire hand (as determined by a timed Allen test), the damaged artery need not be repaired (Gelberman *et al.*, 1977).

Vascular repair follows fracture stabilization when feasible (often by temporary means such as external fixation) in order to provide protection of the repaired vessel. Following revascularization, compartment pressures should be checked and fasciotomy performed as indicated.

Severe mangling upper extremity injury may be best served by early amputation in many cases. However, upper extremity prosthetics provide a poor substitution for upper extremity function and most reasonable attempts at limb salvage are worthwhile (Levin *et al.*, 1980; Jupiter, 1991). In addition to the judgement of the treating surgeon (based upon history, examination, radiological studies and operative exploration), consideration of the mangled extremity score has been found useful in determining the appropriateness of primary amputation (Helfet *et al.*, 1990; Slauterbeck *et al.*, 1994).

Operative treatment of a forearm fracture exposes the patient to the risk of iatrogenic nerve injury. A neurapraxic superficial radial nerve palsy can occur as this nerve is retracted laterally along with the brachioradialis when approaching the proximal radius. Rarely, the nerve, and sometimes the nearby radial artery, is damaged during the approach. This can be avoided by routinely visualizing and protecting these structures. Central to both the anterior (Henry) and dorsal (Thompson) approaches to the radius is protection of the posterior interosseous nerve. This nerve enters through the supinator muscle and, unless protected by maximal supination of the forearm during subperiosteal elevation of the supinator muscle during the volar approach and direct visualization during the dorsal approach, it is at risk of being damaged.

Sympathetic maintained pain (causalgia, reflex sympathetic dystrophy) is a rare, but extremely disabling sequela of extremity trauma which is occasionally seen following forearm fracture. Mitchell *et al.*'s, (1864) initial characterization of this entity associated severe, chronic burning pain and hypersensitivity (allodynia and hyperpathia) with evidence of autonomic nervous dysfunction (trophic, vasomotor and sudomotor changes) in a limb in which a peripheral nerve had been injured by a ballistic injury. However, because these features are commonly seen in injured limbs without identifiable nerve lesions, the term 'sympathetic maintained pain' and its equivalents are applied to chronic pain syndromes of the extremity with autonomic nervous dysfunction, regardless of the nature of the injury (Raj *et al.*, 1992).

Treatment of this syndrome has focused on early, aggressive therapy as well as interruption of the abnormal sympathetic activity by direct block or systemic pharmacological means. Nerve blocks, such as a stellate ganglion block for sympathetic maintained pain of the upper extremity, can provide excellent short-term relief, allowing aggressive therapy and possibly facilitating recovery. Pharmacological therapy (with alpha-adrenergic blockers such as phenoxybenzamine) can be effective, but are associated with undesirable side-effects.

Recent experience has demonstrated that if the sympathetic maintained pain is associated with a definable nerve lesion, an operative approach consisting of a combination of repair, reconstruction and/or lysis of the involved nerve and rotation of a local muscle flap intended to enhance the blood supply to the area of injury and minimize scarring, can be efficacious (Jupiter *et al.*, 1994a).

Infection

At one time there was concern regarding the risks of introducing infection in conjunction with operative treatment of a forearm fracture. However, with current operative techniques and implants (including perioperative antibiotic prophylaxis), infection following operative treatment of closed fractures is uncommon (Anderson *et al.*, 1975; Chapman *et al.*, 1989). Furthermore, recent reports have cited an acceptably low rate of infection in open forearm fractures treated by immediate plate and screw fixation (0–3%) (Moed *et al.*, 1986; Chapman *et al.*, 1989; Jones, 1991; Duncan *et al.*, 1992; Lenihan *et al.*, 1992). Adequate debridement and copious irrigation are requisite with this approach and should be repeated frequently until a clean, healthy soft tissue bed is achieved. Surgical wounds are closed primarily and traumatic wounds by delayed primary closure following adequate wound definition by serial debridement.

When infection occurs, its eradication is not necessarily dependent upon implant removal. As long as all bone fragments and soft tissues are well vascularized, stable internal fixation will facilitate wound care and help maintain length and alignment, as well as range of motion and overall function, without hindering treatment of the infection. Following successful eradication of the infection (with organism-specific antibiotics and local wound care), the wound can be irrigated and closed.

If bone debridement results in a substantial gap, this can be filled with autogenous cancellous bone graft from the iliac crest or other sites at the time of wound closure (Figure 7.4). Bone loss is also common in gunshot fractures. Formerly treated by corticocancellous grafts from the iliac crest and other methods of limited success (Spira, 1954; Nicoll, 1956; Grace and Eversmann, 1976), most recently, early cancellous bone grafting in conjunction with stable plate fixation has been found to be effective (Anderson *et al.*, 1975; Chapman *et al.*, 1989).

Non-union

Closed treatment of both bone forearm fractures in adults has been associated with a substantial rate of non-union (12% in the study by Knight and Purvis, 1949). Non-union was associated with poor reductions in which minimal apposition of fracture fragments had been obtained and held. Thus, open reduction with or without internal fixation has been advocated for inadequate or unstable reductions since early in this century.

Early attempts at internal fixation (including sutures, wires, screws, small plates, and intramedullary wire or pins) provided minimal fracture stabilization, and prolonged cast immobilization was required (Knight and Purvis, 1949). All the disadvantages of operative treatment (additional soft tissue trauma, devascularization of fracture fragments and the risks of infection and iatrogenic neurovascular injury) were thus added to the problems of prolonged immobilization. Among forearm fractures undergoing operative treatment (i.e. fractures unreducible by closed means), the non-union rate remained high, and functional outcomes were dismal (Knight and Purvis, 1949).

In 1957, Smith and Sage published the disappointing results of early attempts at intramedullary fixation of forearm fractures with a variety of pins and nails. Use of Kirschner wires, associated soft tissue injury and delayed treatment were all associated with an increased risk of non-union. Caden (1961) recorded similarly poor results with Rush pin fixation.

After studying cadaver radii, Sage (1982) introduced a nail with proximal and distal angulations and a triangular cross-section in an attempt to improve restoration of radial bow and rotational stability. Complications included nail protrusion, splitting of the cortex, rupture of the extensor pollicis longus tendon, synostosis and radial nerve palsy. Some 11.1% of fractures failed to unite within 6 months, in many cases due to technical errors and complications. Marek (1961) reported 100% union utilizing a square nail. However, he defined union clinically, had a very short follow-up in many cases, and described many fractures as requiring well over 6 months to heal. Street (1986) also used a square nail and reported 7% non-union despite use of a closed nailing technique under fluoroscopic guidance in many cases. He continued to have problems related to nail insertion, including distraction at the fracture site. Despite continued interest in intramedullary nailing and

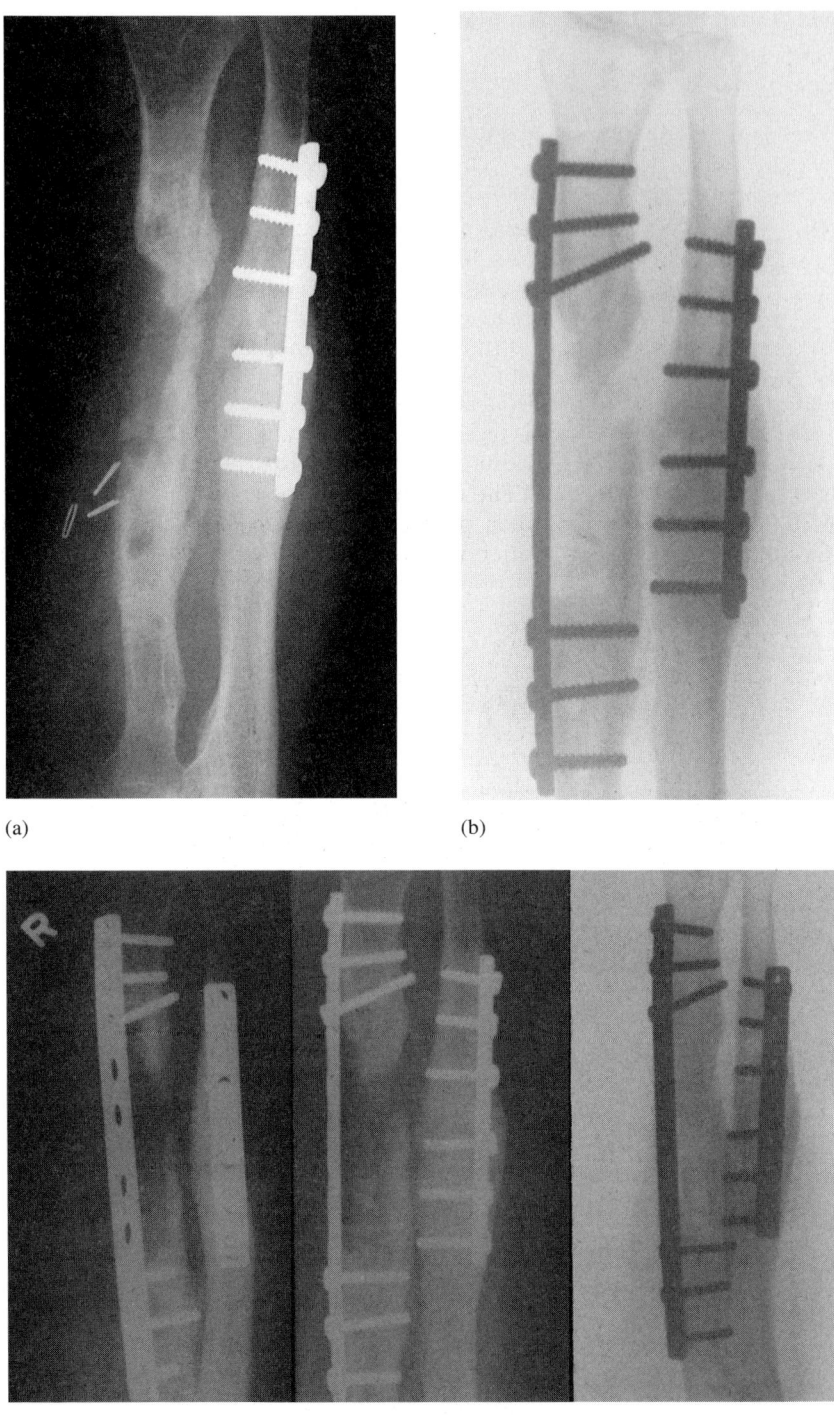

(a)

(b)

(c)

Figure 7.4 A 27-year-old male presented with chronic infection involving his radius. He had had both open bone forearm fractures treated by open reduction and internal fixation, and subsequently the radius was debrided and an external fixator applied. (a) The presenting radiograph reveals the defect of the radius. (b) Following extensive debridement and wound coverage, the defect was bridged with a plate bent off the bone at the level of the defect. The defect was filled with autogenous iliac crest cancellous graft. (c) The non-union healed without recurrence of infection. Excellent forearm rotation was achieved

the periodic introduction of new nail designs, it remains a technically demanding procedure, prone to difficulties and complications.

Improvements in plate and screw fixation have nearly eliminated non-union in forearm fractures. Initial improvements were noted with the use of larger plates such as the Egger's slotted plates (Jinkins *et al.*, 1960; Baker *et al.*, 1969). In 1964, the report of Burwell and Charnley noted that non-union was far less common when plates at least 3.5 inches in length were utilized. With the addition of interfragmentary compression by the AO/ASIF group (Müller *et al.*, 1965), the only problem fractures remained those with substantial bone loss or comminution (Naiman *et al.*, 1970; Dodge and Cady, 1972; Anderson *et al.*, 1975; Grace and Eversmann, 1980; Teipner and Mast, 1980; Chapman *et al.*, 1989). Primary bone grafting of such fractures has further reduced the risk of non-

union (Anderson *et al.*, 1975; Chapman *et al.*, 1989). Also of note, Chapman *et al.* (1989) found that the use of 3.5 mm dynamic compression plates did not increase the non-union rate as compared with traditional 4.5 mm plates (Figure 7.5).

Current recommendations include the use of 3.5 mm dynamic compression plates, applied in the compression mode to appropriate transverse and oblique fractures. In general, fixation to a minimum of 8 cortices (4 bicortical screws) on either side of the fracture is requisite although 6 (3 bicortical screws) may be adequate for some simple transverse fractures. In cases with comminution or bone loss, 10 or 12 hole plates should be utilized in conjunction with immediate autogenous iliac crest cancellous bone grafting.

The current rate of non-union is less than 2% when proper technique is utilized in compliant patients (Chapman *et al.*, 1989). Patients with

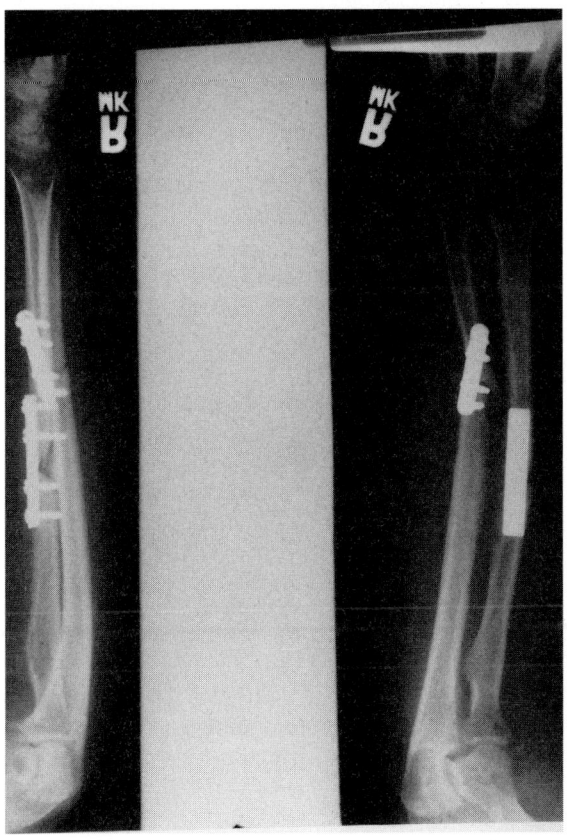

(a)

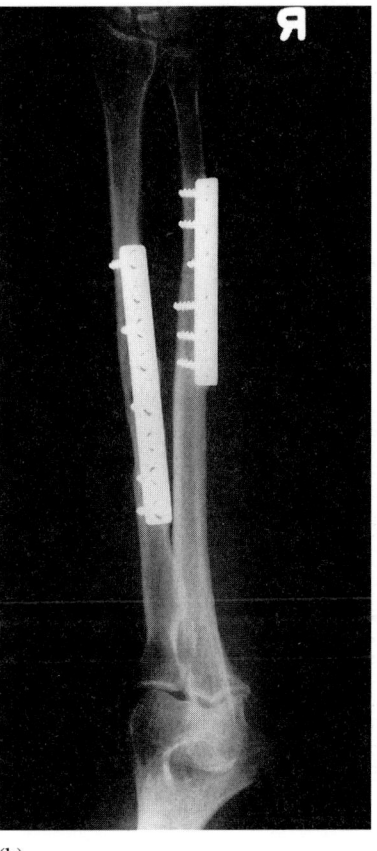

(b)

Figure 7.5 A 40-year-old woman who had failure to heal a fracture of both her radius and ulna due to grossly inadequate plate fixation. (a) The presenting radiographs reveal deformity and non-union with inadequate plate fixation. (b) Both ununited bones were stabilized with longer and more sturdy plates, with rapid union being the result. Full forearm rotation was achieved

stable plate and screw fixation are now mobilized almost immediately postoperatively without an apparent decrease in rates of union (Grace and Eversmann, 1980; Chapman *et al.*, 1989). Non-unions are ascribed to technical errors such as the use of plates of inadequate size (e.g. semitubular plates) or length, inadequate reduction, and failure to bone graft comminuted and open fractures (Hadden *et al.*, 1983; Stern and Drury, 1983; Chapman *et al.*, 1989; Ross *et al.*, 1989; Langkamer, 1991).

Malunion

Following the dramatic success of modern plate and screw fixation of forearm fractures, malunion has proved to be one of the most important determinants of functional outcome. Historically, a good reduction was regarded as one in which apposition of the fracture fragments of both bones without visually obvious deformity had been achieved. Substantial angulation and rotation of fragments was commonplace. Limitation of supination and pronation resulted from impingement between the malunited radius and ulna, rotational malunion, cross-union, and narrowing and contracture of the interosseous membrane, in addition to the stiffness resulting from prolonged immobilization (Figure 7.6).

The problem of rotational malunion was emphasized by Evans (1945) who noted that inadequate or unstable rotational reduction was a source of severe limitation of supination and/or pronation. Patrick (1946) emphasized the role of angular malunion in the limitation of rotational range of motion following forearm fractures. In severe angular malunion, especially when the apices of angular malunion are convergent, radioulnar impingement may occur. Subsequent studies of the effect of angular deformity on pronation and supination in cadaver forearms (performed by the advocates of closed treatment) demonstrated progressive loss of forearm rotation with increasing angular deformity (Sarmiento *et al.*, 1975; Matthews *et al.*, 1982; Tarr *et al.*, 1984). The authors conclude that limitation of rotational movement is acceptable (less than 20%) in closed reductions with 10° or less of combined deformity of the radius and ulna (essentially a near-perfect reduction which is very difficult to achieve).

The importance of restoring the normal anatomical radial bow was first noted by advocates of closed treatment and intramedullary nailing of

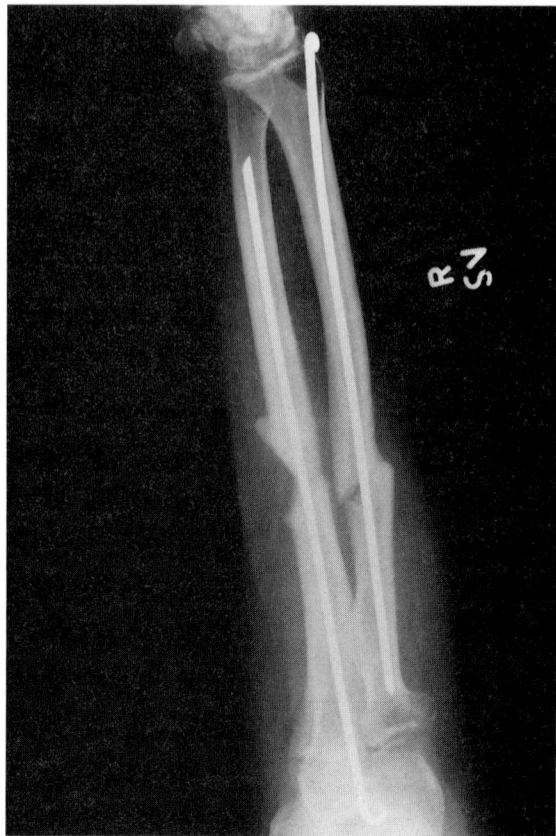

Figure 7.6   A both bone forearm fracture in a 56-year-old woman was treated with unstable intramedullary rod fixation. A rotational malunion developed

forearm fractures (Patrick, 1946; Sage, 1982). More recently, Schemitsch and Richards (1992), reporting on the effect of malunion following plate and screw fixation, demonstrated an important relationship between restoration of the normal magnitude and location of the maximal radial bow and functional outcome. The effects of prolonged immobilization on limitation of forearm motion were minimized in their study by the use of modern plate fixation in all patients with early postoperative mobilization in most. Accounting for the effects of associated soft tissue injury, ipsilateral fracture or complication on forearm rotation, they demonstrated that failure to restore the location and magnitude of the radial bow to within 4–5% of that of the normal arm was associated with greater than 20% loss of forearm rotation. Grip strength was also reduced in malunited fractures. They suggest that attention to restoration of the radial bow (including comparison with the uninjured forearm in difficult cases) with appropriate contouring of the plate should help to maximize functional outcomes.

Synostosis

Post-traumatic synostosis of the radius and ulna is a severely disabling complication estimated to occur in approximately 2% of both bone forearm fractures (Vince and Miller, 1987). It was first recognized in cadaveric dissections in the mid-nineteenth century as a so-called 'vicious union' of the radius and ulna which was regarded as resulting from malreduction with narrowing of the interosseous space or even cross-contact between radial and ulnar fracture fragments with resultant cross-union (Gurdjian and Smathers, 1945). Recent documentation of the local and systemic risk factors for the development of synostosis (including high-energy traumatic injury with soft tissue damage, fracture comminution, dislocation of adjacent joints and/or wide displacement of fracture fragments, prolonged immobilization and

associated head injury, multitrauma or burns) has illustrated the numerous similarities between this entity and heterotopic ossification (Botting, 1970; Watson and Eaton, 1978; Breit, 1983; Garland and Dowling, 1983; Maempel, 1984; Vince and Miller, 1987; Bauer *et al.*, 1991; Jupiter, 1991). Post-traumatic radioulnar synostosis is more common with fractures of the radius and ulna at the same level (Anderson *et al.*, 1975; Breit, 1983) and delayed internal fixation (Botting, 1970). It occurs more commonly in the proximal and mid-forearm than in the distal forearm (Vince and Miller, 1987).

Suggested treatment consists of resection of the synostosis with interposition of various materials intended to discourage recurrence (silastic, muscle or fat) (Yong-Hing and Tchang, 1983; Maempel, 1984; Vince and Miller, 1987) (Figure 7.7). Resection is associated with a risk of damage to neurovascular structures, especially in the proximal third of the forearm (Vince and Miller, 1987). The overall reported recurrence rate following resection is approximately 30% (Vince and Miller,

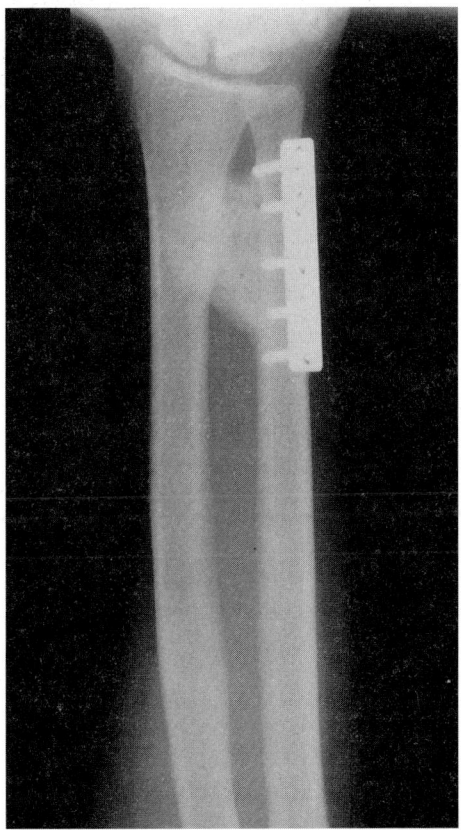
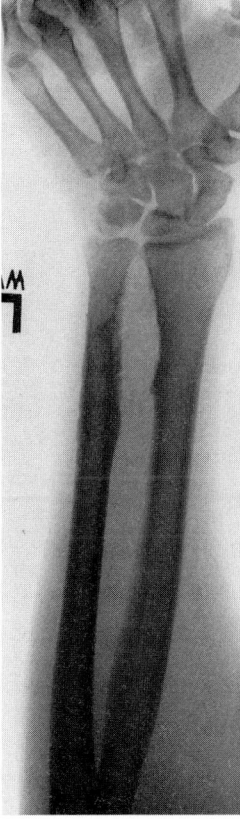

Figure 7.7 Following a fracture of the distal third of the ulna, a complete bony synostosis developed between the ulna and radius in a 32-year-old taxi driver. (a) The anteroposterior radiographs reveal the bony synostosis. (b) A resection of the bony block along with an interposition of free fat resulted in recovery of 50% of forearm rotation

(a)   (b)

1987). Noting the similarities between post-traumatic radioulnar synostosis and heterotopic ossification, postoperative radiation treatments have been attempted with some success in preventing recurrence (Failla *et al.*, 1989; Abrams *et al.*, 1993; Cullen *et al.*, 1994). To this end, non-steroidal anti-inflammatory agents (such as indomethacin) and frequent, early range of motion exercises may also be useful (Jupiter, 1991).

The timing of surgery is critical. With regard to heterotopic ossification of the hip and elbow, delayed intervention is most commonly advised to allow for the maturation of the new bone in the hope that this will decrease recurrence rates. However, the accepted measures of bone maturity (serum alkaline phosphatase level, radiography and bone scanning) are of limited reliability, and excessive delay can lead to contraction of soft tissues with resultant limitations in maximal recovery of range of motion and function (Volkmann, 1881; Jupiter, 1991; Cullen *et al.*, 1994). The use of postoperative radiation may improve the success of early resection (Failla *et al.*, 1989; Abrams *et al.*, 1993; Cullen *et al.*, 1994).

The difficulties of successful treatment of synostosis emphasize the importance of preventing this complication. Surgeon-related risk factors include violation of the interosseous space either by surgical exposure (Vince and Miller, 1987; Bauer *et al.*, 1991) or via a screw of excessive length (Vince and Miller, 1987; Ayllon-Garcia *et al.*, 1993), and placement of bone graft on the interosseous membrane (Vince and Miller, 1987). Proper, stable internal fixation with early motion should help limit the occurrence of synostosis to patients with substantial risk factors such as brain injury.

Refracture

As a result of improvements in implant metallurgy and design, the question has arisen: is routine implant removal necessary? Initially the AO/ASIF recommended removal of all implants following fracture healing (Müller *et al.*, 1965). Because the removal of forearm plates has been associated with a risk of refracture (through either the old fracture site or a screw hole), in addition to the risks of a second operation, the indications for forearm plate removal have been discussed frequently in the literature (Hidaka and Gustilo, 1984; De Luca *et al.*, 1988; Labosky *et al.*, 1990;

Rumball and Finnegan, 1990; Bednar and Grandwilewski 1992; Mih *et al.*, 1994).

The risk of refracture following plate removal is believed to result from a combination of incomplete healing and the osteoporosis which occurs under a plate as a result of some combination of disruption of the vascular supply to the bone and stress shielding. Risk factors for refracture following plate removal include fracture comminution or inability to gain compression of fracture fragments (De Luca *et al.*, 1988; Rumball and Finnegan, 1990), implant size (Chapman *et al.* 1989) noted that refracture was less likely following removal of 3.5 mm than 4.5 mm plates), implant removal earlier than 1 year post-injury (Hidaka and Gustilo, 1984; Rumball and Finnegan, 1990), radiolucency beneath the plate (De Luca *et al.*, 1988) and inadequate protection following plate removal (Hidaka and Gustilo, 1984; Rumball and Finnegan, 1990).

At this time, it is recommended that forearm plates remain in place unless: (1) they cause local symptoms (e.g. tenosynovitis); or (2) the patient is an athlete returning to high-energy activities (in which case the ends of the plates might be expected to act as stress risers and increase the risk of fracture). On the basis of the existing literature, the risk of refracture following plate removal can be expected to be minimal if fractures are fixed with 3.5 mm compression plates, the plates are not removed until at least 1 year following the original injury (perhaps longer in cases in which the fracture was comminuted or in which interfragmentary compression was not achieved), and patients are protected in a brace for 4–6 weeks and avoid high-energy activities for at least 3 months.

Radioulnar dissociation (Galeazzi, Monteggia and bipolar fracture dislocations)

A fracture dislocation of the forearm results in an extremely unstable skeletal dissociation. In addition to complete skeletal dissociation of the proximal and distal portions of the limb (via fracture in one component of the forearm skeleton and dislocation from the adjacent joint in the other), radioulnar dissociation also occurs. As a result, reduction of these fractures is difficult to maintain and they are prone to malunion with resultant radioulnar joint incongruity leading to severe loss of forearm rotation (Speed and Boyd, 1940;

Smith, 1947; Penrose, 1951; Hughston, 1957; Pavel *et al.*, 1965; Bado, 1967; Reckling and Cordell, 1968; Bruce *et al.*, 1974; Reckling, 1982) (Figure 7.8).

The radius and ulna are interconnected at the distal radioulnar joint by the triangular fibrocartilage ligamentous complex, in the mid-portion by the interosseous membrane and at the proximal radioulnar joint by the annular and quadrate liga-ments. Traumatic disruption of these structures occurs in three common patterns: (1) the so-called Galeazzi fracture is a fracture of the radial dia-physis (usually the distal third) in association with dislocation of the distal radioulnar joint; (2) the Monteggia fracture and its variants represent a fracture of the proximal ulna associated with prox-imal radioulnar joint disruption; and (3) bipolar forearm fracture dislocation in which complete

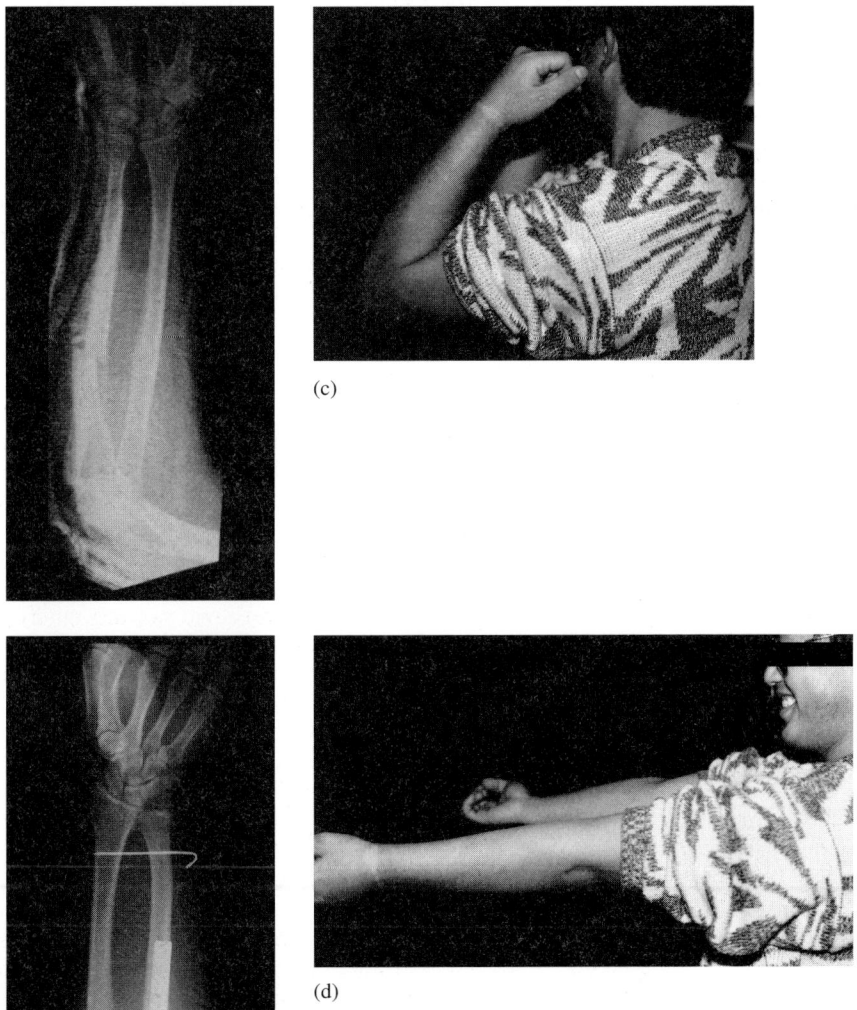

(a)

(b)

(c)

(d)

Figure 7.8 A 24-year-old labourer sustained a major trauma to his left arm when 1000 lb of lumber fell onto him while unloading a truck. (a) The presenting radiograph reveals displaced fractures of the radial head and ulnar diaphysis. In addition, there was an impacted fracture of the distal radial metaphysis and disruption of the interosseous membrane and distal radioulnar joint. (b) Stability to the floating radius was achieved by the placement of a silicone radial head prosthesis, pinning of the distal radius to the ulna, and plate fixation of the ulnar diaphysis. (c) Active flexion and extension of the elbow was begun within days following operative treatment. (d) Full elbow flexion and extension resulted along with restoration of nearly full forearm rotation

dissociation of the radius (floating radius) occurs as a result of disruption of the interosseous membrane and both proximal and distal radioulnar joints as a result of either fracture or dislocation. Stable anatomical reduction is essential for the healing of these structures with maintenance of radioulnar congruity. Failure of these ligaments to heal can result in radioulnar instability, and healing with subluxation can result in an extremely disabling loss of forearm rotation (Boyd and Boals, 1969; Kraus and Horne, 1985; Moore *et al.*, 1985a; Jupiter *et al.*, 1994b).

Reconstruction of the soft tissue constraints (e.g. fascial loop reconstruction of the annular ligament in Monteggia fracture dislocation (Speed and Boyd, 1940)) is unnecessary (Boyd and Boals, 1969). Treatment focused upon accurate anatomical reduction and stable plate and screw fixation has produced excellent results in the treatment of these fractures (Kraus and Horne, 1985; Moore *et al.*, 1985a).

Failure to recognize radioulnar dissociation in a fracture of the forearm represents a common source of complications. Radiographs should visualize the entire length of both bones and the wrist and elbow joints. High-energy injury mechanisms and isolated fractures of one forearm bone are more likely to be associated with radioulnar disruption. However, radioulnar dissociation can occur in any type of forearm fracture, including both bone fractures (Speed and Boyd, 1940; Bado, 1967; Mikic, 1975) and even isolated radial head fractures (the Essex-Lopresti lesion) (Curr and Coe, 1946; Essex-Lopresti, 1951; McDougall and White, 1957).

Clinical and anatomical investigations have determined a number of clues indicating distal radioulnar joint disruption which can be detected on radiographs: (1) fracture of the ulnar styloid at its base; (2) widening of the distal radioulnar joint space; (3) dislocation of the radius relative to the ulna seen on a true lateral radiograph; and (4) shortening of the radius beyond 5 mm relative to the distal ulna under a constant applied load (Moore *et al.*, 1985b; Schneiderman *et al.*, 1993). Proximally the radiohumeral ulnar joint is dislocated if a line through the radial shaft and head does not contact the capitellum in all positions of flexion/extension (McLaughlin, 1959; Guistra *et al.*, 1974).

In the so-called Galeazzi fracture dislocation, a fracture through the distal diaphysis of the radius and a dislocation of the ulnocarpal joint occur in addition to distal radioulnar joint disruption. Reduction of the extremely unstable distal radial fragment cannot be maintained by external means or intramedullary fixation due to the deforming forces of the brachioradialis, pronator quadratus, thumb extensors and weight of the hand (Hughston, 1957). Anatomical reduction and plate and screw fixation using a plate with six screws is optimal (Kraus *et al.*, 1985; Moore *et al.*, 1985a). Following plate fixation, the distal radioulnar joint is usually stable and the forearm can be mobilized immediately postoperatively (Kraus *et al.*, 1985). If the distal radioulnar joint remains unstable, provision of additional stability is required. This can be achieved either by fixation of a large ulnar styloid fracture fragment if present, or by transfixion of the distal ulna to the radius with one or two 0.062 inch Kirschner wires left in for 4 weeks (Mikic, 1975). In either case, the forearm is immobilized in a long-arm cast for 4 weeks while the soft tissues heal.

In Monteggia fractures, stable anatomical reduction of the ulna is sufficient for soft tissue healing, and early mobilization is recommended. Instability or incomplete reduction of the radial head is most commonly a result of ulnar malunion (Boyd and Boals, 1969) (Figure 7.9). Inability to relocate the radial head despite anatomical reduction of the ulna suggests soft tissue interposition (Reckling and Cordell, 1968; Boyd and Boals, 1969; Reckling, 1982).

Stable ulnar fixation requires placement of the plate on the dorsal, tension, surface of the ulna. Comminuted ulna fractures are prone to flexion at the fracture site (with resultant radial head dislocation) when plated on the lateral side (Figure 7.9).

If a missed radial head dislocation is recognized early, it may be possible to revise the ulnar reduction or perform and osteotomy of the ulna and reduce the radial head. In long-standing dislocations, radial head resections may be necessary.

Posterior interosseous nerve palsy has been reported with Monteggia fracture dislocations as the nerve is tented over the dislocated radial head (Spinner *et al.*, 1968; Stein *et al.*, 1971; Jessing, 1975). Complete spontaneous recovery can be expected following stable anatomical reduction. If the radial head remains unreduced, a tardy palsy of the radial head can develop (Lichter and Jacobsen, 1975; Austin, 1976).

Injuries causing disruption of the interosseous membrane and complete radioulnar dissociation (the 'floating' radius) are now more commonly recognized (Edwards and Jupiter, 1988; Bock *et al.*, 1992; Eglseder and Hay, 1993; Jupiter *et al.*, 1994b). The so-called Essex-Lopresti injury (distal

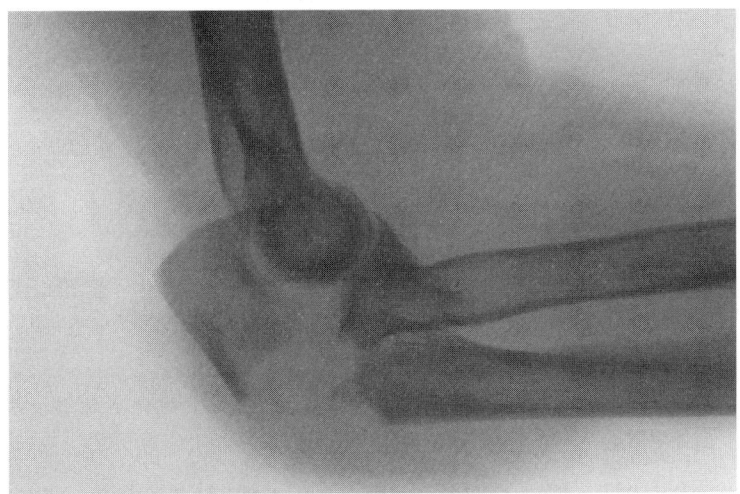

(a)

Figure 7.9   A 60-year-old woman had a posterior Monteggia fracture treated by intramedullary fixation. (a) Lateral radiograph of the initial fracture. (b) A longitudinal fixation pin failed to provide adequate stability, leading to a flexion deformity at the fracture site. (c) The fracture was realigned and stabilized with a dorsally applied contoured 3.5 mm LC–DCP. Full flexion was achieved with a flexion contracture of 20°. Full forearm rotation was achieved

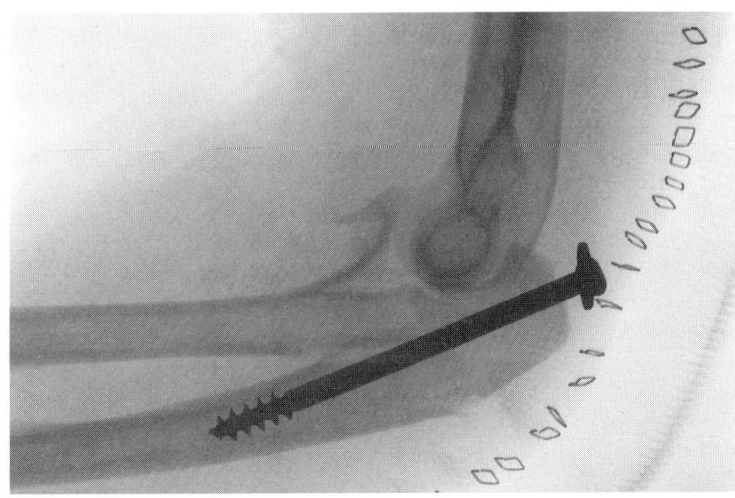

(b)

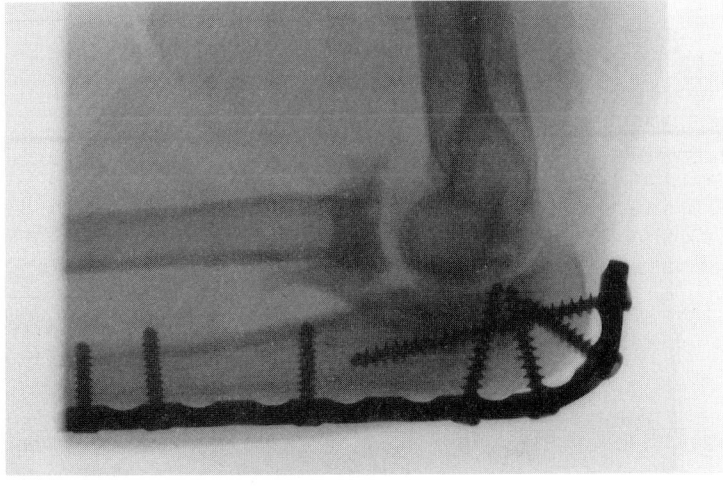

(c)

radioulnar joint dislocation in association with radial head fracture) was the first such injury to be described (Curr and Coe, 1946; Essex-Lopresti, 1951; McDougall and White, 1957), and many floating radius injuries continue to be reported as Essex-Lopresti variants (Bock *et al.*, 1992; Eglseder and Hay, 1993). In our opinion, the more inclusive term 'bipolar fracture dislocation', coined by Odena (1952) to describe complex forearm injuries with disruption of both proximal and distal radioulnar articulations, provides a means by which to categorize all such injuries with complete radioulnar dissociation (floating radius).

With bipolar fracture dislocations, as with Galeazzi or Monteggia fracture dislocations, one of the most important determinants of outcome is recognition of the full extent of injury. The injuries are often complex, but the treatment principles are the same as for other forearm fractures: stable anatomical reduction with adequate plate and screw fixation (Eglseder and Hay, 1993; Jupiter *et al.*, 1994b) (Figure 7.8). Restoration of anatomical length is particularly important in these injuries. Radial head fractures must be reduced and fixed or resected and replaced by a silicone prosthesis because excision without prosthetic replacement will lead to non-anatomical healing of the interosseous membrane in a shortened position and severe loss of rotation (Hotchkiss *et al.*, 1989; Jupiter *et al.*, 1994b).

Summary

This chapter has illustrated that with a sound understanding of the need for anatomical reduction, stable plate and screw fixation and early mobilization of forearm skeletal injuries, avoidance of complications becomes dependent upon complete injury definition and clinical vigilance. Prompt recognition of compartment syndrome and radioulnar dissociation will lead to early appropriate treatment and improved outcomes, avoidance of intraoperative violation of the interosseous space will minimize radioulnar synostosis, and appropriate selection, operative timing and postoperative protection of patients undergoing forearm plate removal should reduce refracture rates.

References

Abrams, R.A., Simmons, B.P. and Brown, R.A. (1993) Treatment of posttraumatic radioulnar synostosis with excision and low-dose radiation. *J. Hand Surg.*, **18A**, 703–7

Anderson, L.D., Sisk, T.D., Tooms, R.E. and Park, W.I. III. (1975) Compression-plate fixation in acute diaphyseal fractures of the radius and ulna. *J. Bone Joint Surg.*, **57A**, 287–97

Austin, R. (1976) Tardy palsy of radial nerve from a Monteggia fracture. *Injury*, 7, 202

Ayllon-Garcia, A., Davies, A.W. and Deliss, L. (1993) Radioulnar synostosis following external fixation. *J. Hand Surg.*, **18B**, 592–4

Bado, J.L. (1967) The Monteggia lesion. *Clin. Orthop.*, **50**, 71–6

Bagley, C.H. (1926) Fracture of both bones of the forearm. *Surg. Gynecol. Obsttet.*, **42**, 95–102

Baker, G.I., Burkhalter, W.E., Barclay, W.A. and Eversmann, W.W. (1969) Treatment of forearm shaft fractures by long slotted plates. *J. Bone Joint Surg.*, **51A**, 1035

Bauer, G., Arand, M., Mutschler, W. (1991) Post-traumatic radioulnar synostosis after forearm fracture osteosynthesis. *Arch. Orthop. Trauma Surg.*, **110**, 142–5

Bednar, D.A. and Grandwilewski, W. (1992) Complications of forearm-plate removal. *Can. J. Surg.*, **35**, 428–31

Bock, G.W., Cohen, M.S. and Resnick, D. (1992) Fracture-dislocation of the elbow with inferior radioulnar dislocation: a varient of the Essex-Lopresti injury. *Skel. Radiol.*, **21**, 315–17

Bolton, H. and Quinlan, A.G. (1952) The conservative treatment of fractures of the shaft of the radius and ulna in adults. *Lancet*, **1**, 700–5

Botting, T.D.J. (1970) Post-traumatic radio-ulnar cross-union. *J. Trauma*, **10**, 16–24

Boyd, H.B. and Boals, J.C. (1969) The Monteggia lesion. *Clin. Orthop.*, **66**, 94–100

Breit, R. (1983) Post-traumatic radioulnar synostosis. *Clin. Orthop.*, **174**, 149–52

Brostrom, L.A., Stark, A. and Svartengren, G. (1990) Acute compartment syndrome in forearm fractures. *Acta Orthop. Scand.*, **61**, 50–53

Brown, P.W. (1970) The time factor in surgery of upper extremity peripheral nerve injury. *Clin. Orthop.*, **68**, 14–21

Bruce, H.E., Harvey, J.P. and Wilson, J.C. (1974) Monteggia fractures. *J. Bone Joint Surg.*, **56A**, 1563–76

Burwell, H.N. and Charnley, A.D. (1964) Treatment of forearm fractures in adults with particular reference to plate fixation. *J. Bone Joint Surg.*, **46B**, 404–24

Buxton, J.D. (1925) Discussion of the treatment of fractures of the forearm, excluding fractures of the olecranon and those of the lower end of the radius of the Colles type. *J. R. Soc. Med.*, **19**, 17–30

Buxton, J.D. (1939) Treatment of closed fractures of the radius and ulna. *B.M.J.*, **2**, 795–9

Caden, J.G. (1961) Internal fixation of fractures of the forearm. *J. Bone Joint Surg.*, **43A**, 1115–21

Carrell, W.B. (1938) Fractures of both bones of the forearm excluding those at the elbow joint and wrist joint. *Surg. Gynecol. Obstet.*, **66**, 506–11

Chapman, M.W., Gordon, J.E. and Zissimos, A.G. (1989) Compression plate fixation of acute fractures of the diaphysis of the radius and ulna. *J. Bone Joint Surg.*, **71A**, 159–69

Compere, E.L. (1948) The treatment of fractures of both bones of the forearm. *Surg. Clin. N. Am.*, **25**, 48–58

Cowie, R.J. (1956) Fractures of the forearm treated by open reduction and plating. *Br. J. Surg.*, **44**, 263–6

Cullen, J.P., Pellegrini, V.D., Miller, R.J. and Jones, J.J. (1994) Treatment of traumatic radioulnar synostosis by excision and postoperative low-dose irradiation. *J. Hand Surg.*, **19A**, 394–401

Curr, J.F. and Coe, W.A. (1946) Dislocation of the inferior radioulnar joint. *Br. J. Surg.*, **34**, 74–7

De Luca, P.A., Lindsey, R.W. and Rowe, P.A. (1988) Refracture of bones of the forearm after the removal of compression plates. *J. Bone Joint Surg.*, **70A**, 1372–6

Dodge, H.S. and Cady, G.W. (1972) Treatment of fractures of the radius and ulna with compression plates. A retrospective study of one hundred and nineteen fractures in seventy-eight patients. *J. Bone Joint Surg.*, **54A**, 1167–76

Duncan, R., Geissler, W., Freeland, A.E. and Savoie, F.H. (1992) Immediate internal fixation of open fractures of the diaphysis of the forearm. *J. Orthop. Trauma.*, **6**, 22–31

Eaton, R.G. and Green, W.T. (1975) Volkmann's ischemia: a volar compartment syndrome of the forearm. *Clin. Orthop.*, **113**, 58–64

Edwards, G.S. and Jupiter, J.B. (1988) Fractures with acute distal radioulnar joint dislocation: the Essex-Lopresti lesion revisited. *Clin. Orthop.*, **234**, 61–69

Eglseder, W.A. and Hay, M. (1993) Combined Essex-Lopresti and radial shaft fractures: case report. *J. Trauma*, **34**, 310–12

Eliason, E.L., Brown, R.B. and Kaplan, L. (1937) Fractures of the forearm – except Colles'. *Am. J. Surg.*, **38**, 511–25

Engkvist, O., Wahren, L.K., Wallin, G. *et al.* (1985) Effects of regional intravenous guanethidine block in post-traumatic cold intolerance in hand amputees. *J. Hand Surg.*, **10B**, 145–50

Essex-Lopresti, P. (1951) Fractures of the radial head with distal radioulnar dislocation. *J. Bone Joint Surg.*, **33B**, 244–7

Evans, E.M. (1945) Rotational deformities in the treatment of fractures of both bones of the forearm. *J. Bone Joint Surg.*, **27**, 373–9

Failla, J.M., Amadio, P.C., Morrey, B.F. (1989) Post-traumatic proximal radio-ulnar synostosis: results of surgical treatment. *J. Bone Joint Surg.*, **69A**, 1208–13

Freelander E. (1986) The relationship between cold intolerance and cutaneous blood flow in digital replantation patients. *J. Hand Surg.*, **11B**, 15–19

Garland, D.E. and Dowling, V. (1983) Forearm fractures in the head-injured adult. *Clin. Orthop.*, **176**, 190–6

Geissler, W.B., Fernandez, D.L. and Graca, R. (1990) Anterior interosseous nerve palsy complicating a forearm fracture in a child. *J. Hand Surg.*, **15A**, 44–7

Gelberman, R.H. and Blasingame, J.P. (1977) The timed Allen test. *J. Trauma*, **21**, 477–9

Gelberman, R.H., Nunley, J.A., Koman, L.A. *et al.* (1982) The results of radial and ulnar arterial repair in the forearm. *J. Bone Joint Surg.*, **64A**, 383–7

Gelberman, R.H., Urbaniak, J.R., Bright, D.S. and Levin, L.S. (1978b) Digital sensibility following replantation. *J. Hand Surg.*, **3A**, 313–19

Gelberman, R.H., Gould, R.N., Hargens, A.R. and VandeBerg, J.S. (1983) Lacerations of the ulnar artery: hemodynamic, ultrastructural, and compliance changes in the dog. *J. Hand Surg.*, **8A**, 306–9

Gelberman, R.H., Garfin, S.R., Hergenroeder, P.T. *et al.* (1981) Compartment syndromes of the forearm: diagnosis and treatment. *Clin. Orthop.*, **161**, 252–61

Gelberman, R.H., Zakaib, G.S., Mubarak, S.J. *et al.* (1978a) Decompression of forearm compartment syndromes. *Clin. Orthop.*, **134**, 225–9

Giustra, P.E., Killoran, P.J., Furman, R.S. and Root, J.A. (1974) The missed Monteggia fracture. *Radiology*, **10**, 45–7

Grace, T.G. and Eversmann, W.W. (1976) The management of segmental bone loss associated with forearm fractures. *J. Bone Joint Surg.*, **58A**, 283–4

Grace, T.G. and Eversmann, W.W. Jr. (1980) Forearm fractures: treatment by rigid fixation with early motion. *J. Bone Joint Surg.*, **62A**, 433–8

Gurdjian, E.S. and Smathers, H.M. (1945) Peripheral nerve injury in fractures and dislocations of long bones. *J. Neurosurg.*, **2**, 202–19

Hadden, W.A., Reschauer, R. and Seggl, W. (1983) Results of AO plate fixation of forearm shaft fractures in adults. *Injury*, **15**, 44–52

Helfet, D.L., Howey, T., Sanders, R. and Johansen, K. (1990) Limb salvage versus amputation: preliminary results of the mangled extremity severity score. *Clin. Orthop.*, **256**, 80–6

Henry, A.K. (1973) *Extensile Exposure*, 2nd edn. Edinburgh: Churchill Livingstone

Hicks, J.H. (1961) Fractures of the forearm treated by rigid fixation. *J. Bone Joint Surg.*, **43B**, 680–7

Hidaka, S. and Gustillo, R.B. (1984) Refracture of bones of the forearm after plate removal. *J. Bone Joint Surg.*, **66A**, 1241

Hotchkiss, R.N., An, K., Sowa, D.T. *et al.* (1989) An anatomic and mechanical study of the interosseous membrane of the forearm: pathomechanics of proximal migration of the radius. *J. Hand Surg.*, **14A**, 256–61

Hughston, J.C. (1957) Fracture of the distal radial shaft. Mistakes in management. *J. Bone Joint Surg.*, **39A**, 249–64

Jessing, P. (1975) Monteggia lesions and their complicating nerve damage. *Acta. Orthop. Scand.*, **46**, 601–9

Jinkins, W.J. Jr, Lockhart, L.D. and Eggers, G.W.N. (1960) Fracture of the forearm in adults. *South. Med. J.*, **53**, 669–79

Jones, J.A. (1991) Immediate internal fixation of high-energy open forearm fractures. *J. Orthop. Trauma*, **5**, 272–9

Jupiter, J.B., Seiler, J.G. and Zienowicz, R. (1994a) Sympathetic maintained pain (causalgia) associated with a demonstrable peripheral-nerve lesion. *J. Bone Joint Surg.*, **76A**, 1376–84

Jupiter, J.B. and Kleinert, H.E. (1988) Vascular injuries in the upper extremity. In *The Hand*, vol. 3. (R. Tubiana, ed.) pp. 593–611, Philadelphia: Saunders

Jupiter, J.B., Kour, A.K., Richards, R.R. *et al.* (1994b) The floating radius in bipolar fracture-dislocation of the forearm. *J. Orthop. Trauma*, **8**, 99–106

Jupiter, J.B. (1991) *Heterotopic Ossification About the Elbow.* Instr. Course Lect., vol. XL. AAOS

Jupiter, J.B. (1991) Nerve injury associated with devascularizing trauma. In *Operative Nerve Repair and Reconstruction* (R.H. Gelberman, ed.) Philadelphia: Lippincott

Kline, D.G. and De Jong, B.R. (1968) Evoked potentials to evaluate peripheral nerve injuries. *Surg. Gynecol. Obstet.*, **127**, 1239–48

Knight, R.A., Purvis, G.D. (1949) Fractures of both bones of the forearm in adults. *J. Bone Joint Surg.*, **31A**, 755–64

Kraus, B. and Horne, G. (1985) Galeazzi fractures. *J. Trauma*, **25**, 1093–5

Labosky, D.A., Cermak, M.B. and Waggy, C.A. (1990) Forearm fracture plates: to remove or not to remove? *J. Hand Surg.*, **15A**, 294–301

Langkamer, V.G. and Ackroyd, C.E. (1991) Internal fixation of forearm fractures in the 1980s: lessons to be learnt. *Injury*, **22**, 97–102

Lenihan, M.R., Brien, W.W., Gellman, H. *et al.* (1992) Fractures of the forearm resulting from low velocity gunshot wounds. *J. Orthop. Trauma*, **6**, 32–5

Levin, L.S., Goldner, R.D., Urbaniak, J.R. *et al.* (1990) Management of severe musculoskeletal injuries of the upper extremity. *J. Orthop. Trauma*, **4**, 432–40

Lichter, R.L. and Jacobsen, T. (1975) Tardy palsy of the posterior interosseous nerve with a Monteggia fracture. *J. Bone Joint Surg.*, **57A**, 124–5

Maempel, F.Z. (1984) Post-traumatic radioulnar synostosis. A report of two cases. *Clin. Orthop.*, **186**, 182–5

Magnuson, P.B. (1922) Mechanics of treatment of fractures of the forearm. *J.A.M.A.*, **78**, 789–94

Marek, F.M. (1961) Axial fixation of forearm fractures. *J. Bone Joint Surg.*, **43A**, 1099–114

Matthews, L.S., Kaufer, H., Garver, D.F. and Sonstegard, D.A. (1982) The effect on supination-pronation of angular malalignment of fractures of bone bones of the forearm. An experimental study. *J. Bone Joint Surg.*, **64A**, 14–17

McDougall, A. and White, J. (1957) Subluxation of the inferior radioulnar joint complicating fracture of the radial head. *J. Bone Joint Surg.*, **39B**, 278–87

Mclaughlin, H.L. (1959) *Trauma*. Philadelphia: Saunders

Mih, A.D., Cooney, W.P., Idler, R.S. and Lewallen, D.G. (1994) Long-term follow-up of forearm bone diaphyseal plating. *Clin. Orthop.*, **299**, 256–9

Mikic, Z.D. (1975) Galeazzi fracture-dislocations. *J. Bone Joint Surg.*, **57A**, 1071–80

Mitchell, S.W., Morehouse, G.R. and Keen, W.W. (1864) *Gunshot Wounds and Other Injuries of Nerves*. Philadelphia: Lippincott

Moed, B.R. and Fakhouri, A.J. (1991) Compartment syndrome after low velocity gunshot wounds to the forearm. *J. Orthop. Trauma*, **5**, 134–7

Moore, T.M., Klein, J.P., Patzakis, M.J. and Harvey, J.P. Jr. (1985a) Results of compression plating of closed Galeazzi fractures. *J. Bone Joint Surg.*, **67A**, 1015–21

Moore, T.M., Lester, D.K. and Sarmiento, A. (1985b) The stabilizing effect of soft tissue constraints in artificial Galeazzi fractures. *Clin. Orthop.*, **194**, 189–94

Morris, A.H. (1974) Irreducible Monteggia lesion with radial-nerve entrapment. *J. Bone Joint Surg.*, **56A**, 1744–6

Mubarak, S.J., Owen, C.A. and Hargens, A.R. (1978) Acute compartmental syndromes: diagnosis and treatment with the aid of the wick catheter. *J. Bone Joint Surg.*, **60A**, 1091

Müller, M.E., Allgower, M. and Willenegger, H. (1965) *Technique of Internal Fixation of Fractures*. New York: Springer-Verlag

Naiman, P.T., Schein, A.J. and Siffert, R.S. (1970) Use of ASIF compression plates in selected shaft fractures of the upper extremity. A preliminary report. *Clin. Orthop.*, **71**, 208–16

Nicoll, E.A. (1956) The treatment of gaps in long bones by cancellous insert grafts. *J. Bone Joint Surg.*, **38B**, 70

Nunley, J.A. and Urbaniak, J.R. (1980) Partial bony entrapment of the median nerve in a greenstick fracture of the ulna. *J. Hand Surg.*, **5A**, 557–9

Nylander, G., Nylander, E. and Lassvik, C. Cold sensitivity after replantation in relation to arterial circulation and vasoregulation. *J. Hand Surg.*, **12B**, 78–81

Odena, I.C. (1952) Bipolar fracture-dislocation of the forearm. *J. Bone Joint Surg.*, **34A**, 968–76

Omer, G.E. Jr. (1974) Injuries to neres of the upper extremity. *J. Bone Joint Surg.*, **56A**, 1615–24

Omer, G.E. Jr. (1982) Results of untreated peripheral nerve injuries. *Clin. Orthop.*, **163**, 15–19

Page, C.M. (1956) An operation for the relief of flexion contracture in the forearm. *J. Bone Joint Surg.*, **38B**, 70

Parks, A. (1951) The treatment of established Volkmann's contracture by tendon transplantation. *J. Bone Joint Surg.*, **33B**, 359

Patrick, J. (1946) A study of supination and pronation, with especial reference to the treatment of forearm fractures. *J. Bone Joint Surg.*, **28B**, 737–48

Pavel, A., Pitman, J.M., Lance, E.M. and Wade P.A. (1965) The posterior Monteggia fracture. A clinical study. *J. Trauma*, **5**, 185–99

Penrose, J.H. (1951) The Monteggia fracture with posterior dislocation of the radial head. *J. Bone Joint Surg.*, **33B**, 65

Prosser, A.J. and Hooper, G. Entrapment of the ulnar nerve in a greenstick fracture of the ulna. *J. Hand Surg.*, **11B**, 211–12

Raj, P., Calodney, A., Jainisse, T. and Cannella, J. (1992) Reflex sympathetic dystrophy. In *Skeletal Trauma. Fractures, Dislocations, Ligamentous Injuries*, vol. 1. (B. Browner, J.B. Jupiter, A. Levine and P. Trafton, eds) pp. 471–99, Philadelphia: Saunders

Reckling, F.W. (1982) Unstable fracture-dislocations of the forearm (Monteggia and Galeazzi lesions). *J. Bone Joint Surg.*, **64A**, 857–63

Reckling, F.W. and Cordell, L.D. 1968 Unstable fracture-dislocations of the forearm: the Monteggia and Galeazzi lesions. *Arch. Surg.* **96**, 999–1007

Ross, E.R.S., Gourevitch, D., Hastings, G.W. *et al.* (1989) Retrospective analysis of plate fixation of diaphyseal fractures of the forearm bones. *Injury*, **20**, 211–14

Rumball, K. and Finnegan M. (1990) Refractures after forearm plate removal. *J. Orthop. Trauma*, **4**, 124–9

Sage, F.P. (1982) Medullary fixation of fractures of the forearm: a study of the medullary canal of the radius and a report on 50 fractures of the radius treated with a pre-bent triangular nail. *J. Bone Joint Surg.*, **64A**, 857–63

Sarmiento, A., Cooper, J.S. and Sinclair, W.F. (1975) Forearm fractures. Early functional bracing – a preliminary report. *J. Bone Joint Surg.*, **51A**, 297–304

Sarmiento, A., Ebramzadeh, E., Brys, D. and Tarr, R. (1992) Angular deformities and forearm function. *J. Orthop. Res.*, **10**, 121–33

Schemitsch, E.H., Richards, R.H. (1992) The effect of malunion on functional outcome after plate fixation of fractures of both bones of the forearm in adults. *J. Bone Joint Surg.*, **74A**, 1068–78

Schneiderman, G., Meldrum, R.D., Bloebaum, R.D. *et al.* (1993) The interosseous membrane of the forearm: structure and its role in Galeazzi fractures. *J. Trauma*, **35**, 879–85

Schuind F., Andrianne, Y. and Burny, F. (1991) Treatment of forearm fractures by Hoffman external fixation. *Clin. Orthop.*, **266**, 197–204

Seigel, D.B. and Gelberman, R.H. (1991) Peripheral nerve injuries associated with fractures and dislocations. In *Operative Nerve Repair and Reconstruction* (R.H. Gelberman, ed.) pp. 619–33. Philadelphia, Lippincott

Slauterbeck, J.R., Britton, C., Moneim, M.S. and Clevenger, F.W. (1994) Mangled extremity severity score: an accurate guide to treatment of the severely injured upper extremity. *J. Orthop. Trauma*, **8**, 282–5

Smith, F.M. (1947) Monteggia fractures. An analysis of twenty-five consecutive fresh injuries. *Surg. Gynecol. Obstet.*, **85**, 630–40

Smith, H. and Sage, F.P. (1957) Medullary fixation of forearm fractures. *J. Bone Joint Surg.*, **39A**, 91–8

Speed, J.S. and Boyd, H.B. (1940) Treatment of fracture of the ulna with dislocation of the head of the radius (Monteggia fracture). *J.A.M.A.*, **115**, 1699

Spinner, M., Freundlich, B.D. and Teicher, J. (1968) Posterior interosseous nerve palsy as a complication of Monteggia fractures in children. *Clin. Orthop.*, **58**, 141–5

Spira, E. (1954) Bridging of bone defects in the forearm with iliac graft combined with intramedullary nailing. *J. Bone Joint Surg.*, **36B**, 642–6

Stein, F., Grabias, S.L. and Deffer, P.A. (1971) Nerve injuries complicating Monteggia lesions. *J. Bone Joint Surg.*, **53A**, 1432–6

Stern, P.J. and Drury, W.J. (1983) Complications of plate fixation of forearm fractures. *Clin. Orthop.*, **175**, 25–9

Strandness, D.E. Jr, ed. (1969) *Collateral Circulation in Clinical Surgery.* Philadelphia: Saunders

Street, D.M. (1986) Intramedullary forearm nailing. *Clin. Orthop.*, **212**, 219–30

Sunderland, S. (1978) *Nerve and Nerve Injuries*, 2nd edn. Edinburgh: Churchill Livingston

Tarr, R.R., Garfinkel, A.I. and Sarmiento, A. (1984) The effects of angular and rotational deformities of both bones of the forearm. *J. Bone Joint Surg.*, **66A**, 65–70

Teipner, W.A. and Mast, J.W. (1980) Internal fixation of forearm fractures: double plating verses single compression (tension band) plating – A comparative Study. *Orthop. Clin. North Am.* **11**, 381–91

Tile, M. and Petrie, D. (1969) Fractures of the radius and ulna. *J. Bone Joint Surg.*, **51B**, 193

Trumble, T., Seaber, A.V. and Urbaniak, J.R. (1987) Patency after repair of forearm arterial injuries in animal models. *J. Hand Surg.*, **12A**, 47–53

Tsuge, K. (1975) Treatment of established Volkmann's contractures. *J. Bone Joint Surg.*, **57A**, 925

Vince, K.G. and Miller, J.E. (1987) Cross-union complicating fractures of the forearm. Part I – adults. *J. Bone Joint Surg.*, **69A**, 640–53

Volkmann, R. (1881) Die ischaemischen Muskellähungen und-Kontrakturen. *Zentralbl. Chir.*, 51–801

Watson, F.M. Jr and Eaton, R.G. (1978) Post-traumatic radio-ulnar synostosis. *J. Trauma*, **18**, 467–8

Whipple, A.O. and St. John, F.B. (1917) A study of one hundred consecutive fractures of the shafts of both bones of the forearm with the end-results in ninety-five. *Surg. Gynecol. Obstet.*, **25**, 77–91

Wolfe, J.S. and Eyring, E.J. (1974) Median-nerve entrapment within a greenstick fracture. *J. Bone Joint Surg.*, **56A**, 1270–2

Wood, M.B. (1987) Upper extremity reconstruction by vascularized bone transfers: results and complications. *J. Hand Surg.*, **12A**, 422–7

Yong-Hing, K. and Tchang, S.P.K. (1983) Traumatic radio-ulnar synostosis treated by excision and a free fat transplant. A report of two cases. *J. Bone Joint Surg.*, **65B**, 433–5

8

Fractures of the forearm in children

K. Graham

Fracture of the forearm bones is one of the most common fractures in childhood and is the most frequent cause of admission to our paediatric fracture unit, accounting for 37% of all emergency admissions. Forearm fractures occur in variable frequency throughout childhood, reflecting the changing activities of a growing child. They are rarely seen before walking age, and child abuse or a fragility syndrome should be suspected if the history is inappropriate in a pre-walker. Once a toddler is on his or her feet, the world is waiting to be explored but the world, including the toddler's own home, can prove to be a hazardous place. Until the art of climbing is mastered, most toddlers' forearm fractures are sustained in minor falls. The energy transfers are low and the fractures are usually simple buckle or greenstick patterns. Preschool children are typically full of curiosity, and the ability to run, jump and climb is well developed before the cortical centres for self-preservation. Hence, there is a sharp rise in the incidence of forearm fractures in children, and occurring at this age they are more likely to be displaced.

School-age children add the hazards of sport to the other causes of forearm fractures. When a new 'craze' such as skateboarding or inline skating appears, 'epidemics' of forearm fractures may occur before the need for protective clothing and proper instruction and supervision of children is recognized. Children of varying ages are often exposed to the hazards of the internal combustion engine before they have the maturity and skills to deal with speed. Whether it is snowmobiling in Canada or riding trial motorcycles in Ireland, the result tends to be the same, and children sustain serious injuries and are sometimes killed (Henderson *et al.*, 1987).

Children also sustain forearm fractures as a result of unguarded access to farm machinery in rural areas and industrial machinery at home or their parents' place of work. Happily these injuries are almost a thing of the past as a direct result of protective legislation.

Boys are more likely than girls to sustain forearm fractures, and the left side is more frequently injured than the right, probably because of the protective role of the non-dominant limb in falls onto the outstretched hand. Forearm fractures and upper limb fractures in general are a seasonal phenomenon: twice as many fractures are sustained in the spring and summer as in the winter.

Types and natural history

Injury patterns in childhood

Children's bones are different biomechanically from those of adults and the patterns of failure under load are also different. Immature bone is more elastic than mature bone and has a thicker, stronger periosteum. Fracture comminution and displacement are less severe in children than in adults. Specific patterns of failure such as plastic bowing, torus fractures and greenstick fractures are commonly seen in children (Figure 8.1). Physeal injuries are also specific to growing children and fractures of the distal forearm may result in growth arrest or progressive deformity. More frequently, growth aids surgeons in the management of forearm fractures in that the active remodelling process may improve an initially disappointing result.

Healing is more rapid in children, and stiffness secondary to immobilization is uncommon. Hence internal fixation of forearm fractures is performed mainly to hold a reduction and is

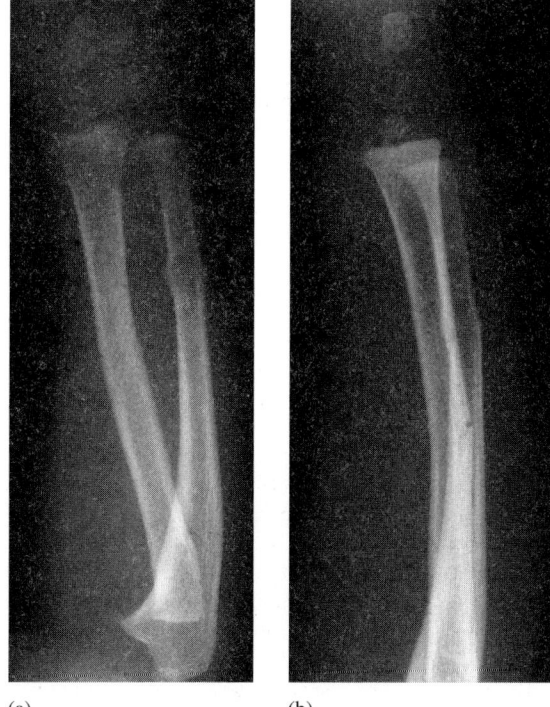

(a)   (b)

Figure 8.1   (a) AP and (b) lateral X-rays of a toddler's forearm showing the patterns of failure of immature bone. In the midshaft of the radius there is a greenstick fracture, and at the junction of the middle and distal thirds of the ulna there is a torus or buckle fracture

usually supplemented with a cast, early mobilization being of little benefit. This dictates a rather different approach to internal fixation from what would be appropriate in the management of forearm fractures in the adult.

The remodelling process varies enormously according to the age of the child and the fracture site and displacement. Some knowledge of these principles is essential to the logical management of forearm fractures in children. Remodelling is more rapid and complete in younger children with active physes. In the forearm, the distal physes are much more active than the proximal physes. Hence residual displacement of a fracture of the radial neck may result in a permanent loss of forearm rotation, but there are very few poor results after distal radial physeal injuries.

Remodelling is rapid and usually complete in the distal metaphysis of the radius and ulna. Friberg (1979) has reported almost 1° per month of correction of residual angulation deformity of the distal radial epiphysis. Remodelling is,

however, slow and unreliable in the diaphysis. The process is one of the smoothing of bumps and jagged ends. Alignment and rotation are scarcely affected. The surgeon who relies on remodelling as a substitute for good fracture care will frequently be both disappointed and embarrassed.

Classification of forearm fractures in children

Classification of forearm fractures in children should proceed in an orderly fashion, starting with the anatomical site. At either end of the radius and ulna, *epiphyseal injuries* are seen and are classified according to the standard Salter–Harris system, which is illustrated in Chapter 2. The more complex systems are not helpful in day-to-day fracture assessment and management. There are a small number of injuries which do not fit the standard Salter–Harris classification, but some will not fit any recognized classification and it is unhelpful to devise new categories for one-off patterns of injury.

Metaphyseal injuries are even more common than physeal injuries and obviously are again seen at either end of the radius and ulna.

The *diaphyseal fractures* are conveniently divided into proximal, middle and distal thirds. Frequently, fractures of both bones are located in different thirds which may or may not be contiguous. This type of fracture pattern often indicates indirect trauma of a twisting nature.

Fractures are usually transverse or short oblique. Long spiral patterns are uncommon. Comminution and segmental fractures are also uncommon. Pathological fractures result from either focal bone abnormalities, e.g. fracture of the distal radius through a simple bone cyst, or from generalized disorders such as osteogenesis imperfecta.

Dislocation of the radioulnar articulations in association with forearm fractures must be recognized and managed appropriately. The Monteggia injury is quite common in children but the Galeazzi pattern is rare, the reverse of the adult pattern.

Monteggia lesions can be classified using the Bado system, according to the direction of radial head dislocation. This is illustrated in Chapter 2. Type 1 has anterior dislocation, type 2 posterior dislocation and type 3 lateral dislocation of the radial head. Type 4 is a rare combination of angulated diaphyseal fractures of the proximal radius and ulna, in association with anterior dislocation of the radial head. This is such a rare injury that a

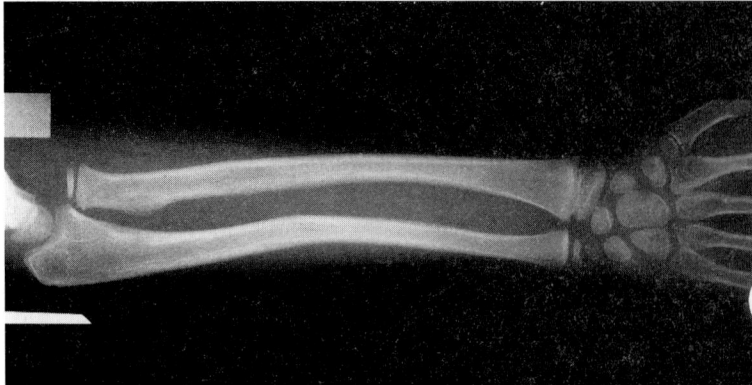

Figure 8.2 Bado type 1 Monteggia fracture dislocation. The ulnar fracture is angulated but has good periosteal stability. Reduction of the radial head dislocation and the ulnar fracture were straightforward .

separate group may be inappropriate (Figure 8.2) (Olney and Menelaus, 1989). There is also an important group of 'Monteggia-equivalent' injuries, including fracture through the radial neck in place of the radial head dislocation and plastic bowing of the ulna in place of a greenstick or complete fracture of the ulna.

Galeazzi fracture dislocations are rare in children and 'Galeazzi-equivalent' injuries may be more common than the classical lesion seen in adults. These have been classified by Letts into four groups according to the site and direction of the radial fracture and according to whether the injury to the distal ulna is a dislocation of the distal ulna or a separation through the distal ulnar physis (Figure 8.3) (Letts and Rowhani, 1993).

In our clinic, about 60% of forearm fractures are in the distal metaphyses or epiphyses, 20% are in the distal third of the diaphyses, 10% are in the middle third, 5% are in the proximal third and 5% are Monteggia or Galeazzi lesions.

Fracture displacement is important. Incomplete fracture patterns include torus or buckle fractures, a characteristically metaphyseal pattern of failure in compression. In the diaphyses, plastic bowing and greenstick fractures are seen. These may be angulated but not displaced because they have strong periosteal hinges and good intrinsic stability. Plastic bowing is not always recognized for what it is, a fracture with deformity. Clinical deformity must be recognized and corrected, if necessary by completing the fracture.

The most convenient description of fracture displacement is in terms of angulation, displacement and rotation. Angulation can be measured on X-ray in degrees and displacement as a percentage of the diameter of the bone at that point. The third plane of displacement of forearm fractures is rota-

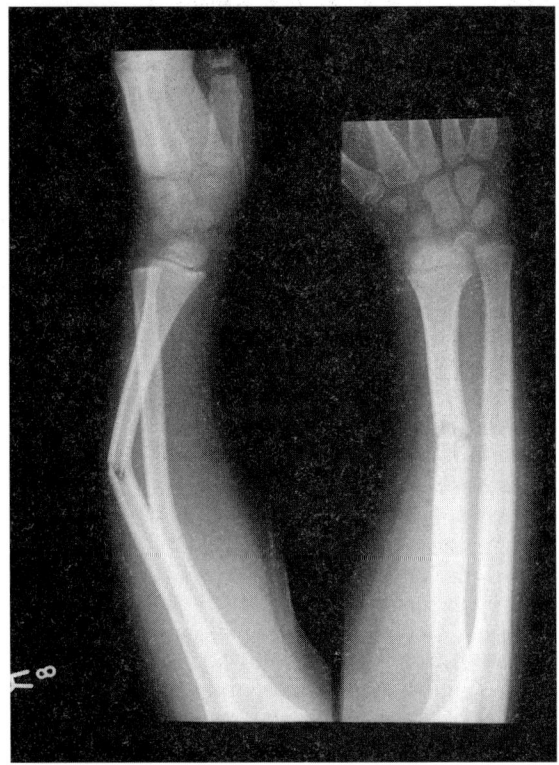

Figure 8.3 Galeazzi fracture dislocation with greenstick fracture of the radius. Closed reduction of both the distal radioulnar dislocation and the forearm fracture was straightforward

tion. This can be difficult to appreciate radiologically and there are no reliable means of measurement. Evans (1951) and Rang (1983) have promoted the bicipital tuberosity view as a guide to the rotational alignment of the proximal radius, but at best this can detect only changes of more than 30° of malrotation. We agree with Kay *et al.* (1986) that tuberosity views in younger children are not reliable. A sudden change in the diameter of the bone or in the normal radial bow are more helpful signs of malrotation, but again these are not quantifiable. Studies that attempt to correlate malrotation at union with final outcome in terms of forearm rotation are flawed because of the lack of a reliable measure of rotation.

The concept of 'axis deviation' has recently been applied to forearm fractures and may be a convenient method in which angular deformity and fracture position are correlated to a single variable. This may provide a better guide to the acceptability of fracture position and therefore management (Younger *et al.*, 1994).

Investigation and assessment of forearm injuries in children

Good management of forearm injuries in children is based on the standard history, examination and investigations. The history may suggest or exclude non-accidental injury or a fragility syndrome, most commonly osteogenesis imperfecta. The details of the injury history may give a guide as to the possibility of associated injuries and the severity of the forearm fracture. High-energy injuries such as a road accident or a fall from an extreme height are more likely to have associated soft tissue injuries, severe swelling and possibly neurovascular injury.

Clinical examination of the injured extremity and the general examination are also important. The first examination should establish the neurovascular state, assess tenderness, deformity, range of motion and the integrity of the adjacent joints. In a frightened child who is in pain, this can be difficult to do efficiently and without adding to the child's distress. Early administration of analgesia and examination of the child on the parent's knee can be helpful.

Neurovascular examination should establish the state of the peripheral pulses and the adequacy of capillary circulation. The single most important symptom of an impending compartment syndrome is excessive pain, and the most important physical sign is pain with passive finger extension. The integrity of the major nerve trunks can be established by a few simple tests of movement and sensation.

X-ray examination is clearly vital but is more likely to be of optimum value if the history and examination are dealt with first. Analgesia and provisional splintage should be considered before X-rays. Standard good-quality anteroposteror and lateral views of the forearm, including the elbow and the wrist, will be required in all cases. If an isolated fracture of the ulna is seen, then a properly centred lateral view of the elbow is required to exclude a Monteggia lesion. Similarly, an isolated radial fracture requires a true lateral examination of the wrist. The possibility of an ipsilateral supracondylar fracture of the humerus should have been excluded on clinical grounds. More complex investigations or imaging are rarely required.

Associated injuries

The majority of forearm fractures in children are simple injuries, uncomplicated by associated injuries in the upper limb or more remotely. There are a number of associated injuries, the management of which may impact on the management of the forearm fracture and should therefore be considered.

Forearm

In the forearm, associated injuries are to the soft tissues. Open fractures of the forearm are uncommon in children, the incidence in our unit varying from 3% to 6%. The majority are Gustilo type I injuries and there is a widespread tendency to treat these less rigorously than other open fractures. This can be a disastrous error as the literature reports a small but steady number of major complications including pyogenic osteomyelitis, gas gangrene and amputation resulting from small puncture wounds communicating with forearm fractures in children (Fee *et al.*, 1977). Gustilo type II and III open injuries are seen in our unit about once per year. Type II injuries have been successfully managed by open reduction through extensions of the wounds and internal or external fixation (Figure 8.4). Restoration of the soft tissue envelope may require a combination of delayed primary suture and split skin grafting. Type III injuries require similar stabilization followed by flap cover, usually free tissue transfer of a muscle flap and split skin grafting.

Neurovascular injuries complicating forearm fractures in children are also uncommon. However, in some units, including our own, forearm fractures have overtaken supracondylar fractures of

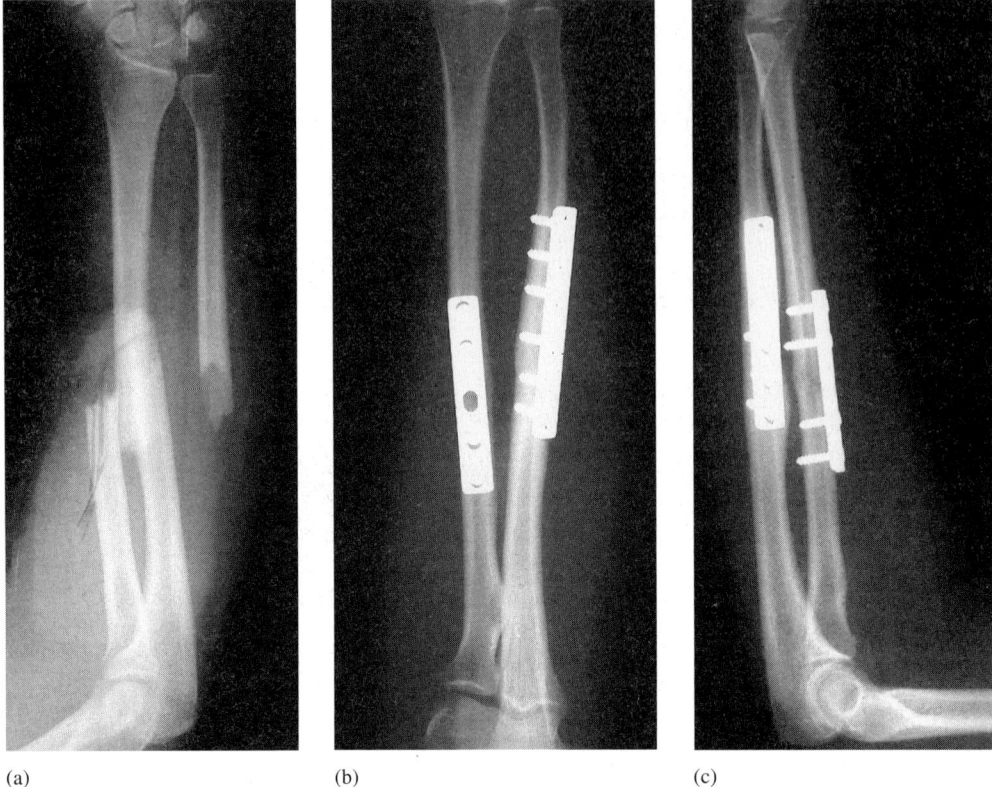

(a)   (b)   (c)

Figure 8.4   (a) Type II open fractures of the midshaft radius and ulna in a 13-year-old girl who was thrown from her horse. At open debridement, both proximal bone ends were heavily contaminated with organic matter. (b, c) The result 3 months after open reduction and internal fixation with delayed primary closure of the wounds was excellent

the distal humerus as the leading cause of fracture-related ischaemic contracture in children. We have seen two patients with ischaemic contractures following forearm fracture. One was transferred to our care because of a major closed head injury. A displaced middle third diaphyseal fracture of both forearm bones was managed in an above-elbow cast. The cast was split because of finger swelling on the fourth day in the intensive care unit but compartment pressures were not measured. The patient recovered consciousness after a protracted period, by which time there was evidence of a Volkmann's ischaemic contracture. The second child had a segmental forearm fracture and a median nerve palsy. A minor degree of ischaemic contracture occurred, the signs of which were masked by the median nerve palsy.

We have seen only one direct arterial injury; this was in a child with an open segmental fracture of the forearm, caused by farm machinery.

Neurological injuries are also uncommon. We have recorded isolated posterior interosseous, anterior interosseous, ulnar and median nerve palsies; the majority are traction injuries in continuity and recover spontaneously.

Arm

The most commonly associated injury is a fracture of the humerus, resulting in the paediatric 'floating elbow'. We recognize two patterns of injury: direct and indirect. The floating elbow which results from indirect forces (a fall onto the outstretched hand) is typically a fracture in the distal third of the forearm and an extension supracondylar fracture of the humerus (Figure 8.5). Direct forces may result in more proximal fractures of both the humerus and forearm, but these are very much less common and are caused by motor vehicle accidents or by machinery (Templeton and Graham, 1995).

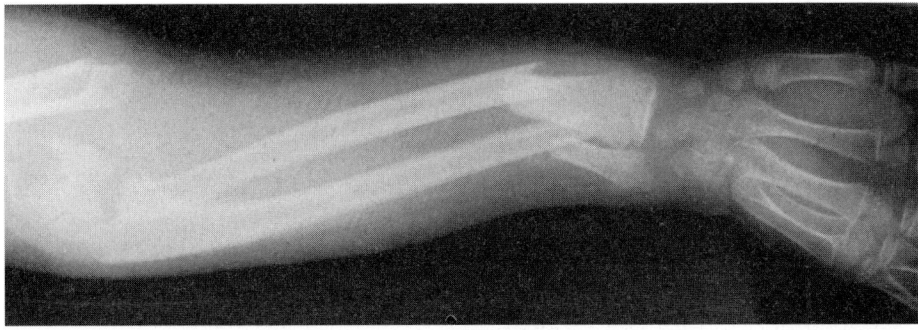

(a)

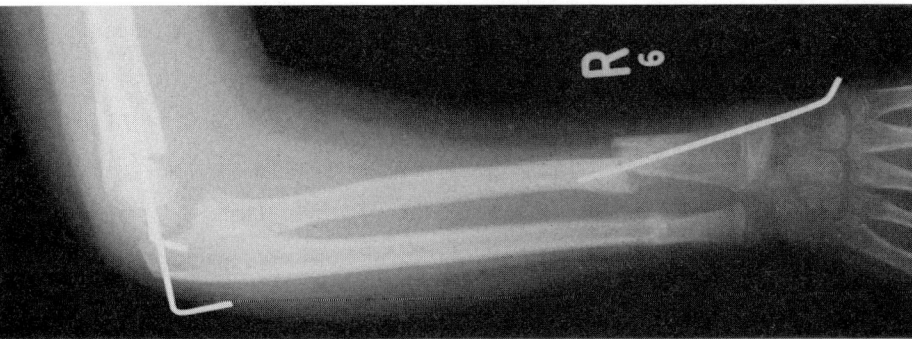

(b)

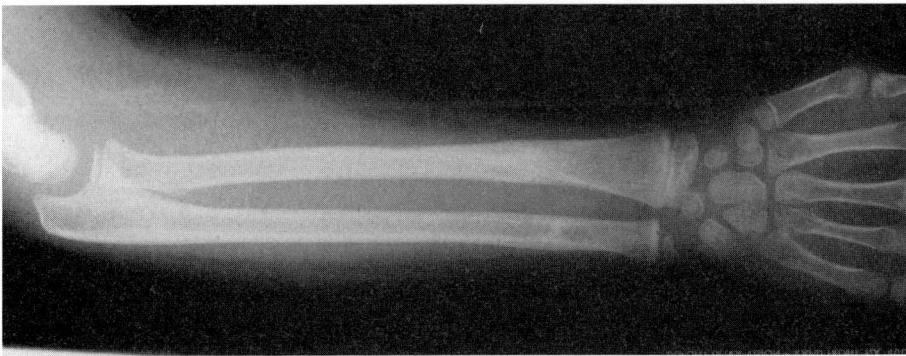

(c)

Figure 8.5   (a) Indirect violence floating elbow, i.e. the result of a fall onto the outstretched hand with displaced fractures of the distal third of radius and ulna and an extension supracondylar fracture. (b) Position 3 weeks after closed reduction and percutaneous Kirschner wire fixation of both the supracondylar fracture and the fracture of the distal radius. (c) Three months later there is sound bony union and early remodelling of both fractures

Forearm fractures and associated fractures in the arm are an indication to consider internal fixation of both fractures. In the indirect violence group, this usually means closed reduction and percutaneus Kirschner wire fixation; in the direct violence group, open reduction and plate osteosynthesis is usually required (Figure 8.6).

Remote

Forearm fractures with remote injuries are seen in three groups of children: those with multiple injuries from road traffic accidents; abused children; and those with bone fragility syndromes. Management is dictated by the clinical picture as a

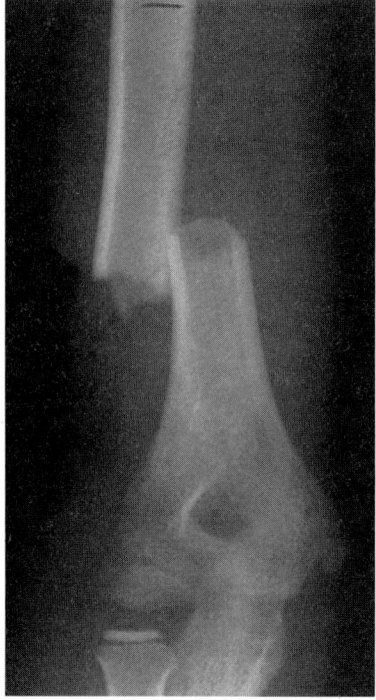

(a)

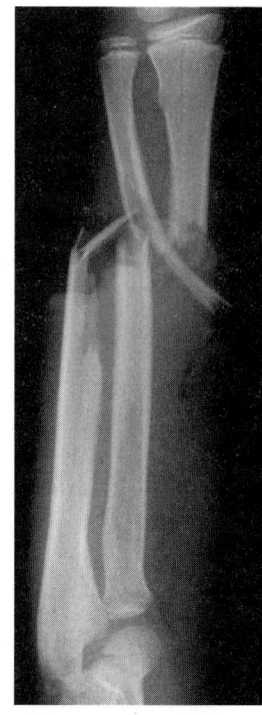

(b)

Figure 8.6 Direct violence floating elbow as a result of an accident involving farm machinery. (a, b) There were type III open fractures of the radius and ulna and a type II open fracture of the ipsilateral distal humeral shaft, with heavy organic soiling. (c) Following debridement, open reduction and plating of the humerus and radius and cerclage wiring of the ulna, non-union of both forearm fractures occurred. (d) Following reconstruction of the soft tissue envelope with a latissimus dorsi muscle flap and split skin grafting, repeat plating and bone grafting were carried out. At 8 months from injury, all fractures were soundly united. There was no evidence of deep infection and function was good with the only deficit being a moderate loss of forearm rotation

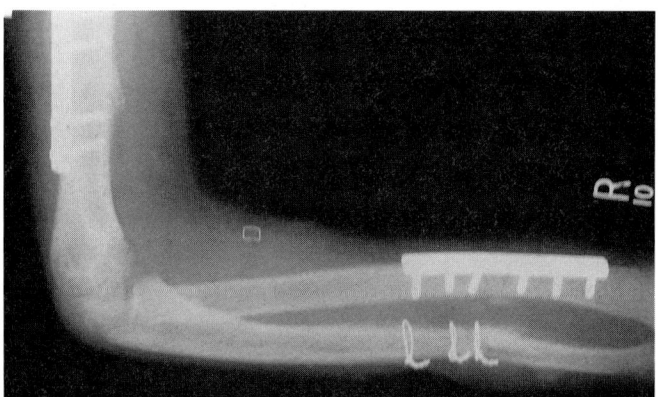

(c)

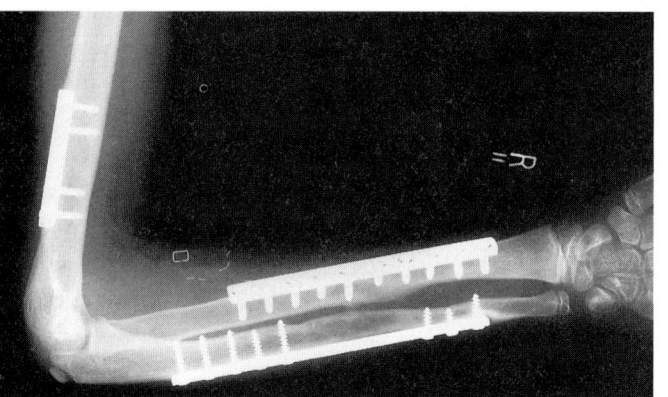

(d)

whole. Associated head injury with a diminished level of consciousness is a risk factor for ischaemic contracture of the forearm. The clinical signs may be masked in the crucial period after injury when reactive swelling is occurring in a closed cast. High elevation and regular monitoring are helpful but the best strategy is to bivalve the cast to permit daily inspection of the limb.

Initial management

The early management of forearm fractures in children is as for any fracture: assess the child for associated injuries, assess the limb for associated injuries and assess the injured forearm. Baseline neurovascular function should be recorded and the limb splinted pending X-rays. Good-quality anteroposteror and lateral X-rays are required, including the elbow and wrist for diaphyseal fractures. Isolated ulnar fractures should usually have separate elbow views to exclude a Monteggia lesion and isolated radial fractures, and wrist views to exclude a Galeazzi lesion.

It can be helpful to assess the stability of a forearm fracture clinically and radiologically, in an effort to plan appropriate management:

Good
X-ray: a mainly uniplanar greenstick angulation deformity, with little displacement and no shortening.
Clinical: reduction to anatomical alignment is easy, stability is good, over-reduction is difficult or impossible.

Poor
X-ray: complex deformity with angulation, displacement, shortening and rotation.
Clinical: unstable reduction, several possible positions noted on fluoroscopic screening including over-reduction or reversal of deformity, difficulty to 'hitch'.

Fair is intermediate between the above extremes; usually one bone has good and one bone has poor periosteal stability.

An analysis of the X-rays should give a reasonable guide as to the management plan. The degree of periosteal stability is the key to closed reduction and maintenance of a satisfactory position until healing. Periosteal stability can usually be inferred from an analysis of the initial X-rays and confirmed by fluoroscopic screening of the frac-

tures with the patient under general anaesthesia. Successful closed treatment requires good or fair periosteal stability in at least one of the fractures (Figure 8.7). If both fractures have good stability, successful closed treatment is virtually assured; if both fractures have poor stability, closed management will be difficult and the likelihood of remanipulation and a less than ideal position at healing are greatly increased (Figure 8.7). Open reduction may be required and should be considered early rather than late, particularly for older children. The possibility of open or closed fracture management should be discussed with the parents and permission sought for either procedure. There is little point in returning the child to the ward with an incomplete, unstable reduction which everyone but the parents know is doomed to failure.

Management options

Non-operative

Techniques of forearm fracture management in children can be considered as a pyramid. At the base is casting of undisplaced and minimally displaced fractures for pain relief and to prevent displacement. On the next level is closed reduction of displaced fractures and casting or bracing. Towards the top of the pyramid is open reduction and internal fixation. Finally, at the top of the pyramid is external fixation. The vast majority of forearm fractures are managed by plaster cast application. Hence a reliable safe casting technique and a follow-up service are essential. All of the other options will be required from time to time, including open reduction and fixation by a variety of techniques, and occasionally external fixation.

Plaster can bring almost instant pain relief and comfort to a frightened child with a forearm fracture. On the other hand, familiarity should not breed contempt. The only compartment syndromes in children with forearm fractures seen in our unit in the past 10 years have been caused by the cast, not by the injury. A fracture can be difficult to assess once the plaster has been applied.

Torus or buckle fractures are managed by cast immobilization to relieve pain and prevent further injury. As these injuries occur mainly in younger children and in the distal metaphyses, 2–3 weeks in a below-elbow cast are all that is required. Follow-up X-rays are rarely necessary but are often done.

Greenstick fractures and plastic bowing injuries should be studied carefully, both clinically and

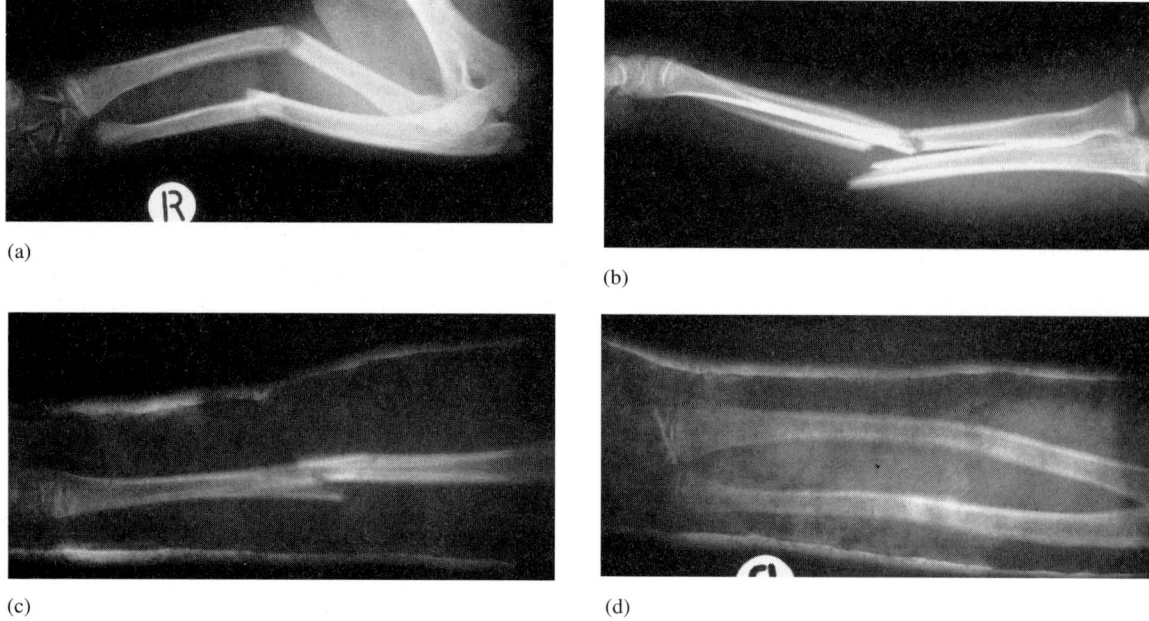

(a)

(b)

(c)

(d)

Figure 8.7   (a, b) Middle third diaphyseal fractures of the right forearm in a 12-year-old girl. The radius has good periosteal stability, but the ulna has poor periosteal stability. The overall grading is fair and conservative management should be attempted. (c, d) Using the almost intact periosteal sleeve of the radius, a good reduction has been secured. Union was achieved in this position and the result was excellent

radiologically. If there is no clinical deformity there is unlikely to be a need for reduction. However, a carefully applied cast is necessary because there is a small risk of progressive deformity. Fractures with deformity may have a combination of angulation and rotation; both components require correction.

The majority of, but not all, *complete fractures* will require reduction and cast immobilization. It is a matter of judgement and experience which children can be dealt with safely and efficiently by outpatient anaesthetic techniques (Bier block, self-administered nitrous oxide) and which should be admitted for general anaesthesia and access to fluoroscopic screening.

Closed reduction and casting

Children who have fractures with greenstick angulation deformity can often be managed as outpatients. However, it is pointless to attempt reduction of unstable diaphyseal fractures in an older child in less than optimum conditions. Difficult reductions are made easier by general anaesthesia, fluoroscopy, skilled assistants and enough time to do the job properly. Junior surgeons should not be expected to reduce and hold the fracture and then apply the cast single handed. In some hospitals these are the conditions which may pertain! 'Chinese finger traps' are a partial substitute for one assistant but not for both. The proximal fragments tend to adopt a position which is dictated by muscle pull on intact skeletal levers. Hence the distal fragments must be positioned to align with the proximal, taking into account rotational factors. An attempt is made to reverse the deforming forces and then retrace the movement of the fractured bone ends. The deformity may first need to be made worse in order to 'hitch' the bone ends, especially for fractures in the distal third of the forearm with a reverse obliquity pattern.

Image intensification is very useful but gives a much better representation of apposition rather than alignment. In most children's fractures, alignment is the key, not apposition. Therefore conventional plate films may be required in the operating theatre in many cases.

By convention, proximal third fractures are positioned in supination, middle third fractures in neutral and distal third fractures in pronation. This serves as a general management guide and is based on the position likely to be adopted by the proximal fragments under the influence of muscle pull. However, each fracture should be individually assessed and some will be found which will not follow this simple guide. Extreme positions of forearm rotation should not be used throughout the period of immobilization but only until the fractures are 'sticky'.

The majority of Monteggia fracture dislocations can be managed by closed reduction, but the dislocation can be unstable and careful postoperative follow-up is required. The most stable position is flexion of the elbow just above 90° and supination of the forearm.

Closed reduction and percutaneous fixation

Some fractures are reducible but either very unstable or so liable to late displacement that fixation should be considered. Percutaneous Kirschner wire fixation can be a useful option but should not be overused. The main indications are an isolated, completely displaced, distal metaphyseal fracture of the radius (Gibbons *et al.*, 1994) and fractures

of the distal third of the forearm in association with supracondylar fractures of the humerus (the 'floating elbow') (Figure 8.5) (Templeton and Graham, 1995). This technique is also of use in displaced radial neck fractures in children (Bernstein *et al.*, 1993).

Closed reduction and intramedullary fixation

Closed intramedullary fixation of diaphyseal fractures can be performed using fluoroscopic screening and special flexible rods. This specialized technique gives excellent results with minimum morbidity in experienced hands. The problem is that if the indications are strictly observed (fractures which are reducible but very unstable) then it is difficult to maintain expertise to use this as an occasional technique (Figure 8.8) (Amit *et al.*, 1985; Lascombes *et al.*, 1990). For most surgeons, a limited-exposure open technique is probably preferable. The need for external cast immobilization after IM nailing of forearm fractures is controversial. With flexible, small-diameter rods it should be used routinely.

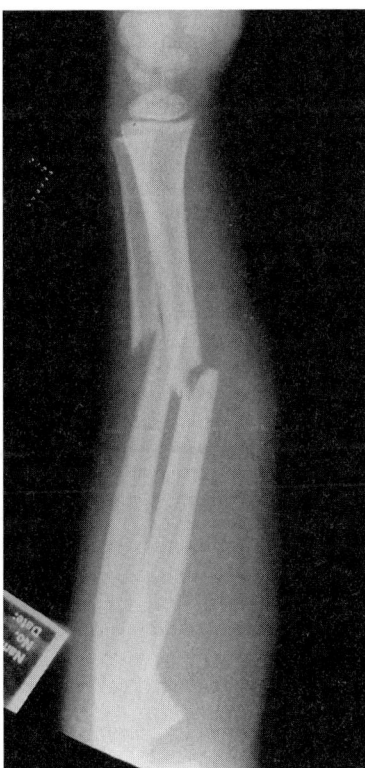 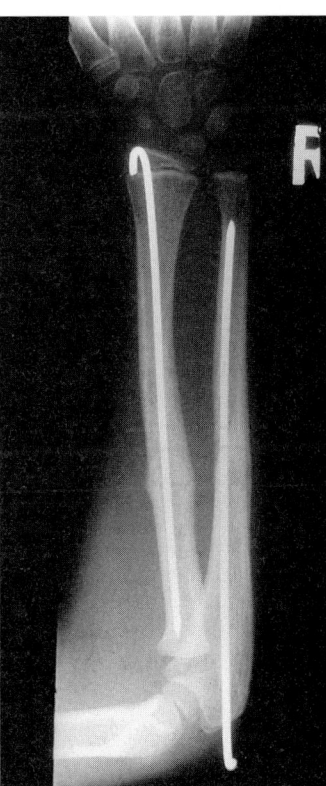

Figure 8.8 Closed intramedullary nailing of midshaft forearm fractures. This is an elegant technique with excellent published results. However, a 'minimally open' technique is easier for the occasional operator

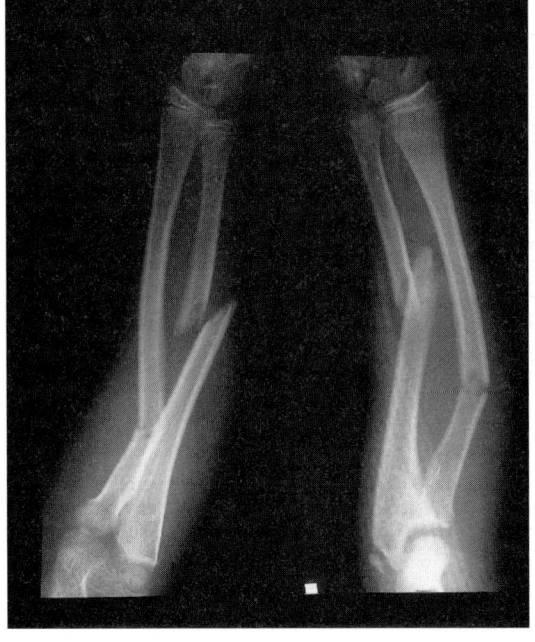

(a)

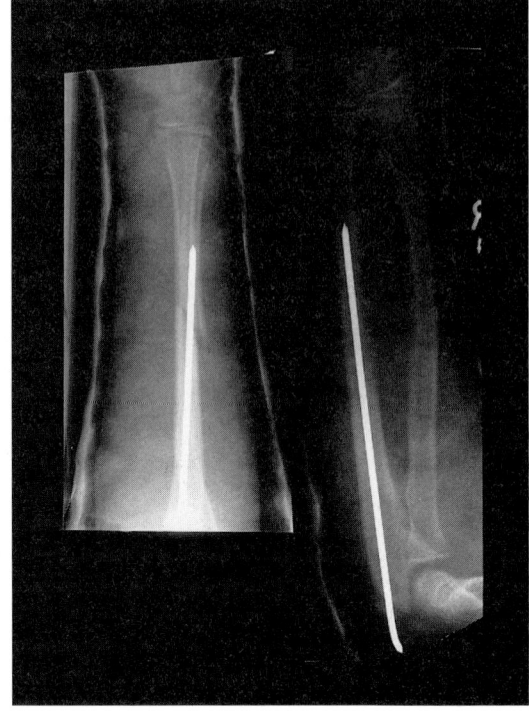

(b)

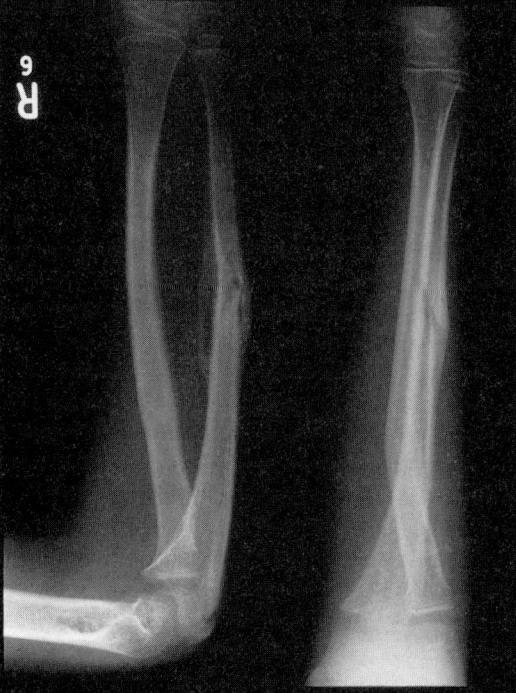

(c)

Figure 8.9 (a) Type I open fracture of the right ulna and closed fracture of the radius. The stability of the radial fracture was good; that of the ulna was poor. (b) Following debridement, retrograde rodding of the ulnar fracture was performed using a Steinmann pin. The Steinmann pin was removed after 4 weeks. (c) The appearances at 8 weeks post-fracture are seen. The union of the radial fracture is more sound than that of the ulna. Sometimes fixation of only one fracture is acceptable

Open reduction and internal fixation

The indications for open reduction are relative, usually involving failure to obtain a reduction or a reduction which is very unstable. Other relative indications are older children, irreducible Monteggia or Galeazzi fracture dislocations, associated fractures of the humerus and open fractures.

There is probably no role for open reduction without fixation: the surgeon who chooses this course must be completely confident of his or her ability to hold the fracture in perfect position in cast, otherwise the fracture, having been opened, should be fixed. Open reduction utilizes approaches similar to those used in adult fracture management but the exposure does not need to be so extensive because different priorities are observed in paediatric fracture fixation. Internal fixation is usually supplemented with a cast, and hence rigid internal fixation is not required.

Diaphyseal fractures with different degrees of periosteal stability do not always require fixation of both fractures; the most unstable fracture is fixed and the effect of this on the other fracture is carefully assessed before opening the second fracture (Figure 8.9).

Intramedullary fixation with Kirschner wires, Steinmann pins or purpose-made flexible nails is a very useful technique for the treatment of paediatric forearm fractures. Exposure of the fracture is much less than that required for plating, union is as rapid and satisfactory as following plate osteosynthesis, and the operation for metal removal is usually very simple. Disadvantages include possible violation of the growth plates and less rotational control of the fracture.

The results of plate osteosynthesis in paediatric forearm fractures are usually excellent. Union is rapid and motion usually full. Smaller implants than would be considered appropriate in adult forearm fractures are used because internal fixation is usually supplemented by external cast fixation. The major disadvantages are prominent scars and the need for a second operation to remove metal (Figure 8.10). The risks of infection, cross-union and stiffness are small but must be borne in mind.

External fixation

The indications for external fixation in the primary management of paediatric forearm fractures is limited to a few children with severe open injuries, wound contamination and soft tissue loss. Half pins ranging from 2 to 5 mm in diameter and

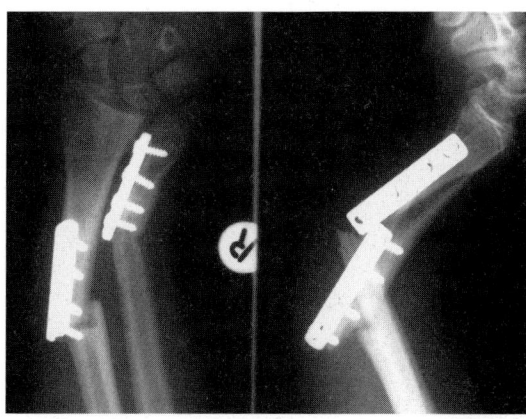

Figure 8.10   Should forearm plates be removed in children? This boy had sound union and full return of function following open reduction and internal fixation of his fractures in the distal third of the right forearm 2 years earlier. He returned to riding his trail bike and sustained refractures just proximal to his plates. Management was by removal of these plates and replating using larger implants. Plates and screws are stress risers and are a concern in active children

monolateral frames are required to cover the full age range (Figure 8.12). Circular frames of the Ilizarov type are useful in reconstructive surgery for post-traumatic problems, the commonest of which is lengthening of the radius following a distal radial physeal injury.

Timing of surgery

The majority of forearm fractures in children are isolated injuries and are not complicated by neurovascular injury or by open soft tissue injuries. The timing of intervention can therefore be relaxed, observing the need to have the child fasting, etc. It is good practice to align severely displaced fractures and to apply temporary splintage in the accident & emergency department when formal reduction is likely to be delayed.

Positioning, preparation and draping

The patient is invariably placed in the supine position. Smaller children are sometimes placed eccentrically across the operating table to permit access to the upper limb for radiographic screening of the fractures with a C arm image intensifier. Fluoroscopy is widely used and good standards of

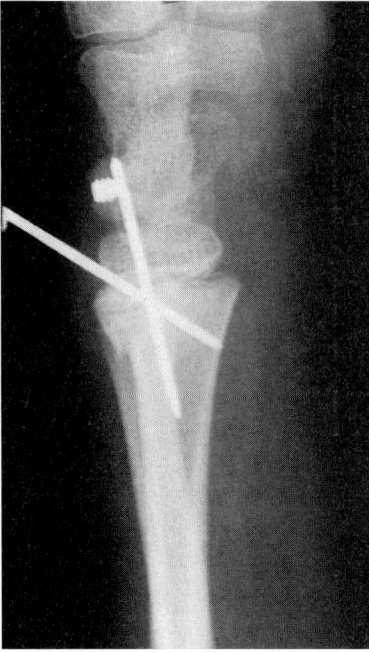 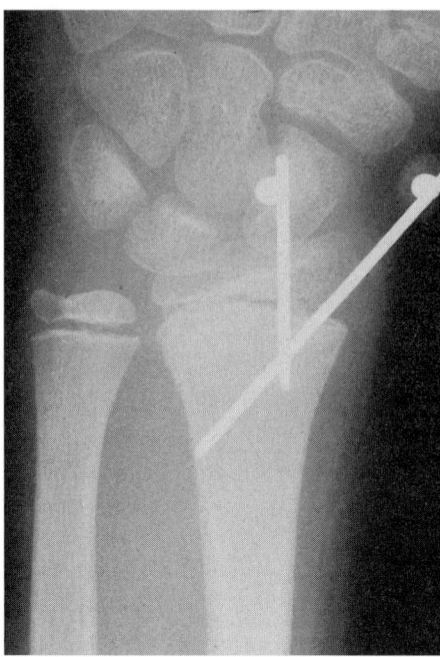

Figure 8.11 The Kapandji technique can be applied to distal radial fractures in children. This child had a floating elbow with a completely displaced supracondylar fracture of the distal humerus and a completely displaced Salter–Harris type II injury of the distal radius. Percutaneous stabilization of the distal radial injury was performed after closed reduction. The Kirschner wires should enter the fracture site from a dorsal and radial direction and engage the proximal cortex. The Jurgen 'pin balls' can be seen on the ends of the wires

practice should be observed to protect patients and staff from radiation.

Many closed reductions and percutaneous fixation procedures are performed using the image intensifier as an operating table. This increases the scatter of radiation but this is offset by a shorter screening time and fewer exposures.

Percutaneous fixation and open procedures require skin preparation of the entire forearm, including the elbow and wrist. Younger children should have the hand prepared to avoid having to exclude a small hand in a bulky towel from the operative field. Tourniquets are not required for percutaneous procedures but are useful for open reductions.

Surgical approaches

These do not differ significantly from those described in Chapter 5 for the management of adult forearm fractures. The growth plate should not be violated by screws, the periosteum should not be elevated within 1 cm of the physis and, because shorter plates are often used, exposures can be more restricted.

The superficial radial nerve and branches are at risk from percutaneous fixation of distal radial fractures and from the short incisions used for intramedullary fixation.

Surgical techniques

Percutaneous Kirschner wire fixation

This is a useful technique after closed reduction when the reduction is very unstable and for segmental injuries such as the 'floating elbow'. The advantage of fixation without opening the fracture is considerable.

Metaphyseal fractures of the distal radius are the main indication for use of the Kapandji intrafocal technique. The majority of distal radial metaphyseal fractures are dorsally displaced. Following closed reduction, the position is checked fluoroscopically. A single smooth Kirschner wire is drilled across the fracture, starting from the dorsal aspect of the wrist and angulated to engage intact metaphysis, proximal to the fracture. If the fracture was radially displaced, a second wire can be added from the radial side; the starting point is through the fracture, and the wire should engage the ulnar side of the radial metaphysis, proximal to the fracture. The fracture is 'buttressed' against redisplacement without violating the growth plates. A sterile

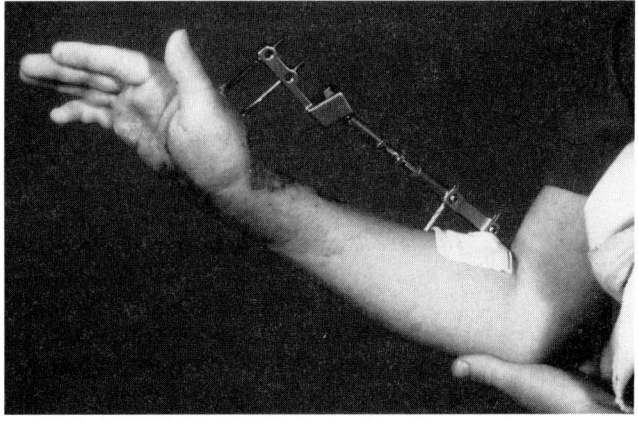

(a)

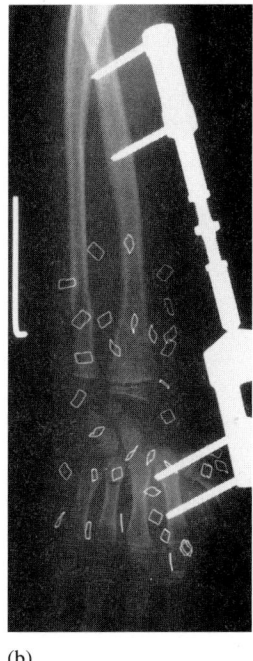

(b)

Figure 8.12 (a) External fixation following type III open fractures of the right distal third radius and ulna in a 5-year-old boy who was ejected from a bus in a motor accident. There was significant soft tissue loss over the fractures and dorsum of the wrist. (b) Monolateral external fixation gave excellent stability until the soft tissue envelope was reconstructed

gauze dressing is applied by piercing it with the wire, and a Jurgen 'pin ball' is attached to prevent wire migration and to facilitate wire removal. The pin ball should not be placed tightly against the skin or ulceration may occur during the period of post-fracture swelling. The wires are cut distal to the pin ball and a suitable plaster cast applied. Wires should be removed after 3 weeks. Complications from percutaneous wires are minimal with this technique because the skin–wire interface is stabilized, reducing infection and avoiding migration (Templeton and Graham, 1995) (Figure 8.11).

Percutaneous Kirschner wire fixation can be used for more distal injuries, for example to the distal radial growth plate, but the fixation wires may then have to cross the physis. There is a small but definite risk of growth disturbance, which would be a disaster in this rapidly growing physis. The risks of this complication need to be justified by the indication (Figure 8.5). Percutaneous Kirschner wire fixation can be used for some favourable distal third diaphyseal injuries but the technique becomes increasingly difficult in hard diaphyseal bone.

Intramedullary fixation

Closed reduction and intramedullary fixation is performed in a sterile field, without a tourniquet, either using the image intensifier as the operating table or with the C arm under a radiolucent table. Special flexible rods and introducers are best but a range of Kirschner wires, Steinmann pins and Rush rods can be used, according to the age of the patient and the availability of the equipment.

The fracture which has the most periosteal stability is reduced and fixed first. The radius is approached through a 1 cm incision over the styloid, identifying and protecting the superficial radial nerve branches. The distal radial growth plate must be avoided and the level of the incision should be first checked with the image intensifier. A 2–3 mm drill hole is made on the lateral side of the distal metaphysis at about 45° to the long axis of the limb. The nail is chosen, the tip bent about 10–20°, and it is inserted using a holder or pliers. A combination of axial pressure and rotation movements are required for insertion, especially to cross the fracture. Insertion of the ulnar rod is through the proximal lateral metaphysis, in order to avoid the apophysis and the ulnar nerve. Rods or wires are cut to the correct length, i.e. short enough not to cause skin irritation and long enough to permit easy removal (Figure 8.8). Even with experience, closed intramedullary fixation can be difficult and it can sometimes be less traumatic to convert to an open procedure (Figure 8.9). Open intramedullary fixation is a much easier procedure for the occasional operator. The choice of implants and techniques of insertion are much the same as described above. The fractures can be exposed and reduced through much smaller incisions than would be required for plate fixation. In older children, retrograde rodding of the ulna has not been associated with any evidence of growth disturbance.

Open reduction and internal fixation

The main differences between plate osteosynthesis in children and adults are related to the differences in fracture biology and the goals of fixation. In adult forearm fractures, union is slow and unreliable. Rigid internal fixation, avoidance of casting and early motion are desirable. In children, union is rapid and predictable; loss of forearm rotation is because of malunion, not the period of cast immobilization. Plate fixation is therefore designed to obtain anatomical union, not early motion. Shorter and lighter implants can therefore be appropriate. In younger children, one-third tubular plates are used and in older children, 3.5 mm DCPs (Figure 8.4). On the other hand, very unstable forms of internal fixation such as crossed Kirschner wires are rarely appropriate. Kirschner wire fixation is useful after closed reduction because union is rapid and predictable. Even after limited open reduction, healing is much slower and the combination of periosteal stripping and unstable fixation may produce delayed union or non-union (Lewallen and Petersen, 1985).

External fixation

A wide range of devices and fixation pins can be used, according to the size of the child, fracture configuration and availability. Simple monolateral fixators with half pins are adequate for most acute fracture problems in children. Usually external fixation of one of the forearm bones will provide adequate stability to manage associated soft tissue injuries. External fixation of the subcutaneous ulna is easier and safer than external fixation of the radius. In the ulna, half pins can be inserted by a closed technique, but in the radius a 'semi-open' technique is preferable, to minimize the risk of neurovascular injury (Figure 8.12).

The versatility of circular frames may be required for some difficult post-traumatic reconstructions such as lengthening and deformity corrections.

Soft tissue problems and management

Open forearm fractures in children are uncommon. The majority are Gustilo type I soft tissue injuries and should be managed by exploration, irrigation and debridement. After debridement, some wounds can be allowed to heal by secondary intention but some will require delayed primary suture or skin grafting. Fracture management is decided according to the fracture pattern and the age of the child. These children are more likely to be older and have displaced, unstable diaphyseal fractures. Open reduction through an extension of the debrided wound and intramedullary or plate fixation may be required. Gustilo type II and III open injuries are rare. Management should be individualized according to the degree of bone and soft tissue injury, soft tissue loss, contamination, associated injuries and age of the child. Stable fracture fixation is important and can be achieved by external or internal fixation. Early restoration of the soft tissue envelope will reduce morbidity and improve outcome. This may require a rotation flap or free microvascular soft tissue transfer (Figure 8.6).

Compartment syndromes are not rare in paediatric forearm fractures. Some are the result of a crush injury and others are a complication of swelling in an unsplit cast. Measurement of compartment pressures can be helpful in deciding when to intervene. Decompression of the volar compartment is accomplished through an anterior Henry approach which leaves an ugly scar but is infinitely preferably to an ischaemic contracture. The extensor compartment is less often affected but should always be considered.

Postoperative management

Postoperative management of forearm fractures is important because early recognition of complications and appropriate intervention can rescue the situation. The management of forearm fractures treated by closed reduction and casting is most important because most forearm fractures are dealt with in this way.

Should the plaster cast be split? All fractures are accompanied by swelling which approximates to the degree of displacement and soft tissue injury. The majority of casts should be split because it is safer to do so. At least some compartment syndromes are the result of swelling in a plaster cast.

What are the disadvantages of splitting the cast? The cast is weakened and it is difficult to make casts child-proof anyway. There may be a slightly increased risk of losing position of the fracture.

When should a split cast be closed? At a patient's first visit to the fracture clinic we X-ray the fracture, and in conjunction with the original films and the findings at reduction, try to anticipate the future stability of the fracture. Most casts can then be closed and reinforced with a layer of fibreglass. A child who is tearful and holding his arm carefully in a sling at the time of discharge from hospital may be climbing on monkey bars in a few days time, when his pain has gone and his confidence has returned.

When should the child be reviewed? Distal forearm fractures, especially epiphyseal injuries, heal very rapidly. If we are to detect and correct loss of position it will be in the first 5 days after injury. On the other hand, diaphyseal fractures remain 'sticky' for 2–3 weeks after injury. It is often easier to correct a minor degree of persistent angulation or recurrent displacement at this stage than within a few days of injury.

When do problems arise? X-rays can be lost or not done, views may be inadequate, or the films may not be looked at. Appointments may not be kept, instructions can be misunderstood, and casts broken or removed.

The 3-week look and remanipulation

Assessing an X-ray 3 weeks from injury can be difficult. There will be signs of healing but often some residual displacement or angulation. Is the position acceptable? It is often more helpful to remove the cast and inspect the arm than to

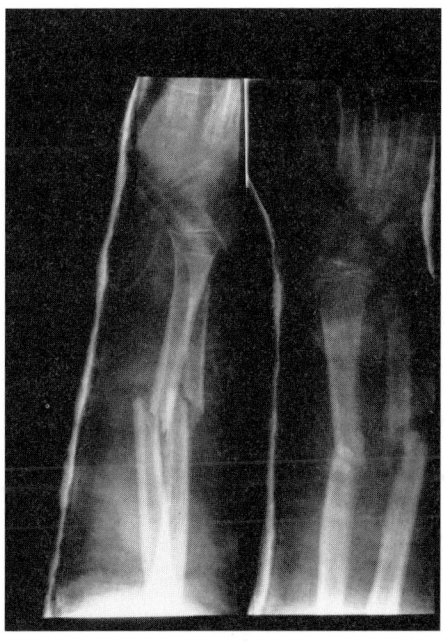
(a)

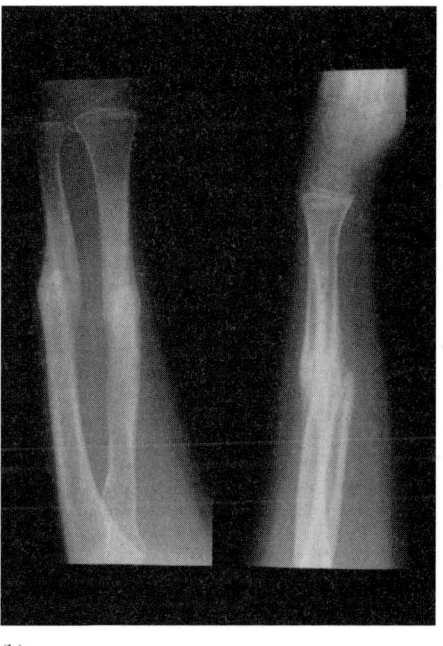
(b)

Figure 8.13   (a) A common appearance of fractures of the forearm 3 or 4 weeks from injury. There is early union with angulation of the fractures and loss of the interosseous space. The fracture was remanipulated and the cast was moulded to prevent sagging of the ulnar fracture. (b) Eight weeks after injury, there was sound bony union with acceptable alignment of both fractures. At 6 months from injury, forearm rotation was completely full and there was no discernible deformity. Rereduction at 2–4 weeks from injury is a viable option rather than proceeding to open reduction for many forearm fractures in which there is loss of position

agonize over X-rays. A different appearance on X-ray may be the result of different positioning in the X-ray department, as well as a loss of fracture position. Most older children will cooperate with a cast change at 3 weeks without too much difficulty. The appearance in terms of clinical deformity and the range of forearm rotation should be examined gently. If there is about 50% of the normal range at 3 weeks, the full range is likely to be regained in time. If the original position was an extreme position of forearm rotation, is this still necessary? The answer is usually no, and subsequent X-rays are much easier to interpret in a position of near-neutral rotation. Fractures that were in acceptable alignment after reduction but which have lost position can often be corrected by a gentle osteoclasis at this stage, without recourse to open reduction (Figure 8.13). About 7% of displaced forearm fractures in children will require a remanipulation for loss of position, and the majority will go on to a satisfactory result (Voto *et al.*, 1990).

The ulnar border is subcutaneous and a posterior sag or bow will be easily seen and considered unacceptable. Moderate loss of the normal radial bow is often not seen because of the muscle cover; if it does not significantly interfere with forearm rotation, it can be accepted.

In very young children with short, chubby forearms, maintaining alignment in above-elbow casts with the elbow flexed to 90° can be impossible. In this case it is safe to extend the elbow and use a carefully moulded straight-arm cast for a few weeks until the fractures are stable (Rang, 1983).

After open reduction, children are discharged when comfort and neurovascular stability permit. An early return to the fracture clinic is advised. The reduction should be stable and is not the main concern. Infection is very unusual after plate osteosynthesis and we only inspect the incision if clinically indicated, because children are usually frightened to have their cast removed in the early postoperative period. Pin site sepsis is a constant risk after percutaneous procedures or external fixation.

Metaphyseal fractures and growth plate injuries require 4–6 weeks of cast immobilization. The time to healing for diaphyseal fractures is much more variable and there is a significant risk of refracture. After 6–8 weeks of cast immobilization, an X-ray out of cast is usually appropriate. The stability of the fracture can be discussed with parents: they will usually indicate if a child can be relied upon to protect the limb against further injury. When there is doubt, the best answer is usually a further 2 or 3 weeks in a below-elbow cast.

Complications

Some complications are more accurately described as a failure of initial management. Many children are discharged from hospital with a fracture in a barely acceptable position in an inadequate cast. At 3 weeks post-injury, the position is clearly unacceptable and the options can then be limited.

Early complications of forearm fractures include neurovascular deficits and compartment syndrome. Isolated injury to the radial or ulnar arteries may occur but are largely unrecognized because distal ischaemia is very rare. Isolated nerve palsies are uncommon but well described. The highest incidence is in association with Monteggia fracture dislocations. Most are traction injuries in continuity and recover after 6–12 weeks. Multiple nerve palsies are sometimes seen as part of a neglected compartment syndrome, when irreversible ischaemic changes in the forearm muscles have already taken place.

The early signs of neurovascular impairment and compartment syndrome can be difficult to distinguish from the normal response of a child with a forearm fracture. Most children are in some pain after a forearm fracture, many are reluctant to fully extend their fingers actively and all dislike passive finger extension. With prevailing early discharge policies, careful cast instructions are necessary for every child and family, along with a contact telephone number in case of difficulties out of hours.

Cast sores are a constant risk in the management of forearm fractures. Good cast technique requires experience, optimum conditions during application and experienced help. The commonest errors are flexing the elbow as the cast is setting, producing a sore at the flexor crease, and over-zealous moulding of the cast with fingers and thumbs. The sign of impending pressure sores is unrelieved pain which is well localized. After a few days the pain may go, when full-thickness skin necrosis has occurred. Parents whose concerns have been too lightly dismissed are unlikely to be happy with this outcome.

The commonest complication of forearm fractures is malunion resulting in a deformity and reduced function. Perfect position is hard to achieve following displaced diaphyseal fractures, and defining what constitutes an acceptable position can be difficult. There is little agreement in the literature or much in the way of scientific information. Clinically obvious deformities should be corrected. This usually means angulation of more than about 10°, especially in the ulna. Bayonet apposition is quite acceptable in children with 2 or more years of growth remaining (Figure 8.13). Rotational deformities are difficult to measure clinically or radiologically and it is therefore difficult to quote acceptable limits. Probably no more than 20° should be accepted.

Age is important. Younger children usually have less displaced fractures and more rapid growth and remodelling. Most open reductions are required in girls over the age of 10 and boys over the age of 12 years (Kay *et al.*, 1986).

The fracture site is important. Fracture remodelling is faster close to the most active physes, i.e. the distal radius and ulna. There are very few poor long-term results after distal epiphyseal and metaphyseal injuries. Diaphyseal remodelling is slow, incomplete and unrealiable.

The direction of the displacement is important. Posterior bowing or sagging of the ulna is very easily seen, but loss of radial bow is less obvious. Angulation of diaphyseal fractures is readily seen on X-ray and should be corrected. Rotational malalignment can be much more subtle to see on X-ray but the clinical effects are much worse. The worst functional deficits are the result of malrotation which has almost no capacity to remodel. It is my impression that rotational deformities of more than about 20° are associated with permanent loss of forearm rotation. Malunion is the result of neglecting a malposition. The 3-week check is vital and if there is a concern the cast should be removed and the forearm checked clinically. At this stage, a remanipulation will usually rescue the situation. At 6–8 weeks, minor degrees of malunion are common and most will remodel (Price *et al.*, 1990). Major deformities will not go away and some should be dealt with by an early drill osteoclasis rather than hoping for a miraculous degree of remodelling (Rang, 1983).

Infection in closed fractures is rare but not unknown. The infection rate following open reduction and internal fixation is very low. External fixation and percutaneous techniques are associated with a significant incidence of pin site sepsis.

Non-union of forearm fractures in children is uncommon (Lewallen and Petersen, 1985). There

has usually been a major open injury, loss of bone and infection. Once the soft tissue envelope has been properly restored and infection controlled, conventional plating and bone grafting is usually successful for non-union associated with modest bone loss. Major bone loss occasionally requires bone transport. Pathological fractures associated with neurofibromatosis can result in 'disappearing bone disease'. Major deficits may require bone transport or vascularized fibular grafts (Figure 8.14).

Cross-union or radioulnar synostosis is infrequent but a major problem when it occurs. The predisposing factors are high-energy, displaced fractures of both forearm bones at the same level, very proximal or distal fractures, convergence of the fractures with loss of interosseous space and open reduction, especially when performed through a single incision. Excision of synostosis has a high recurrence rate and does not always restore full forearm rotation (Kelly and Miller, 1987).

In percentage terms, growth disturbance following forearm fractures is rare but distal physeal injuries are so common that an appreciable number of these problems are seen. In a few children, growth arrest may be predicted because the physeal injury was type 4 or 5 or the forces involved were very great. However a number of innocuous-looking type 2 injuries result in growth arrest and this can occur after variable intervals from the injury. It is desirable but impractical to follow all children with forearm physeal injuries until the risk of growth disturbance has passed.

Proximal growth arrest

Growth arrest following radial neck fractures and epiphyseal injuries is not uncommon. However, the proximal physis contributes much less to longitudinal growth of the radius than the distal physis. Premature closure may be associated with a mild degree of cubitus valgus and an increase in the carrying angle of the elbow. Intervention is not required (Steele and Graham, 1992).

Distal growth arrest

Growth arrest in the distal radius and ulna will result in undergrowth and relative shortening of the forearm. More commonly, an isolated arrest of the radius, or less commonly the ulna, is seen. This is more serious because a progressive deformity may occur because of the disparity in

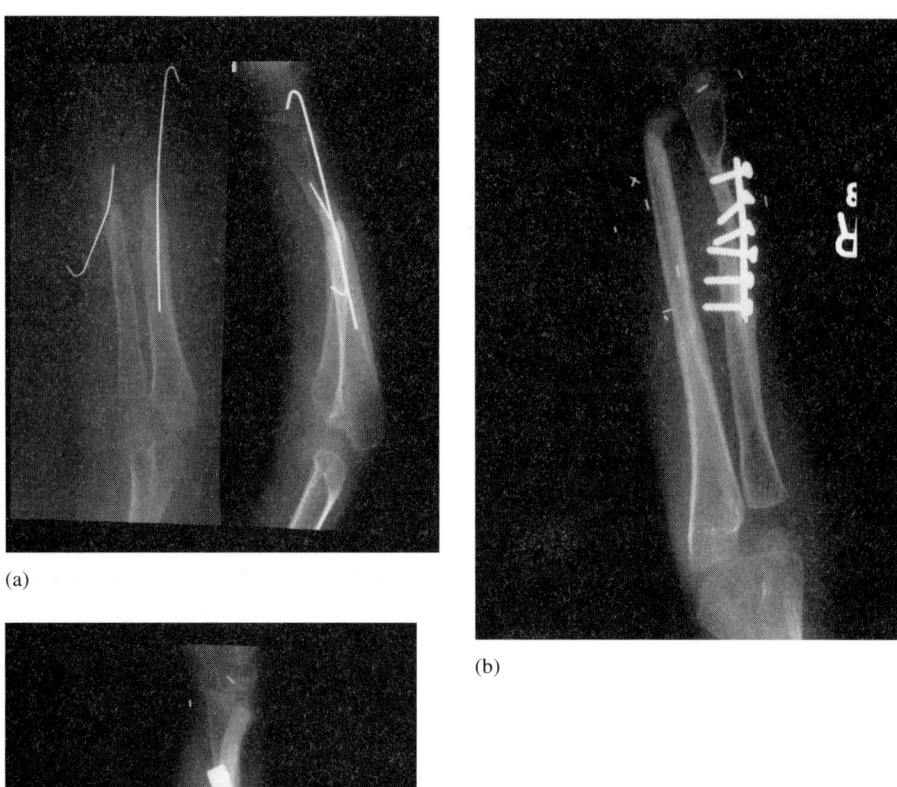

(a)

(b)

(c)

Figure 8.14   (a) Fractures at the junction and middle third right radius and ulna in a 5-year-old boy with multiple café au lait spots. He had been managed by open reduction and inadequate Kirschner wire fixation. (b, c) The fracture of the ulna was managed by resection and free vascularized fibula grafting. Sound union was obtained, but there is deformity at the level between the junction of the fibula graft and the distal ulnar physis. Despite plating and autogenous iliac crest grafting of the radius, non-union persists and fracture of the plate is imminent

growth between the two forearm bones, with disturbance of the distal radioulnar joint.

Growth arrest affects the distal radius more commonly than the distal ulna. The physeal bar should be mapped by tomography, and the surgical options can then be considered in relation to the patient's age, degree of deformity and expected progress of deformity during remaining growth.

The most physiological procedure may be bar excision and fat graft interposition. Completion of the physiodesis of the radius and/or ulna may be appropriate in certain circumstances. Residual deformities at maturity may require distal radial osteotomy, ulnar shortening and occasionally a combination of lengthening and deformity correction with a circular frame.

Author's choice of management

Good closed management is the cornerstone of forearm fracture management in children. When closed management is chosen, attention to detail is essential. These fractures are very common, and at peak times of the year, even well-run fracture clinics can be overwhelmed. We scrutinize the position carefully at 1 and 3 weeks after injury and perform a remanipulation when required.

Epiphyseal separations of the distal radius and ulna

Salter–Harris type I and II: Our preferred option is closed reduction and a below-elbow cast for 3–4 weeks.

Salter–Harris type III and IV: Our preferred option is open reduction and Kirschner wire fixation with a below-elbow cast for 3–4 weeks.

Distal metaphyseal fractures

Our preferred option is closed reduction. For irreducible fractures we use open reduction and Kirschner wire fixation, avoiding the growth plates. For unstable reductions we use percutaneous Kirschner wire fixation by the Kapandji intrafocal technique. Isolated distal radial fractures and floating elbow injuries are the main indications.

Diaphyseal fractures

Our preferred option is closed reduction and casting, with an above-elbow cast for 6–10 weeks, plus a below-elbow cast for 0–4 weeks. Closed reduction and intramedullary fixation is used for easily reducible but unstable injuries; not for the occasional operator. Open reduction and internal fixation is reserved mainly for failures of closed management. In girls over the age of 10 years and boys over 12 years it can be used as the primary option if the fractures are unstable. Intramedullary fixation should be considered before plate osteosynthesis.

Segmental injuries: the floating elbow

Our preferred option is fixation of both fractures. Distal metaphyseal and growth plate injuries can be managed by closed reduction and the Kapandji technique. Distal third diaphyseal fractures can sometimes be managed by closed percutaneous Kirschner wire fixation but open reduction can be necessary. Proximal fractures should be managed by plate osteosynthesis.

Monteggia and Galeazzi lesions

Our preferred option is closed reduction of both the radioulnar joint dislocation and the forearm fracture. Failure of radioulnar joint reduction is managed by open reduction, usually involving removal of soft tissue obstructions. Failure to obtain a stable closed reduction of the forearm fracture requires open reduction and intramedullary fixation or plate osteosynthesis.

References

Amit, Y., Salai, M. Checik M.D. *et al.* (1985) Closed intramedullary nailing for the treatment of diaphyseal forearm fractures in adolescence: a preliminary report. *J. Pediatr. Orthop.*, **5**, 143–6

Bernstein, S.M., McKeever, P. and Bernstein, L. (1993) Percutaneous reduction of displaced radial neck fractures in children. *J. Pediatr. Orthop.*, **13**, 85–8

Evans, E.M. (1951) Fractures of the radius and ulna. *J. Bone Joint Surg.*, **33B**, 548–61

Fee, N.F., Dobranski, A. and Bisla, R.S. (1977) Gas gangrene complicating open forearm fractures. *J. Bone Joint Surg.*, **59A**, 135–8

Friberg, K.S. (1979) Remodelling after distal forearm fractures in children. *Acta Orthop. Scand.*, **50**, 537–46

Gibbons, C.L.M.H., Woods, D.A. Pailthrope, C. *et al.* (1994) The management of isolated distal radius fractures in children. *J Pediatr. Orthop.*, **14**, 207–10

Henderson, S.A., Graham, H.K. and Piggot, J. (1987) Death and serious injury in child motorcyclists. *B.M.J.*, **294**, 1259

Kay, S., Smith, C. and Oppenheim, W.L. (1986) Both bone midshaft forearm fractures in children. *J. Pediatr. Orthop.*, **6**, 306–310

Kelly, V.G. and Miller, J.E. (1987) Cross-union complicating fracture of the forearm. Part II: children. *J. Bone Joint Surg.*, **69A**, 654–61

Lascombes, M.D., Prevot, J., Ligier, J.N. *et al.* (1990) Elastic stable intramedullary nailing in forearm shaft fractures in children: 85 cases. *J. Pediatr. Orthop.*, **10**, 167–71

Letts, M. and Rowhani, N. (1993) Galeazzi equivalent injuries of the wrist in children. *J. Pediatr. Orthop.*, **13**, 561–6

Lewallen, D.G. and Petersen, H.A. (1985) Non union of long bone fractures in children. *J. Paed. Orthop.*, **5**, 135–142

Olney, B.W. and Menelaus, M.B. (1989) Monteggia and equivalent lesions in childhood. *J. Pediatr. Orthop.*, **92**, 19–23.

Price, C.T., Scott, D.S., Kurzner, M.E. and Flynn, J.C. (1990) Malunited forearm fractures in children. *J. Pediatr. Orthop.*, **10**, 705–12

Rang, M. (1983) *Children's Fractures*, 2nd edn. Philadelphia: Lippincott

Steel, J.A. and Graham, H.K. (1992) Angulated radial neck fractures in children: a prospective study of percutaneous reduction. *J. Bone Joint Surg.*, **74B**, 115–8

Templeton, P. and Graham, H.K. (1995) The floating elbow in children. *J. Bone Joint Surg.*, **77B**, 791–796

Voto, S.G., Weiner, D.S. and Leighley, B. (1990) Redisplacement after closed reduction of forearm fractures in children. *J. Paed. Orthop.*, **10**, 85–89

Younger, A.S.E., Tredwell, S.J., Mackenzie, W.G. *et al.* (1994) Accurate prediction of outcome after pediatric forearm fracture. *J. Pediatr. Orthop.*, **14**, 200–6

9

Fractures of the radial head

C. Geel

With the development of methods of secure fixation of small bony fragments, there has developed an increased interest in the operative management of fractures of the radial head. Traditional concepts of non-operative surgical removal of fractured radial heads as the methods of choice are now being challenged (Swanson *et al.*, 1981; Coleman *et al.*, 1987; Geel *et al.*, 1990).

Among the topics that have witnessed greater investigation is the relationship of the radial head fracture with associated soft tissue injury about the elbow and/or forearm. The radial head has long been identified as one of the anatomical restraints to valgus stress about the elbow (Youm *et al.*, 1979; Werner *et al.*, 1982; Morrey and An, 1983; An and Morrey, 1985; Ekenstam and Hagert, 1985; Pribyl *et al.*, 1986; Robbin *et al.*, 1986). In our laboratory we have been able to demonstrate the effect on elbow stability with graded injury to the radial head and medial collateral ligament, as well as the interosseous membrane of the forearm (Werner *et al.*, in press). With sequential repair of these structures, the load transmission across the elbow was quantified with the most significant improvement in stability. Two-thirds of stress resistance are provided by the medial collateral ligament and the anterior capsule. Pribyl *et al.* (1986) noted that the radial head provides 28% of resistance to valgus stress, which is in good agreement with values in the literature (Youm *et al.*, 1979; Werner *et al.*, 1982; Ekenstam and Hagert, 1985; Robbin *et al.*, 1986). The radial head provides stability through the radial capitellum joint forces which resist valgus stress. After excision, the remaining structures must resist this force. Excision of the head may change the centre of rotation of the elbow in the coronal valgus-varus plane. Movement of the centre of rotation towards the medial epicondyle would shorten the moment arm of the medial collateral ligament, effectively increasing the stress within the ligaments. Our own non-published data (Werner *et al.*, in preparation) with sequential dissection and sequential repair of a wedge fracture, the interosseous membrane and the ulnar collateral ligament show that with increasing division of the interosseous membrane as well as the ulnar collateral ligament, a significant displacement occurs while applying the forces through the elbow joint. Further, we are able to demonstrate that with the sequential repair of the structures, the load transmitted through the elbow in neutral, 45° and 90° flexion return to normal. With fracture repair alone the most significant change is achieved.

The role of injury to the interosseous membrane with radial head fracture has also been subjected to study (Taylor and O'Connor, 1964; Palmer and Werner, 1981; Hotchkiss *et al.*, 1989; Geel and Palmer, 1992). Some recently published work by Hotchkiss and co-workers has attempted to clarify the role of the interosseous membrane to longitudinal stiffness of the forearm. Their work suggested that the membrane was responsible for 71% of the longitudinal stiffness after radial head excision (Kapandji, 1987; Palmer, 1989). In addition, the contribution of the triangular fibrocartilage complex of the distal radioulnar joint to the stiffness was evaluated, with the observation that this structure contributes to less than 10% of the longitudinal stiffness (Kapandji, 1987; Palmer, 1989). As the radius pronates, there is an increasing contact at the radiocapitellar joint. Conversely, as the forearm supinates, there is less contact. Less contact may preclude effective radiocapitellar load sharing leading to increased demand on the central band and the entire interosseous membrane.

It has become quite evident from these scientific investigations as well as clinical experience that in some instances the soft tissue injuries associated with radial head fractures will have a profound effect on outcome and influence the therapeutic approach (McDougall and White, 1957; Coleman *et al.*, 1987).

Classification

Mason, in 1954, described and classified fractures of the radial head based solely upon the bony lesions. Recent clinical experiences and cadaver studies focus more on the combination of fracture and soft tissue injury. The classification system developed by the AO/ASIF group identifies radial head fractures alone as well as in combination with proximal ulnar fractures (Müller *et al.*, 1990). By the same token, this classification also lacks the definition of the presence or absence of associated soft tissue lesions. We propose a separate classification using Mason's work to serve as a basis for defining the skeletal injury with four subgroups of lesions. The latter include: subgroup 1 identifying the medial collateral ligament injury; subgroup 2 defining incomplete or complete disruption of the interosseous membrane; subgroup 3 referring to combined medial collateral ligament and interosseous membrane injury; and subgroup 4 defining an ipsilateral proximal ulnar fracture (Table 9.1).

Diagnosis

The first step in the evaluation of these lesions is to rule out the Essex-Lopresti type lesion of the distal radius. The resulting proximal migration of the radial shaft that occurs at the time of injury is *sine qua non* of the Essex-Lopresti radial head fracture and distinguishes it from gradual migration of the radius sometimes seen months or years after radial head excisions (Essex-Lopresti, 1951; Edwards and Jupiter, 1988). Because attention is focused on the more obvious fracture at the proximal end of the radius, the distal radioulnar joint subluxation can be overlooked at the time of the initial examination.

The concomitant lesion of the interosseous membrane is difficult to identify. Newer diagnostic modalities such as magnetic resonance imaging (MRI) and bone scanning to date have not proved reliable. MRI may be suitable for better visualization of the central band of the interosseous membrane, although our early attempts have been somewhat discouraging (Hotchkiss *et al.*, 1989).

However, when treating a patient with a radial head fracture, the index of suspicion must be high that there is an associated soft tissue lesion which might include an elbow dislocation, rupture of the medial collateral ligament, and/or separate distal radioulnar joint problems.

In the radiographic evaluation of the morphology of the radial head fracture, the standard anteroposterior and lateral X-ray views of the elbow should be supplemented with a 45° oblique view (Figure 9.1). This additional view projects the radial head without interference of the proximal ulna. The size and orientation of the fragment, as well as its interaction with the capitellum, is more readily recognized (Heim and Pfeiffer, 1982). The CT scan may not add more information about the fracture and fragment size, but could be helpful in the assessment of additional soft tissue injuries such as avulsion fractures of the collateral ligaments. The value of MRI and CT scans, however, remains under continuous scrutiny.

Table 9.1   A proposed classification of radial head fractures including soft tissue injury

Extra-articular neck fracture		Wedge fracture		Comminuted articular fracture		Soft tissue subgroup	
A1	Simple	B1	Simple	C1	Multiple fragments (<3)	1.	Isolated collateral ligament
A2	<3 fragments	B2	No depression but multifragmentary	C2	Multiple fragments (>3)	2.	Rupture of the interosseous membrane
A3	>3 fragments	B3	Depression and multifragmentary			3.	Combination of 1 and 2
						4.	Ipsilateral proximal ulnar fracture

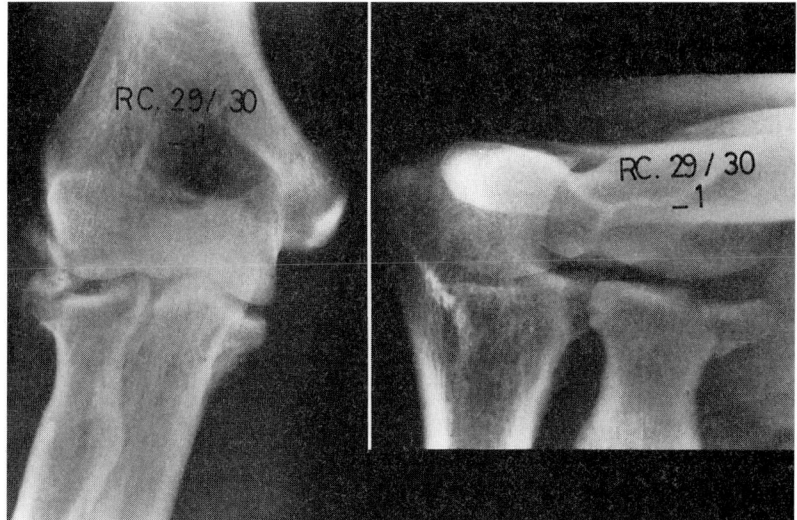

(a)

Figure 9.1 (a) Preoperative X-ray of proximal radial head fracture in anteroposterior and 45° oblique view (wedge fracture). (b) Postoperative X-ray of ORIF proximal radial head fracture

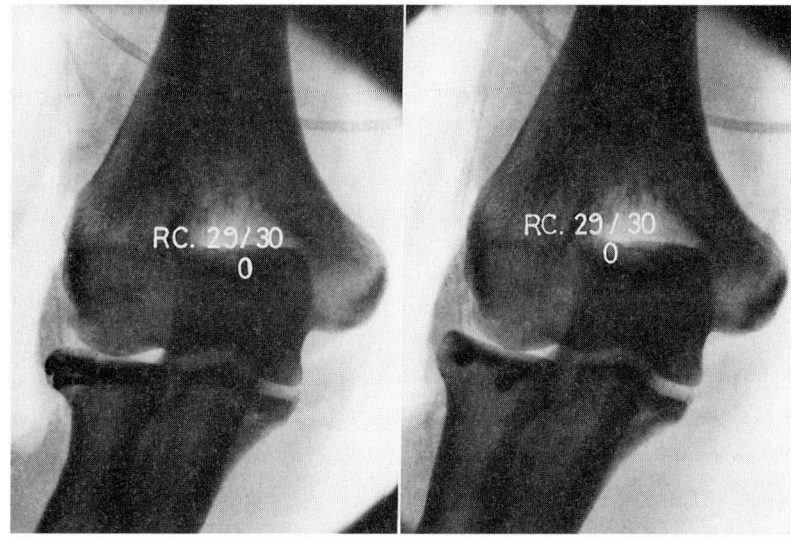

(b)

Treatment options

Treatment decisions are based upon the determination of the fracture morphology, presence or absence of associated injury and the experience of the surgeon.

Isolated fractures are likely to be the result of low-energy trauma. Slightly reduced forearm rotation and a stable elbow joint as well as a pain-free wrist joint are the key findings. The radiographs commonly show a fracture with minimal displacement (less than 2 mm) and an otherwise intact elbow and wrist joint. These are the fractures which are amenable to conservative care.

Based upon the literature and our own results, we define the indication for open reduction and internal fixation of radial head fractures as follows (Morrey *et al.*, 1979; Heim and Pfeiffer, 1982; Curtis and Corley, 1986; Geel *et al.*, 1990; Geel and Palmer, 1992):

● The size of the main fragment must be greater than one-quarter of the articular surface
● The fragment is displaced more than 2 mm

Table 9.2   Indications for ORIF of radial head fractures

Wedge fractures
• Severe dislocation
• >1/4 articular surface
• Additional lesions
    Cartilage
    Proximal ulna
    Collateral ligament
    Distal radioulnar joint

Neck fractures
• Proximal fragment >15° angulated

Impression fracture
• Impression >15° angulated
• Additional lesions
    Cartilage
    Proximal ulna
    Collateral ligament
    Distal radioulnar joint

● There are additional lesions of the cartilage of the capitellum humeri and/or fracture of the proximal ulna as well as a ruptured collateral ligament or distal radioulnar joint disruption (Table 9.2).

For a subcapital transverse fracture, operative intervention is the treatment of choice if the proximal main fragment is displaced as well as tilted more than 15°. Radial head fractures which have united with a radial head tilt of more than 30° are associated with a loss of forearm rotation.

The impression type of fractures with more than 2 mm of depression and a displaced articular surface with additional lesions of the cartilage of the distal humerus (capitellum) are best treated by ORIF combined with a cancellous bone graft underneath the radial head. The same principle for the treatment of tibial plateau fractures applies in cases with depression and/or split fracture combinations. The cancellous bone graft is conveniently obtained from the lateral epicondyle of the humerus.

Non-reconstructible, severely comminuted fractures of the radial head are treated by resection and replacement of the radial head by a prosthesis. The comminuted high-energy type of radial head fractures are frequently associated with distal radioulnar joint disruptions and rupture of the interosseous membrane. We are aware of the fact that the prosthesis system is a controversial topic because there is not one prosthesis system available that has good long-term results. In our opinion, however, the prosthesis should provide enough stability to provide an opportunity for the interosseous membrane to heal. Silicone radial head replacement has been recommended for treatment of proximal migration. The compressive behaviour of the prosthesis suggests that 1 cm of displacement is necessary to achieve the same degree of stiffness as the intact interosseous membrane (Hotchkiss *et al.*, 1989). This may explain the lack of uniform success in preventing normal migration by the use of the prosthesis. Some have suggested that the radial head replacement might be of value in two rather uncommon clinical situations (Morrey *et al.*, 1979, 1981): firstly, for the patients with acute fractures of the radial head and elbow dislocation who have a demonstrable instability of the elbow after the radial head is excised; and, secondly, for patients with acute dissociation of the distal radioulnar joint at the time of the fracture of the radial head, the silastic radial head implant may be of value to stabilize what has been termed a floating radius. The prosthesis would act as a place holder in order to allow the ipsilateral collateral ligaments to heal and to prevent a cubitus valgus. The use of a stiffer implant may improve the elbow's resistance to valgus stress but cannot be expected to restore normal biomechanical function (Pribyl *et al.*, 1986). Edwards and Jupiter (1988) recommend temporary transfixation at the level of the distal radioulnar joint with a pin inserted for cases of primary radial head resection in combination with a collateral ligament rupture. One must be aware of the potential for secondary pin breakage as well as loss of forearm rotation by this tactic.

In children and in adolescents with a skeleton still growing, the same indications apply and an additional factor of the growth plate of the proximal radius has to be considered (Papavasilou and Nenopoulos, 1986). The remodelling process is very helpful in fractures of children. However, restoration of angulation of more than 15° is rarely observed, even a few years after the trauma. Fowles and Kassab (1986) treated fracture angulation of greater than 60° with open reduction and internal fixation.

Schmittenbecher *et al.* (1991) report on their treatment protocol of forearm fractures in children with late results 3–5 years post-injury. In patients with more than 10° of deviation from the normal axis at the end of the initial treatment, 13.9% satisfactory and 6.9% unsatisfactory results were found. Exceeding those 10° were the cases with the most severe late functional restrictions later, requiring 57% secondary reductions and 76.9% secondary operations to restore the function of the forearm.

The only difference is in the choice of implants. Whereas in adults, screws and plates are used, in children, tiny K wires and plaster of Paris post-operatively are common measures taken to protect the elbow joint. Ehrensperger (1990) used poly-dioxanone as a replacement for K wires. The advantage of resorbable pins is that those pins do not have to be removed. The thick periosteal sleeve in adolescents and young adults is amen-able to additional stability if the periosteum can be reattached. The reattachment of the periosteum combined with the increased potential for callus formation and fracture healing has to be consid-ered as the additional instrumentarium for internal fixation in children's fractures.

Operative technique

The patient is positioned supine and we use a rad-iolucent arm board which allows for intraoperative fluoroscopy. The operative exposure is carried out with a pneumatic tourniquet and, before closing the wounds, the tourniquet is deflated to control haemostasis. The surgical approach to the radial head is an extended anterolateral approach. The insertion of the radial collateral ligament onto the lateral epicondyle of the humerus is osteotomized in order to allow extended exposure of the radial head and to provide a secure reattachment of the collateral ligament with a 4.0 mm cancellous screw and washer (Figure 9.2). Before the osteotomy is done this screw is drilled. If necessary, the

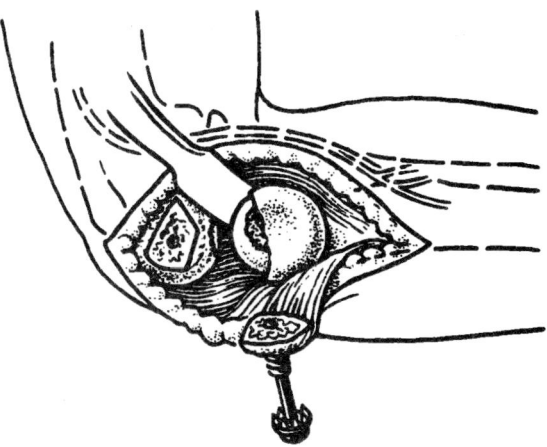

Figure 9.2 Surgical approach to the radial head with osteotomy of the ligamentous attachment of the radial epicondyle of the humerus

osteotomy can also serve as a site for collection of cancellous bone graft to fill any defect in an impacted radial head fracture. This approach allows a better exposure to the entire radial head and it is less likely that the radial nerve is damaged while using a Homan retractor. Preliminary fix-ation and reduction can be obtained with a cerclage wire which is brought around the radial head or neck (Figure 9.3). The two-pointed reduction forceps serves as an excellent reduction tool and prevents impaired pronation and supination. In case of a transverse fracture of the radial head, a mini condylar plate is applied in a buttress mode

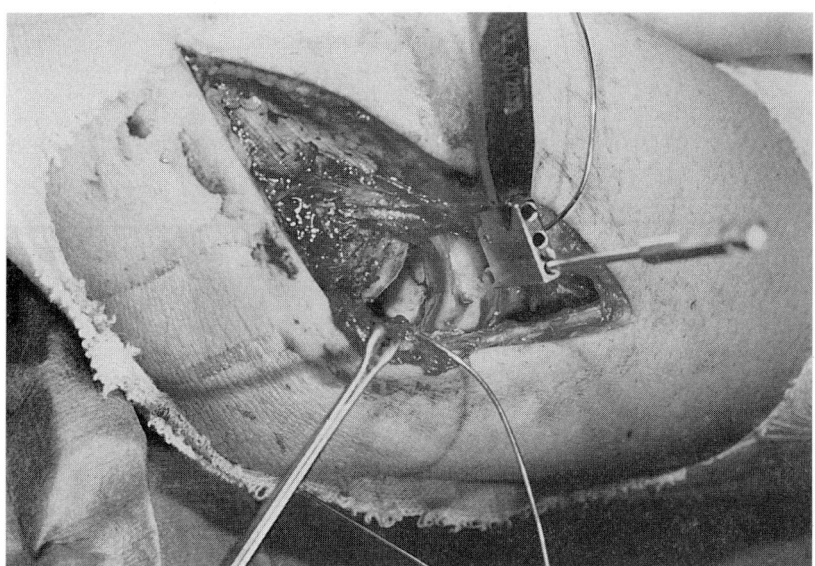

Figure 9.3 Preoperative view of a comminuted proximal radial head fracture with temporary fixation with a cerclage wire. Note the osteotomy of the collateral ligament at the epicondyle humerus

(Figures 9.4 and 9.5). If a plate has to be used in the articular surface, the plate has to be placed in a prepared groove such that pronation/supination are not altered (Figure 9.6).

Very frequently, an osteocartilaginous fragment impinges on the so-called depressed and comminuted radial head fractures. This osteocartilaginous bone fragment originates from the capitellum. Inspection of the capitellum as well as the radial head is important. These fragments have to be removed because they prevent an anatomical reduction of the radial head fracture (Figure 9.7). The outcome of those fractures is probably more related to the cartilage damage at the capitellum than the radial head.

The repair of the ruptured medial collateral ligament of the elbow needs to be addressed with a second incision. The ulnar nerve should be identified, protected and transposed into the adjacent subcutaneous tissues. The collateral ligaments are

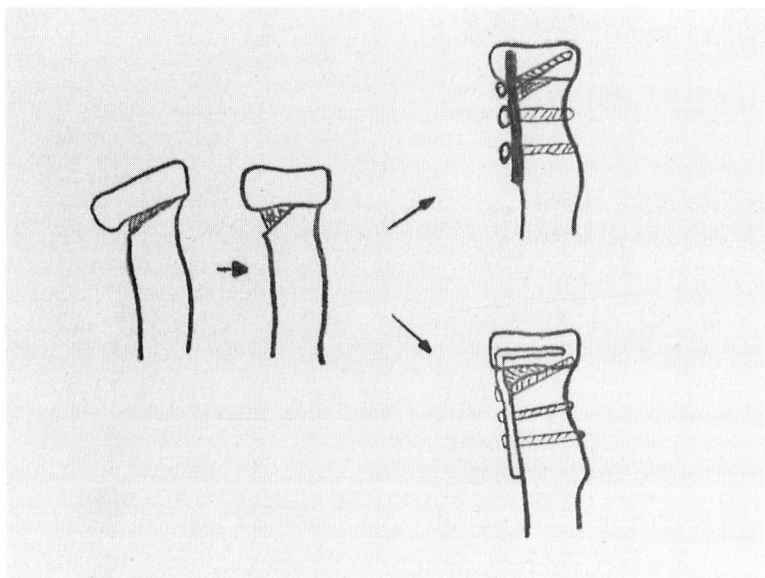

Figure 9.4   Schematic drawing of a transverse proximal radial head fracture and application of plates. Note the placement of the plates in relation to the cartilage

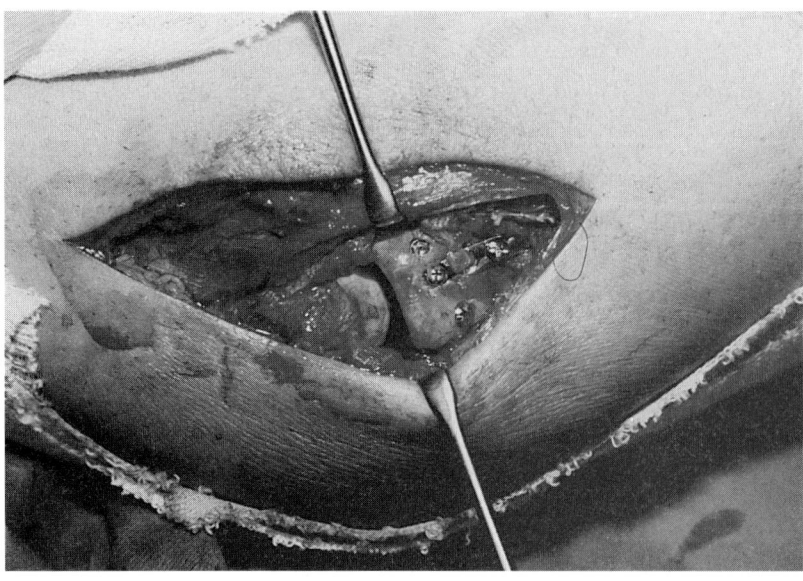

Figure 9.5   Perioperative view of placement of a plate for transverse radial head fractures. The screws as well as the plates are countersunk into the cartilage

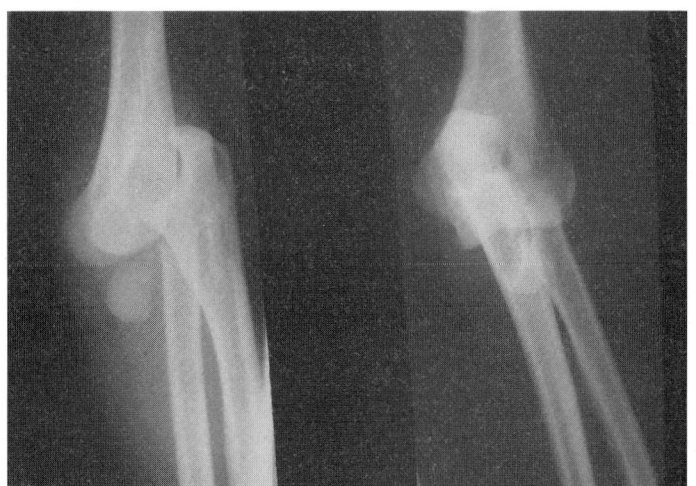

Figure 9.6 (a) Perioperative X-ray of proximal radial head fracture with elbow dislocation. (b) Postoperative X-ray of ORIF of the proximal radial head fracture using a mini condylar plate and proximal screw fixation of the detached radial collateral ligament. (c) One year follow-up after implant removal of proximal radial head fracture (transverse) and implant removal

(a)

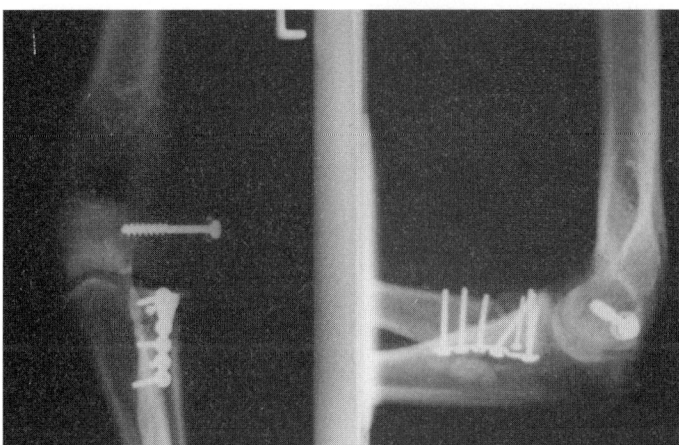

(b)

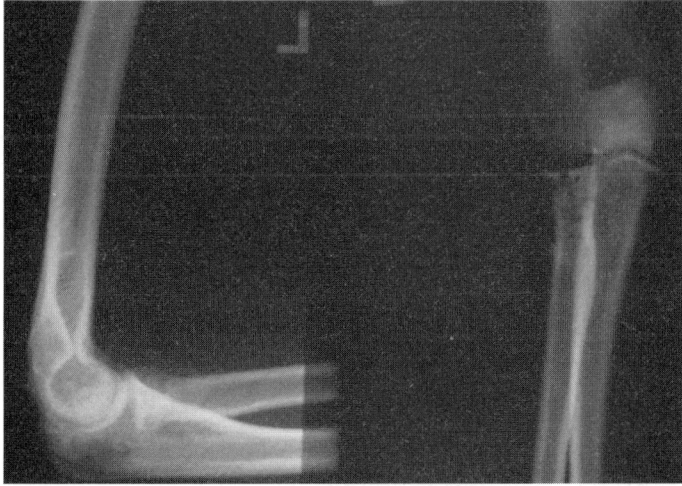

(c)

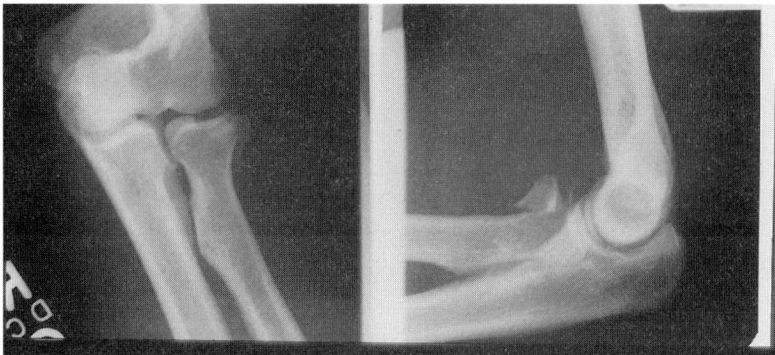

(a)

Figure 9.7 (a) Preoperative view of a depressed comminuted radial head fracture with an intra-articular osteocartilaginous fragment and damage to the capitellum humeri. (b) Immediate postoperative view of the same patient. The osteocartilaginous fragment is reattached with a 1.5 mm lag screw into the condyle, and a proximal 3.5 mm lag screw with washer is used for refixation of the osteotomized insertion of the radial collateral ligament

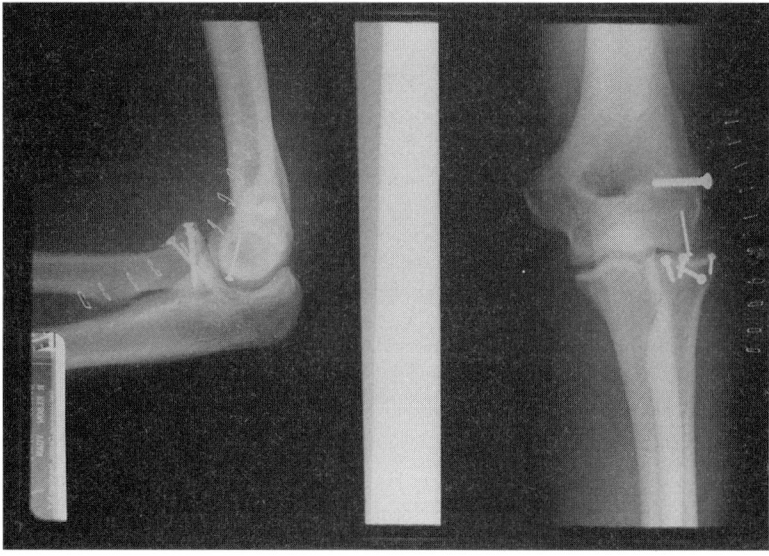

(b)

reattached, preferentially by cancellous screws with washers, or by transosseous suture fixation.

If two incisions are used they must be at least four finger breadths apart and longitudinal in order to prevent any damage to the blood circulation which could result in skin necrosis and wound healing problems, especially in the treatment of an additional ipsilateral proximal ulnar fracture (Heim and Pfeiffer, 1982; Perry and Tessier, 1987; Geel *et al.*, 1990).

During the entire surgery, irrigation with antibiotic solutions is necessary to prevent the cartilage and ligaments from damage. The anterolateral incision is closed with resorbable sutures and includes the joint capsule and the annular ligament to re-establish the ligamentous stability of the joint.

The image intensifier will permit intraoperative control before wound closure to assure the exact positioning of all the screws, their length and the reduction. At the same time, under fluoroscopy, the pronation/supination as well as flexion of the elbow are tested and visualized in order to prevent any subluxation of the two joints and/or interference with the internal fixation.

Postoperative treatment

Postoperatively the patient is placed in a well-moulded posterior splint. In the absence of associated collateral ligament disruption, early range of motion exercise are prescribed. Severe disruption of the medial collateral ligament with documented

elbow instability requires a longer immobilization. The length of this immobilization is an individual decision based upon the amount of damage
to the collateral ligament, the quality of the repair
of the ligaments, and the stability of the internal
fixation of the radial head fracture. In the worst
case scenario, immobilization should not exceed
2–3 weeks postoperatively to prevent any further
stiffness of the elbow joint. Preferably, physical
therapy is started after the wound healing is complete with active range of motion exercises when
out of the posterior splint. The posterior splint
should be removed, at the very latest, 3 weeks
postoperatively. Murphy (1990) used an adjustable supination splint in selected cases in which
traditional mobilization therapy was not produc-
tive in terms of gains in supination. They reported
that the success of the splint varied depending
upon many factors including the type and severity
of the injury, the timing of the intervention, the
patient's age, and the patient's tolerance and compliance with the treatment programme. Use of the
splint is contraindicated in patients with unstable
fractures or with injuries that require surgical
intervention before splinting.

Results

We treated 24 patients with this protocol (Geel
et al., 1990). The follow-up time averaged
12.3 months. No patient developed wrist pain. The
range of motion of the elbow and forearm was
complete in 16 patients and minimally restricted
in eight patients. These eight patients have a
deficit of 5.1° in extension, 14° in flexion, and 8°
in pronation/ supination. For the evaluation, the
functional rating index of the elbow joint modified
after Morrey was utilized (Morrey, 1985; Geel
et al., 1990) (Table 9.3). Sixteen patients scored
98 points and the eight patients with restricted
range of motion scored 93.2 points. Four out of
the eight patients had mild to moderate pain in the
elbow after heavy exercise and labour. Only one
patient subsequently required a radial head excision due to avascular necrosis of the fragments,
and one patient developed heterotopic ossification
in combination with concomitant closed head
injury.

We noted two temporary radial nerve palsies
that resolved at follow-up. At the time of follow-
up, no infection or septic arthritis was reported
and no post-traumatic arthritic changes were
observed.

Table 9.3   Functional rating index

Variable	Points
Motion	
Degree of flexion (0.2 × arc) 150°	30
Degree of extension (0.2 × arc) 10°	2
Degree of pronation (0.2 × arc) 80°	16
Degree of supination (0.2 × arc) 80°	16
Strength	
Normal (80% of opposite side)	12
Mild loss (50% of opposite side)	8
Moderate loss (30% of opposite)	4
Severe loss (limits everyday tasks, disabling)	0
Stability	
Normal	12
Mild loss (no limitation)	6
Grossly unstable	0
Pain	
None	12
Mild (activity normal, no medication)	8
Moderate (with or after activity)	4
Severe (at rest, constant medication)	0
Results	
Excellent	90–100
Good	80–90
Fair	70–80
Poor	<70

Total maximum points: 100
Modified after Morrey *et al.* (1985)

It is noteworthy that the patients in this consecutive series were operated upon within 5 days after
the accident, and 86% of the patients had been
operated upon within 24 hours of their accident.

References

An, K.K. and Morrey, B.F. (1985) Biomechanics of the elbow.
  In *The Elbow and its Disorders* (B.F. Morrey, ed.)
  pp. 43–61, Philadelphia: Saunders
Coleman, D.A., Blair, W.F. and Shurr, D. (1987) Resection of
  the radial head for fracture of the radial head. *J. Bone Joint
  Surg.*, **69A**, 385–92
Curtis, R.J. and Corley, F.G. (1986) Fractures and dislocations
  of the forearm. *Clin. Sports Med.*, **5**, 663–80
Edwards, G.S. and Jupiter, J.B. (1988) Fractures with acute
  distal radioulnar joint dislocation. The Essex-Lopresti lesion
  revisited. *Clin. Orthop.*, **234**, 61–9
Ehrensperger, J. (1990) Osteosynthesis of proximal radius
  fractures using polydioxanone pins. *Z. Unfallchir.
  Versicherungsmed.*, **83**, 84–90
Ekenstam, F. and Hagert, C.G. (1985) Anatomical studies on
  the geometry and stability of the distal radioulnar joint.
  Scand. J. Plast. Reconstr. Surg., **19**, 17–25
Essex-Lopresti, P. (1951) Fractures of the radial head with
  distal radioulnar dislocation. Report of two cases. *J. Bone
  Joint Surg.*, **33**, 244–7

Fowles, J.V. and Kassab, M.T. (1985) Observations concerning radial neck fractures in children. *J. Pediatr. Orthop.*, **6**, 51–7

Geel, C.W. and Palmer, A.K. (1992) Radial head fractures and their effect on the distal radioulnar joint. *Clin. Orthop.*, **275**, 79–84

Geel, C.W., Palmer, A.K., Ruedi, T. *et al.* (1990) Internal fixation of proximal radial head fractures. *J. Orthop. Trauma*, **4**, 270–4

Heim, U. and Pfeiffer, K.M. (1982) *Small Fragment Set Manual: Technique Recommended by the ASIF Group.* New York: Springer-Verlag

Hotchkiss, R., An, K., Sowa, D. *et al.* (1989) An anatomic and mechanical study of the interosseous membrane of the forearm: pathomechanics of proximal migration of the radius. *J. Hand Surg.*, **14A**, 256–261

Kapandji, A. (1987) Bioméchanique du carpe et du poignet. Mise au point. *Ann. Chir. Main*, **6**, 147–169

Mason, M.L. (1954) Some observations on fractures of the head of the radius with a review of 100 cases. *Br. J Surg.*, **42**, 123–32

McDougall, A. and White, J. (1957) Subluxation of the inferior radioulnar joint complicating fracture of the radial head. *J. Bone Joint Surg.*, **69A**, 278

Morrey, B.F. (1985) Functional evaluation of the elbow. In *The Elbow and its Disorders.* pp. 73–91, Philadelphia: Saunders

Morrey, B.F. and An, K.N. (1983) Articular and ligamentous contributions to the stability of the elbow joint. *Am J. Sports Med.*, **11**, 315–9

Morrey, B.F., Asken, L. and Chao, E.Y. (1981) Silastic prosthetic replacement for the radial head. *J. Bone Joint Surg.*, **63A**, 454–58

Morrey, B.F., Chao, E.Y. and Hui, F.C. (1979) Biomechanical study of the elbow following excision of the radial head. *J. Bone Joint Surg.*, **61A**, 63–8

Müller, M.E., Nazarian, S., Koch, P. and Schatzker, J. (1990) *The Comprehensive Classification of Fractures of Long Bones.* Berlin: Springer-Verlag

Murphy, M.D. (1990) An adjustable splint for forearm supination. *Am. J. Occup. Ther.*, **44**, 936–9

Palmer, A.K. (1989) Triangular fibrocartilage complex lesions: a classification. *J. Hand Surg.*, **14A**, 594–606

Palmer, A.K. and Werner, F.W. (1981) The triangular fibrocartilage complex of the wrist anatomy and function. *J. Hand Surg.*, **6A**, 153–62

Papavasiliou, V. and Nenopoulos, S. (1986) Ipsilateral injuries of the elbow and forearm in children. *J. Pediatr. Orthop.*, **6**, 58–60

Perry, C.R. and Tessier, J.E. (1987) Open reduction and internal fixation of radial head fractures associated with olecranon fracture or dislocation. *J. Orthop. Trauma*, **1**, 36–42

Pribyl, C.R., Kester, M.A., Cook, S.D. *et al.* (1986) The effect of the radial head and prosthetic radial head replacement on resisting valgus stress at the elbow. *Orthopedics*, **9**, 723–6

Robbin, M.L., An, K.L., Linscheid, R.L. and Ritman, E.L. (1986) Anatomic and kinematic analysis of the human forearm using high speed computed tomography. *Med. Biol. Eng. Comput.*, **24**, 164–68

Schittenbecher, P.P., Dietz H.G. and Uhl, S. (1991) Late results of forearm fractures in childhood. *Unfallchirurg*, **94**, 186–90

Swanson, A.B., Jaeger, S.H. and LaRochelle, D. (1981) Comminuted fractures of the radial head. The role of silicone implant replacement arthroplasty. *J. Bone Joint Surg.*, **63A**, 1039–49

Taylor, T. and O'Connor, B. (1964) The effect upon the inferior radioulnar joint of excision of the head of the radius in adults. *J. Bone Joint Surg.*, **46B**, 83

Werner, F.W., Geel, C., Fortino, M.D. *et al.* (1998) Biomechanical study of radial head fractures. Publication in preparation

Werner, F.W., Palmer, A.K. and Glisson, R.R. (1982) Forearm load transmission: the effect of ulnar lengthening and shortening. ORS 28th Annual Meeting, New Orleans, Louisiana, January 19–21

Youm, Y., Dryer, R.F., Thambyrajah, K. *et al.* (1979) Biomechanical analyses of forearm pronation–supination and elbow flexion–extension. *J. Biomech.*, **12**, 245–55

10

Fractures of the olecranon

B.J. Holdsworth

The olecranon process is literally 'the head of the elbow' in Greek. It is, strictly speaking, only that part of the upper ulna which lies proximal to the transverse non-articular strip in the trochlear notch, forming about half of the articular surface of that notch. However, in clinical practice the coronoid process often shares in relevant injuries and the adjacent radial head is very often disturbed in association with olecranon fractures. These associated injuries to the proximal radio-ulnar joint may alter the treatment of the olecranon itself and so are included in this discussion. The olecranon has very little soft tissue cover and is hence easily injured by a direct blow. It has powerful muscle attachments which commonly lead to displacement of fragments whether due primarily to direct or indirect forces. It forms an

essential part of the extensor mechanism of the elbow and due to its length increases the moment of the force generated by the triceps muscle. Its shape mirrors the profile of the trochlea of the humerus. Together with the collateral ligaments, and the effect of the extensor and flexor muscles, this shape contributes to the stability of the elbow joint (Figure 10.1).

Prior to the introduction of operative fixation, even displaced olecranon fractures were splinted with the injured elbow in full extension in the hope of reducing the degree of displacement and encouraging union. There is no very well documented prospective trial of the results of this method but the similar technique of initial passive extension followed, after 2 weeks of 'massage', by active movements was said to give excellent

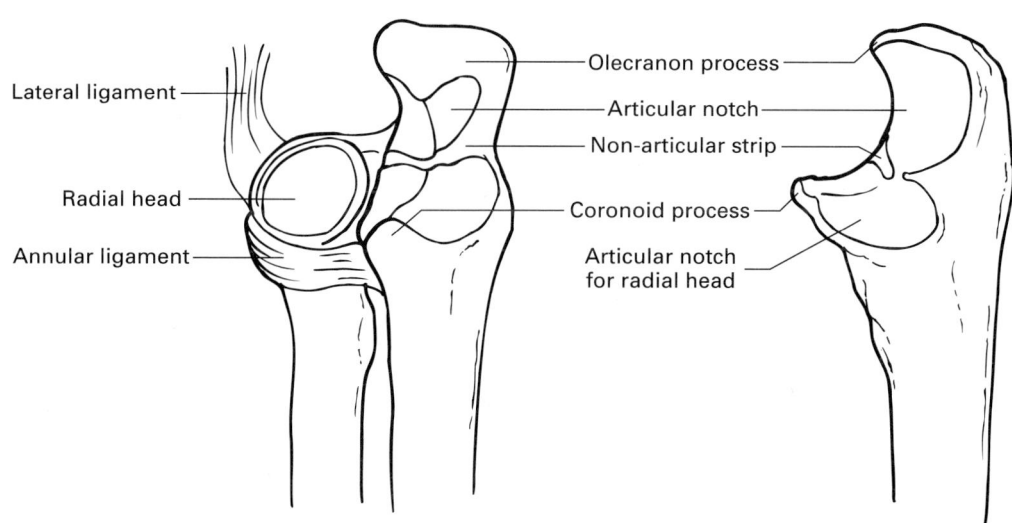

Figure 10.1 The main anatomical details of the olecranon and proximal radioulnar joint

results by Sachs in 1894 (Rockwood and Green, 1991). Presumably this resulted in fibrous union of the displaced olecranon fragment at best, but with less stiffness than follows prolonged splintage.

Fibrous union is compatible with moderately good elbow function, although without the full range of extension. Some reduction of extensor power could be expected (Rowe, 1965). Despite these disadvantages, this method is still useful in extreme old age when very poor bone stock may preclude secure fixation. Only those fractures which are confined to the olecranon process proper and are not associated with major instability of the elbow joint are suitable.

Excision of the shattered fragments in more proximal olecranon fractures has been advocated by several authors and was popularized by McKeever and Buck (1947). The triceps tendon is reattached to the remaining stump of olecranon process or directly to the shaft of the ulna.

Lord Lister's nephew included the following passage in his biography, which is of great interest in terms of the history of fracture treatment:

'Before 1873, Lister had often said to Hector Cameron that he thought the use of silver suture, antiseptically applied, should be extended to fractures of the patella and also to those of the olecranon. The first suitable case, a man with a fractured olecranon, happened to come to Cameron who, as he had no beds at the time in the Glasgow Infirmary, brought the patient to Edinburgh where he was successfully operated upon by Lister. In this, as in all the early cases, the wire was removed once union was sound.

The man obtained a perfectly useful elbow, and wielded the hammer in an iron shipbuilding yard with his former energy. Lister had two other equally good cases of fractured olecranon. One of these patients had consulted eighteen surgeons before coming to him. Now for the first time he cut short the end of the wire which, after the knot had been hammered down flat, was left permanently *in situ*. This caused no inconvenience and had the advantage of allowing the patient to use his limb much sooner than if the wire had been removed.' (Godlee, 1924)

Thus a displaced olecranon was the first fracture that Lister treated by internal fixation using his antiseptic method. That was in 1873 (Le Vay, 1990). Lister was not, however, the first to attempt fixation it seems, and Rigaud was cited as using screws of Swedish iron in 1870 (Berenger-Ferand,

1870). Thus internal fixation of olecranon fractures has a long, if not always uniformly glorious, pedigree.

Many and varied techniques have been suggested since Lister's time, using a wide variety of materials. The commonest have been an assortment of screw types, either intramedullary or oblique transcortical or, more recently, following Pauwels' teaching, a tension band device made from wires or a surface plate. Most recently, various authors have experimented with absorbable substances in the hope of reducing the problem of prominence of metal implants (Hope *et al.*, 1991; Wissing and van der Werken, 1991).

The use of internal fixation gradually became the usual treatment in the majority of displaced olecranon fractures and recent debate has mainly concerned the various methods of internal fixation and their indications.

Classification: types of olecranon fracture

Undisplaced fractures occur both in adults and children, though in the latter they are perhaps more common. In adults, the more brittle bone with its relatively weak periosteum and the greater leverage effect of body weight often lead to considerable fragmentation and displacement. The relatively shorter lever provided by the child's olecranon may be a protective factor. As with patellar fractures, soft tissue retinaculae are, if not torn, quite capable of supporting the broken olecranon during fracture healing even when early gentle movements are allowed.

Displaced olecranon fractures, though not entirely confined to adults, are very unusual in children. Dislocation of the elbow or injury to the distal humeral growth plate are common in childhood and perhaps protect the olecranon from further injury by absorbing the energy of the impact. Of the various classifications suggested in the past, several emphasized the importance of an intact extensor mechanism.

Bakalim and Wilpula (1971) showed in 109 adult cases that only 4% had a small fragment avulsed from the tip of the olecranon. Of the rest, approximately equal numbers had transverse, oblique and comminuted fractures. Twelve per cent were undisplaced. Triceps avulsion without fracture has been described but is thought by most authors to be exceptional.

The Swiss-based AO/ASIF group (Association for the Study of Internal Fixation) led by Müller initially developed a basic universal classification of articular fractures into groups A, B or C. Type A are extra-articular (metaphyseal). Type B involve only part of the width of the joint surface and C the complete articulation. Within each type there were at first only three grades of severity. This had the advantage of its universality but was not quite comprehensive in that rare cases could not readily be fitted into these basic categories.

Based on and expanding their previous type A/B/C and group I/II/III method, a new comprehensive, expanded method has been devised which has few, if any, loopholes. However it has not yet been used in its full form in any relevant publications and is not easy to apply. It does encompass the entire elbow articulation, and the various patterns of associated injury to the radial neck and head are included (Müller *et al.*, 1990).

One of the most widely quoted classifications has been that of Colton (1973) who divided adult fractures, in terms of increasing complexity, into four groups (Figure 2.11):

Colton classification of olecranon fractures

Group 1: Transverse — avulsion of a single fragment, whether small or large

Group 2: Oblique — the fracture line starts as an oblique shear

type 2A — simple two-part fractures

types 2B–D — multifragmentary, degree of impaction determines b, c or d. (Occasionally further complicated by a sagittal split)

Group 3: 'High Monteggia' or transolecranon dislocations — a fracture at the level of the coronoid, with dislocation of the radial head, resulting in gross instability (in the varus/valgus plane)

Group 4: Complex or unclassifiable injuries — High-energy injuries with additional damage to distal humerus or the forearm (so-called side-swipe or baby-car injuries)

— Stress fractures in sportsmen and women
— Epiphyseal lesions of various types are described in children.

This relatively simple classification will be used in this chapter.

Mechanism of injury

Group 1

True avulsion of triceps by indirect force does occur but is unusual, with or without a small fragment of bone. If the lateral expansions of triceps are also torn, displacement and loss of extensor function is inevitable. It is more commonly seen in elderly patients who fall on the outstretched arm.

Group 2

Oblique fractures are sometimes said to be due to hyperextension, but as pointed out by Colton (1973), this would not easily explain the frequent presence of a centrally depressed triangular fragment. Colton explained that a sudden bending force applied to a bone held taut by simultaneous contraction of the elbow flexors and extensors might better explain the observed patterns. This author is not aware of any experimental work that might prove the exact pattern of forces involved. Recent texts attribute most fractures to a combination of direct and indirect forces. Direct blows to the elbow are commonly involved, as shown by marked skin abrasions.

Group 3

Transolecranon fracture dislocations are probably due to a direct force acting just distal to the flexed elbow, causing huge shear forces. The type where a fracture occurs obliquely in the sagittal plane together with radial head disruption is due to a very marked valgus force as in falling with the elbow held semi-flexed and the hand externally rotated. In these cases, instability is increased even further by tearing the medial collateral ligament.

Group 4

The severely pulverized elbow seen in side-swipe injuries is undoubtedly due to huge direct force.

Natural history

We are aware of only one carefully analysed series of olecranon fractures treated by simple observation with no active treatment. Twenty-two patients, with a variety of fracture patterns, attended the Birmingham Accident Hospital and were reviewed after an average of 26 months. The patients either refused or were not fit for operation. They were immobilized for a maximum of 10 days and were then encouraged in active movement. Results compared well with cases treated by operation, even when the same criteria for grading were applied (Parker *et al.*, 1990).

Conservative treatment

Lou (1949) reviewed a series including 11 patients with displaced fractures treated by plaster without reduction and found only one patient requiring late correction by excision of a fragment blocking extension. Results for eight patients were described as excellent, though without precise criteria for this grading, and two patients had 'minor troubles'.

Perkins (1936) pointed out that many patients suffering from olecranon fractures are frail old ladies for whom surgery may not always be advisable. He advised, in such cases, a sling at right angles and encouragement of early active movement, and commented: 'As a rule the patient has regained, for her, normal use in the arm in a few weeks'.

In the author's experience, occasional patients who were too unfit to be offered surgery or who refused to contemplate it are known to have regained a good range of flexion with active extension. The range of extension inevitably is not full in the presence of severely displaced fractures, but the fibrous union that forms is quite strong and may allow good function (Figure 10.2).

Therefore, if internal fixation or excision of fragments cannot be carried out for overwhelming

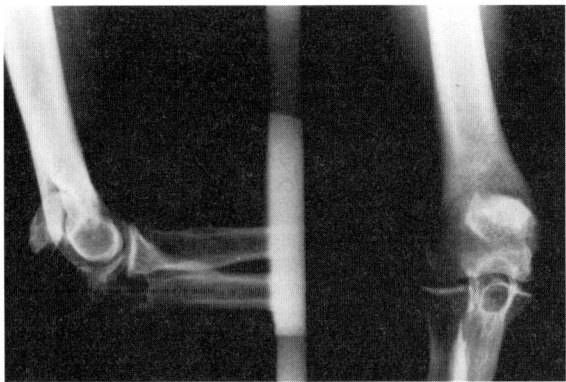

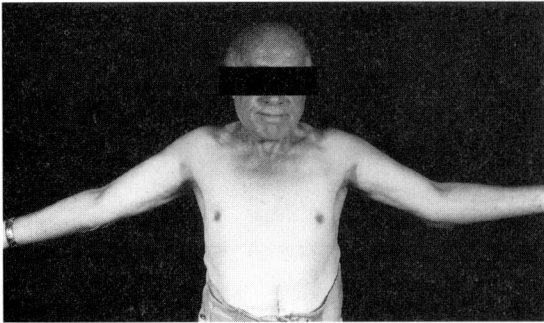
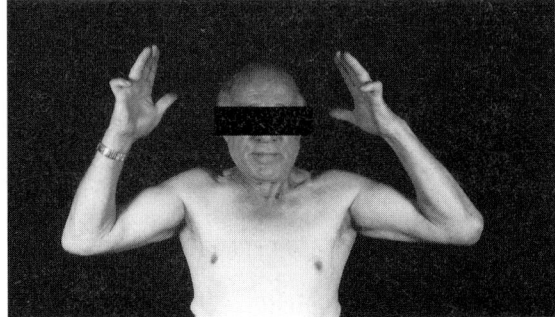

Figure 10.2   This man injured his arm in childhood. He worked all his life down a coal mine, only aware of slight problems with straightening his arm. His presenting problem was entirely unrelated to the old olecranon fracture. Note that his fracture did *not* render his elbow unstable or the outcome would have been much worse. When the option of non-operative treatment is chosen, great care must be taken to ensure that the elbow remains stable, or rapidly progressive arthritis will ensue

clinical reasons, it is not necessary to immobilize the elbow for prolonged periods in hopes of achieving bony anatomical union, but it is better to aim for early return of perhaps somewhat limited function by allowing gentle active movements as soon as pain allows. Total failure to regain active elbow extension has not been observed by the authors, though it certainly can occur, as described by Lister and more recently recorded by Waldram and Porter (1987), 26 years after a fracture in childhood.

Investigations

Anteroposterior and lateral radiographs suffice in the majority of simple fractures, but when complex displaced fractures are encountered then oblique films may help in delineating the smaller fragments. Always anticipate more fragments than are obvious on the radiographs. Comparative films of the other elbow may help if planning a reconstruction following this rare type of fracture. In children, when an atlas of developmental views at various ages may not be available or when rare epiphyseal abnormalities are suspected, again comparative films may help. In this case they must be taken from the same direction and with the normal arm in the same posture as the injured limb.

More expensive investigations, for example CT or MRI scanning, are neither helpful, indicated or necessary for the vast majority of olecranon fractures.

Associated injuries

These can be considered as local or general.

Local

Overall, both in children and adults, about 20% of olecranon fractures are associated with some other bony injury in the same elbow. Varying degrees of disruption of the proximal radioulnar joint are far from rare and must be always borne in mind when deciding on appropriate treatment.

Fractures of the radial head, when multifragmentary, are associated with a major degree of elbow instability necessitating special care in internal fixation. Avulsion fracture of the coronoid process is also significant for the same reason.

General

General injuries found in association are legion. Most series of olecranon fractures include several multiply injured patients. These patients are often quite incapable of any cooperation in early rehabilitation, which mitigates against a full recovery of elbow function. Head injuries are of particular significance and this is not only due to the inability of the patient to cooperate in active movements. There seems to be a humoral effect, possibly mediated via the hypothalamus, resulting in a tendency to myositis ossificans, especially in the elbows (and hips) of these patients. The outcome is often a very severely compromised joint, requiring late arthrolysis to restore a useful range of movement.

Among 52 olecranon fractures, six had other significant injuries in the same elbow. Five patients suffered multiple injuries and only one patient, with a head injury, developed significant myositis ossificans (Holdsworth and Mossad, 1984).

Initial management

The emergency management should be to dress any open wound with a sterile antiseptic pad after taking Polaroid photographs of the wound in the accident department. This dressing should only be removed in the main operating room under full asepsis when formal debridement and internal fixation are being carried out. A systemic broad spectrum antibiotic is given, as well as tetanus prophylaxis if the patient is not already protected. The arm can be rested on pillows or perhaps on a thickly padded, slightly flexed splint. Radiographs are taken, as described above. In complex fractures or multiple injuries, a formal treatment plan, including drawings of all internal fixation devices, should be made at this stage.

Timing of surgery

There is no doubt that the elbow joint has a 'window' of optimal timing for surgery, though its exact edges are difficult to specify. In general, the first week after injury is the best time to perform any operations. Beyond that, myositis ossificans becomes increasingly likely to result from surgery until perhaps 3 months post-injury.

Any open wound should be thoroughly debrided within 6 hours. Closed fractures are not quite so urgent, but in our experience, surgery within 36 hours has been associated with no case of myositis, whereas after 7 days this becomes an increasingly common complication.

A tourniquet is usual, sealed at its lower edge. The exact limit of tourniquet time is argued but when very severe fractures are being reassembled a maximum of 2 hours should be adhered to. In practice, this operating time will rarely be exceeded in the treatment of olecranon fractures, but if necessary the tourniquet may be released at 2 hours for 10 minutes, and then briefly reinflated.

Positioning

The use of general anaesthesia is usual. The standard patient position is simply supine with the arm across the chest. In more complex fractures which can be difficult to reassemble, it may be preferred to have the patient in the true lateral position with the arm resting over a short, well-padded rest. This will, of course, depend on the other injuries, which might not allow such a posture.

In the usual, across-chest position, the arm can be supported by an assistant or by a sterile bandage which, after application of sterile drapes, is first tied to the wrist and then secured to the base of the operating table by a technician.

A tourniquet is applied high on the upper arm and carefully sealed using waterproof adhesive tape to avoid any spirit-based preparation fluids leaking under the cuff. Unless this precaution is rigorously observed then it is advised that only aqueous fluids are used for skin sterilization. The risk otherwise is of a 'burn' which at first appears as superficial blistering but later breaks down to form an ulcer, resulting in ugly and permanent scarring in the axilla.

The hand is first prepared then held by an assistant using a sterile stockinette. The rest of the arm is then prepared and two waterproof towels are clipped around the arm just below the tourniquet. The stockinette is rolled up and an extremity drape applied. After cutting an aperture in the stockinette, Betadine-impregnated adhesive film is applied over the point of the elbow.

Surgical approaches

The incision is usually fairly straight, slightly to the radial side of the point of the elbow, thus avoiding damage to the ulnar nerve. The presence of substantial open wounds may necessitate slight adjustment of the position of the incision. Sharply curved wounds are best avoided, as skin flaps, which have already been traumatized in the original accident and are often grazed, may become hypoxic. The author's habit is to locate but not to denude the ulnar nerve. When treating complex fractures, it helps to pass a nylon tape gently round the nerve at the medial epicondylar groove to act as a 'flag' throughout the operation. It is otherwise easy to forget the exact site of the nerve while concentrating on the niceties of internal fixation with potentially disastrous results for the nerve, which rapidly takes on the same colour as the wound in general and tends to disappear. Prophylactic transposition of the nerve is *not* advised as it results in unnecessary disturbance to the blood supply.

Experience shows that routine transposition is unnecessary. It is always advisable to make a careful written note of the position of the nerve after any elbow operation to avoid possible damage during later procedures.

Treatment methods and surgical technique

Undisplaced fractures

'Undisplaced' in this context means less than 2 mm of gap, no step at the articular surface and an intact extensor apparatus. The term 'sub-aponeurotic' emphasizes that the periosteum can function as a tension band.

There is no argument that *undisplaced* olecranon fractures in children or adults do very well with early gentle active range of movement exercises. Resisted extension should not be attempted for 3 weeks.

If some splintage is thought essential, then a simple back slab, leaving the front of the elbow totally free, is strongly advised. The splint is loosely held in place above and below the joint to accommodate inevitable swelling.

Displaced simple fractures

The historical methods recommended for treatment of displaced simple fractures, whether transverse or oblique with no disruption of the radioulnar joint, have been many and various. Almost any strong material capable of resisting distraction has been advocated for internal fixation. The following methods are still recommended for displaced fractures.

Excision of the fragments and reattachment of the triceps tendon to the ulnar shaft

This method has had many strong advocates. It requires the presence of an intact coronoid process and stability of the elbow as a whole. McKeever and Buck (1947) gave as indications for this technique, very old patients, old un-united fractures (presumably when pain and/or weakness were marked), comminuted fractures, and those fractures which did not involve the trochlear notch. Up to 80% of the olecranon was sometimes excised without compromising the stability of the elbow.

Gartsman *et al.* (1981) retrospectively reviewed the notes of 107 patients and clinically re-examined 29, of whom 15 had been fixed and 14 had had the fragment excised. This small group was judged to be typical of the overall series in terms of age, type of fracture and type of treatment. Based on the occurrence of far more complications in the fixed group (13 of 54 as opposed to 2 in the 53 who had excision) and their failure to show any functional advantage in the fixed group, they concluded that excision was a preferable treatment. However, the methods of fixation were far from uniform, the majority being screwed and only 11 tension band wired. All their patients were immobilized for 3 weeks, which would no longer be regarded as correct technique. Despite the obvious weaknesses of such a retrospective review of a non-homogeneous group, the point was well made that technical difficulties are not rare in internally fixed olecranon fractures. Excision seems a relatively easy technique with which to restore acceptable elbow function.

The advantages of excision of fragments are said to be:

- It is an easier technique, especially in the presence of isolated comminuted olecranon fractures
- There are fewer postoperative skin problems
- It is said to be a comfortable technique with few patients having difficulty regaining motion
- The possibility of joint incongruity due to malunited fragments is lessened, removing one possible source of post-traumatic osteoarthritis.

Internal fixation

The contrasting advantages claimed for internal fixation are:

- Restoration of normal anatomy should restore full power of extension by maintaining the lever arm through which the extensor apparatus works
- The ulnar nerve is not exposed to minor knocks, and need not be transposed
- Stable fixation allows earlier active movement which may allow recovery of a greater range of movement.

It is also applicable when stability of the elbow joint has been destroyed, for example by involvement of the coronoid process or by the disruption of the radial head.

In these circumstances, stable fixation of the olecranon may restore stability to the joint. Persistent instability is a potent cause of rapidly progressive post-traumatic osteoarthritis which is less likely following anatomically precise and stable fixation.

A technique which may simplify fixation, without totally abandoning the bony insertion point of the triceps, was that of Barford who advised removal of interpolated small fragments and reshaping the tip of the olecranon to fit the remaining shaft (personal communication, Colton, 1973) (Figure 10.3).

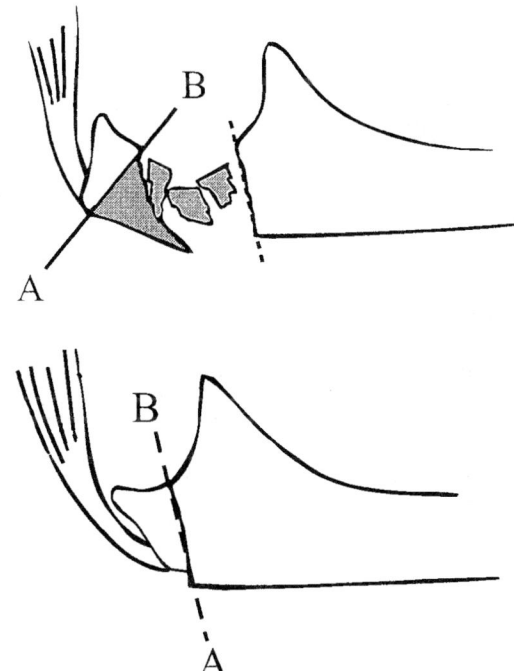

Figure 10.3   The method of partial excision attributed to Barford, useful in elderly patients in whom the bone may not support formal internal fixation devices

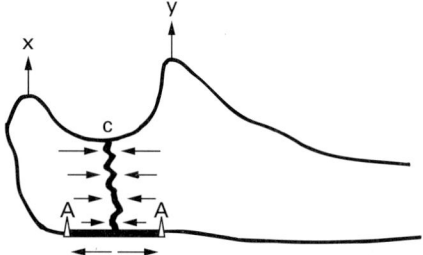

Figure 10.4   The tension band principle. All tensile distractive force at the outer surface of the bone is absorbed by the superficial prosthesis. At the same time, compressive forces are created across the fracture surface thus stabilizing the fracture. It is essential that the fixation device be inextensible, the anchorage points (A) must be rigid and the bone must be capable of resisting compressive forces of considerable magnitude

The tension band principle

The most common recently advised methods of internal fixation have in common the application of the tension band principle. The emphasis of the usefulness of this technique in orthopaedic applications is credited to Pauwels (1965). The basis of this principle is that when a non-elastic fixation device is fixed rigidly at the surface of the reduced fracture, all distracting force due to muscle activity is absorbed by the fixation device. An equal and opposite compression force is generated which increases the coaption of the fragments and discourages any loss of reduction. This mechanical principle has been widely applied in internal fixation of fractures throughout various parts of the body with considerable success (Figure 10.4). The commonest form of tension band is made by using wires as detailed below.

By wiring alone   The main indication for wiring alone is in simple isolated fractures, whether oblique or transverse, and even in Colton types 2b, i.e. with minimal comminution. Some authors have demonstrated good results with wires alone or sometimes with a single additional screw, even in mildly comminuted fractures of the olecranon (Deliyannis, 1973; Holdsworth and Mossad, 1984; Wolfgang *et al.*, 1987). The presence of sagittal splits in the olecranon makes fixation more difficult, and some form of transverse support to the tension band must be added. It is quite possible to combine small interfragmentary screws in this situation with a tension band to achieve a sound fix.

By plates and screws   Plates and screws, if placed on the tension aspect of the ulna, can also act as a tension band, with the extra advantage of increased resistance to lateral flexion (Horner *et al.*, 1989). Stable fixation may be achieved even in transolecranon dislocations (Colton type 3), in which the radial head otherwise tends to redislocate. Wiring alone is *not* advised in the presence of instability, whether due to marked comminution or to dislocation of the radial head. A plate should always be used in these cases. In rare side-swipe injuries, plates and tension wiring may sometimes be combined.

Postoperative management

Plaster of Paris splints are *not* advised following internal fixation of elbow fractures. They are unnecessary and can occasionally be dangerous as considerable swelling is usual. When it occurs within the confines even of a back slab it carries a risk of Volkmann's ischaemic contracture. A light dressing is applied to the wound, held gently in place by plaster wool and a very loose bandage applied from the hand up, with the elbow flexed. Even then it may be best to release these by cutting part of the bandages to expose the cubital fossa. The upper limb is elevated within a thick foam sling suspended from an intravenous drip stand.

The suction drain is removed after 24 hours and active elbow movements are encouraged. Few patients need the formal attentions of a physical therapist. For these few who are perhaps too timid to dare to move at first, then active assisted movements are permitted. It is important to avoid at all costs any occasional forceful manipulation of a recalcitrant elbow as this encourages interstitial haemorrhage, leading rapidly to myositis ossificans. The first clue to this is increasing pain in the presence of a falling range of movement. Paradoxically, the best treatment if this does occur is to desist from all movement for a few weeks, combined with anti-inflammatory medication.

Complications

Peroperative complications are very unusual. Damage to the ulnar nerve is possible and must be guarded against. Technical failures can occur either through inexperience or particularly with complex fracture patterns. It is important to take radiographs in the operating room and to study them closely at the time. Particular attention needs to be paid to the position and length of any wires or screws, and to the reduction. It is disappointing and embarrassing to the operator to read in the notes 'X-rays satisfactory', only to realize later, when the patient has regained consciousness, that some basic flaw in technique was not spotted and corrected at the time. The proximal radioulnar joint needs to be specifically checked for any residual malalignment.

The common postoperative complications are loosening of Kirschner wires, which we believe can usually be avoided by adequate burial of the wires deep to the whole triceps, and transient paraesthesiae in the ulnar nerve. Despite its subcutaneous site, the ulna is not particularly prone to infection, though patients with a previous tendency to olecranon bursitis are at risk .

The problem of wire loosening

Many surgeons have reported complications with loosening of wires. Macko and Szabo (1985) found symptomatic prominence of Kirschner wires in 16 of 20 patients, of whom four had skin breakdown over the wires. In 12 of 16, the wires had not been properly seated at operation. Wolfgang *et al.* (1987) also confirmed that true migration of Kirschner wires was much less common than faulty initial placement, and thought the problem was largely due to faulty technique.

Several authors have devised modifications to the original AO technique of tension band wiring in order to reduce the incidence of the Kirschner wires backing out. The simplest of course is to omit them, but this reduces the stability of fixation. Cheese-wiring of the figure of eight loop is then more likely, as illustrated by Danis (1966). Special perforated pins were devised by Netz and Stromberg (1982) who showed that these pins were very successful. Eight of their first 10 cases still had no problems despite the pins being *in situ* at 7 years. Netz pins have also been shown to be very effective by Larsen and Lyndrup (1987).

An alternative way to anchor the longitudinal wires to the figure of eight is to create a 'pig-tail' curl in the end of the Kirschner wires, preferably deep to the triceps tendon, and then to pass the figure of eight through the loops so formed (Montgomery, 1986). An easy way to make the staple ends in the Kirschner wires, or the pig-tail curl if preferred, is to use the turnkey from the AO

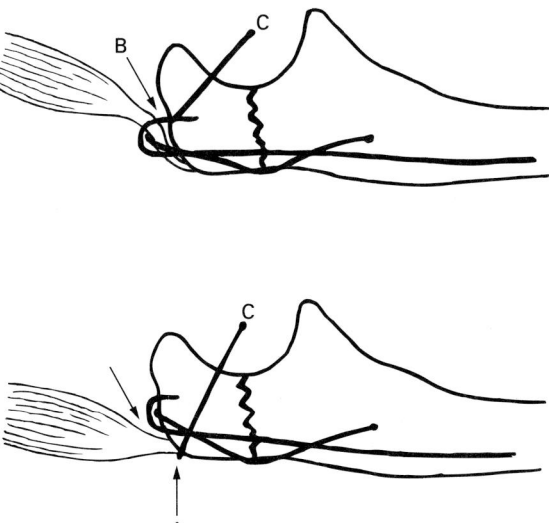

Figure 10.5 All wires must be under the triceps tendon. The effect on the lever arm of the triceps can be seen by the difference between lines CA and CB. The upper diagram explains why wires may be pulled out if this precaution is not observed

wire tightener as pointed out by Findlayson and Nicol (1983). It is essential first to pre-drill a track for each Kirschner wire, then to back out the wire a little or it could prove impossible to seat the curled wire properly.

Both the excursion of triceps to full extension and its moment of action will be limited by any wire not buried deep to triceps. Otherwise a tug-of-war develops between the triceps and the resistance of the wires to sliding. As the tips of the wire drill a hole slightly greater than the diameter of the wire, loosening of any wire remaining superficial to the triceps tendon is inevitable (Holdsworth and Mossad, 1984) (Figure 10.5).

In addition, it has been shown in a cadaver study that any wires protruding more than a few millimetres beyond the upper surface of the olecranon prevent full elbow extension by blocking entry of the olecranon into the olecranon fossa.

Reduced range of movement

Persistent elbow stiffness, though not commonly severe after olecranon injuries, is found in very old and frail patients or occasionally even in young patients if associated with ectopic calcification (myositis ossificans). The latter is prone to occur if surgery is delayed beyond about 5 days and is much more common in association with head injuries. This is not fully explained but may be due to the thrashing about of unconscious limbs or in part due to hypothalamic factors affecting calcium balance. When feasible, such patients should be given a course of an effective anti-inflammatory agent preferably combined with an antipeptic ulcer drug (e.g. ranitidine or misoprostol). This treatment is strongly advised in the rare late cases when severe stiffness requires formal capsulectomy and removal of excess bone to restore elbow extension (arthrolysis). In general, the range of movement regained is optimal if active movements are practised immediately following removal of the vacuum drain at 24 hours, and no intermittent passive movements are allowed. The role of continuous passive movement is not established in primary treatment. It is very useful following arthrolysis.

Non-union

The surprising lack of symptoms, noted by several writers, in those few patients who refuse surgery for their primary fracture has been referred to above. This may explain why symptomatic non-union of the olecranon is a rarity. Obviously if no fibrous union occurs or it is later torn through, then symptoms are much more likely. The fact that gravity alone will extend the elbow has been said to explain the lack of complaint in some old ladies with non-union after non-intervention. Total loss of extensor power is highly unusual.

The more common scenario is that of a sturdy young manual worker who has disrupted a primary repair and has continuing discomfort. Despite the niggling discomfort often described with wires, failure of bony union is more common in the author's experience when longitudinal screws are used as fixation whether alone or in combination with a tension band loop. This was confirmed by Helm and colleagues (1987) who found a considerably higher rate of separation at the fracture surface when screws were compared with tension band wiring. This is in keeping with the biomechanical experimental work of Fyfe *et al.* (1985).

The simple presence of a small gap on radiographs at 6 months should not be taken as an indication to intervene if the patient has no complaints. Consolidation can occur gradually over a long period.

If persistent complaints of weakness and pain are found, together with an increasing gap on X-ray, then operation with bone grafting and stable internal fixation using a plate and screws is advised. The exact incidence of this complication is unclear, but it was not observed in their series by Holdsworth and Mossad (1984), and is perhaps in the region of 2–4%.

Author's preferred technique

'Accuracy is the keystone of success in joint fractures.'

Zuelzer, 1951

Transverse fractures (Colton type 1): tension band wiring

The great majority of these are closed or are only grade 1 open fractures (i.e. small, sharp and usually clean cut) which may be treated as a closed fracture. Tension wires as illustrated give excellent results (Figure 10.6).

Various technical points have been mentioned in the text but it is worth re-emphasizing them for clarity. The Kirschner wires should at first be over-drilled by 2 cm depth then backed out to allow final impaction though the triceps tendon and well into the olecranon. The wires should be 1.6 mm as finer ones are too flimsy to provide

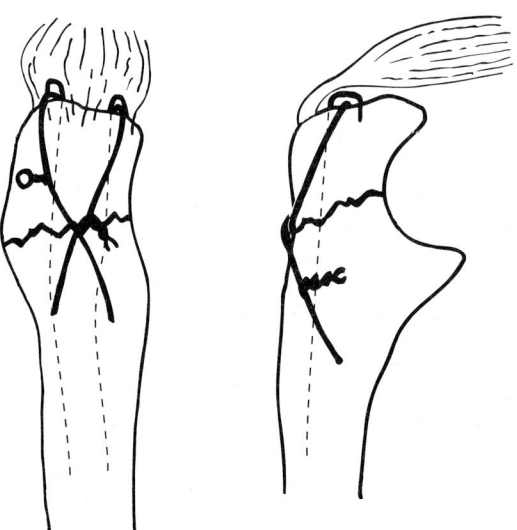

Figure 10.6 The author's preferred method of producing a tension band fixation for isolated simple olecranon fractures. Please refer to the text for important practical points of technique. Note the deep placement of wires under triceps. The cross-over point of the figure of eight only works correctly if it overlies the fracture

rotatory stability. It is essential that both sides of the figure of eight are tightened. Avoid the common beginner's mistake of having two loops for tightening in the same side of the 'eight'. Continue twisting the loops only for as long as the knot can be lifted away from the surface of the bone using the pliers, or the wire will snap. In order to achieve deep burial of the Kirschner wires, the ends should be fully bent into a staple end and cut obliquely. The special pliers in the AO wiring kit can be used, which automatically create a neat right-angled bend (Figure 10.7). A second bend is then created which completes the staple shape.

The author has not used Netz pins but has no objection to them nor to pseudo-Netz pins made by creating a pig-tail loop in the Kirschner wire as described. It would be advisable to pass the loops carefully through the tendon of triceps before threading the figure of eight wire to avoid reduced triceps action as explained above.

The transverse drill hole used to anchor the distal loop of the figure of eight should be well away from the superficial surface of the shaft of the ulna to prevent cutting out and possibly to prevent the tendency for the fracture to open at the articular surface as the wires are tightened. It should be placed approximately as far distal to the fracture as the fracture is from the tip of the olecranon. Use stout wire for the figure of eight as the forces within it are huge.

Oblique fractures (Colton type 2a, b and c): tension band wiring

For the great majority the same technique applies. If the fracture is very oblique with a shallow surface spike, additional security can be achieved with supplementary interfragmentary screws perpendicular to the fracture line.

Mildly comminuted oblique fractures (Colton type 2d)

With these fractures it may be necessary to apply a plate. The forces acting against this plate are principally tensile, which a one-third tubular plate can well withstand. These plates are not so bulky and are perhaps better tolerated than the dynamic compression plate (DCP). For heavy, well-built patients, a DCP would be preferred.

Transolecranon fracture dislocation (Monteggia type, Colton group 3)

It is essential not to risk failure of the fixation device by attempting tension band wiring in these cases. The shearing force up through the fracture plane and across the wires is too much for them to resist. Occasional disastrous loss of fixation has occurred which, if not appreciated straight away, leads to malunion or non-union with the attendant risk of arthritis. This is all the more likely if it has

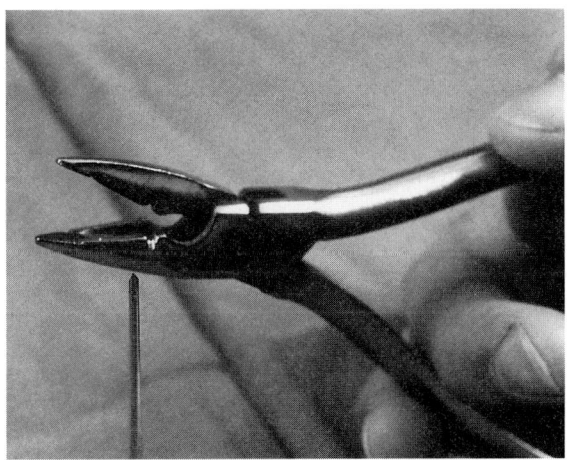

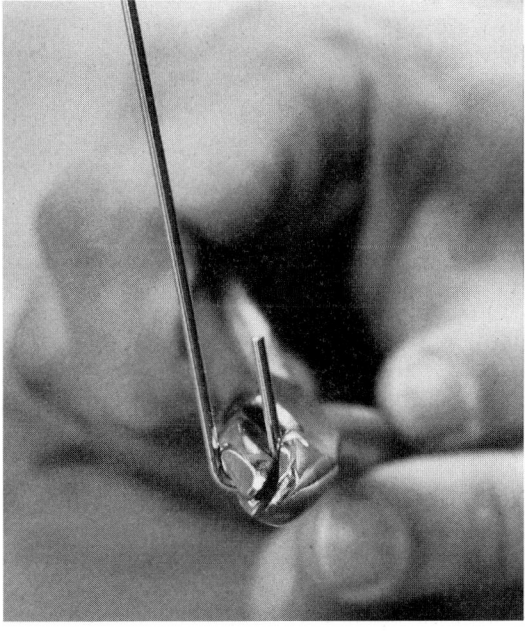

Figure 10.7 The use of the pointed pliers in making staple ends to ensure sound embedding into the olecranon

proved necessary to excise a comminuted radial head. In these cases the additional strength of a 3.5 mm DCP is well worth any slight discomfort caused by its subcutaneous position.

Complex fractures

The more severe complex fractures in which the proximal ulna is effectively exploded provide a major challenge to even the most intrepid surgeon and should be approached with great care. It is usually possible, by meticulous technique, to build the fragments back gradually onto the surviving shaft and apply a plate shaped to curve around the olecranon onto which the bone can be rebuilt. In these circumstances a strong type of plate, for instance a 3.5 mm DCP, is needed, and semi-tubular plates are not advised.

Whenever possible, the proximal radioulnar joint should also be reconstructed. It is not often feasible to rebuild a comminuted radial head and the fragments may need to be excised. They should not be discarded but can be useful as grafts to assist in reconstruction of the olecranon and in particular in reconstructing a coronoid process, without which profound instability otherwise persists. Excision of the radial head results in increased valgus stress on the olecranon and it may be useful, even if only temporarily, to insert a silastic radial head as a spacer. Most authorities do not strongly recommend these prostheses for long-term use as they tend to fragment, but in our experience they fulfil a useful temporary role (Stoffelen and Holdsworth, 1994). It is worth emphasizing once more that the use of excision of the olecranon cannot be extended to these more distal injuries.

Olecranon fractures in children

These are unusual and rarely displaced. Landin and Danielsson (1986) studied a large series of children's fractures from Sweden. They found that elbow fractures comprised 7% of all fractures and that of 589 elbow fractures, 7% involved the olecranon. A fifth of olecranon fractures (as in adults) were not isolated but associated with other local elbow fractures. They were commoner in boys and twice as common in summer as in winter.

It must be realized that the olecranon ossification centre does not appear until the age of 10 years. Before that age, only clinical examination is likely to reveal damage in the immature cartilaginous olecranon process. Special care must be exercised to avoid missing an associated displacement of the radial head (Hume, 1957). This can always be assessed by the alignment of the radial shaft which points towards the capitellar ossification centre, present from the age of 1 year onwards. Due to the pure swivel nature of the proximal radioulnar joint, it is not possible by radiographic malpositioning to disturb this relationship.

The olecranon process often ossifies from several centres and may be mistaken for a fracture. Persistence of the growth plate has been recorded even into adult life and again has led to confusion.

Colton (1973) pointed out the relative rarity of displaced fractures of the olecranon in children. It is thought at least in part to be due to the shortness of the immature process. All series of children's fractures agree that displaced fractures are unusual and comprise no more than 10%. The recommended treatment for such fractures is by a simple figure of eight wire, avoiding Kirschner wires and so not penetrating the growth plate. A brief period of resting in a back slab or sling can then be followed by active movement. Absorbable but strong sutures are an alternative to wire, which avoids the need for a second procedure for removal. Hope *et al.* (1991) in a randomized prospective trial of children's elbow fractures used polyglycolic acid pins with good overall results and no cold abscesses. However there were only two displaced olecranon fractures in his series of 24 elbows. There were more problems in the group in which wires were used, and this was statistically significant.

Matthews (1981) reviewed 49 children with olecranon fractures at up to 3 years following injury. Twenty-two had associated fractures, some of which were very severe with correspondingly poor outcomes. He concluded that plaster is contraindicated due to the risk of Volkmann's ischaemia and that the olecranon fracture is merely one factor in often very major insults to the elbow.

References

Bakalim, G. and Wilpulla, E. (1971) Fractures of the olecranon II. Excision of the fragment and reinsertion of triceps tendon in comminuted fractures. *Ann. Chir. Gynaecol.*, **60**, 102–4

Berenger-Ferand, L.J.B. (1870) *Traité de L'immobilisation Directe des Fragments Osseux dans les Fractures.* Paris

Colton, C.L. (1973) Fractures of the olecranon in adults: classification and management. *Injury*, **5**, 121–9

Danis, A. (1966) Nouvelle technique de l'ostéosynthèse de olécrâne. *Acta Orthop. Belg.*, **32**, 872–8

Deliyannis, S.N. (1973) Comminuted fractures of the olecranon treated by the Weber-Vasey technique. *Injury*, **5**, 19–24

Findlayson, D. and Nicol, A. (1983) Bends made easy. *Injury*, **15,** 145

Fyfe, I.S., Mossad, M.M. and Holdsworth, B.J. (1985) Methods of fixation of olecranon fractures: an experimental mechanical study. *J. Bone Joint Surg.,* **67B**, 367–72

Gartsman, G.M., Sculco, T.P. and Otis, J.C. (1981) Operative treatment of olecranon fractures: excision or open reduction with internal fixation. *J. Bone Joint Surg.,* **63A,** 718–21

Godlee, R.J. Bt. (1924) *Lord Lister,* 3rd edn. pp. 481–2, Oxford

Helm, R.H., Hornby, R. and Miller, S.W.M. (1987) The complications of surgical treatment of displaced fractures of the olecranon. *Injury,* **18**, 48–50

Holdsworth, B. J. and Mossad, M. M. (1984) Elbow function following tension band fixation of displaced fractures of the olecranon. *Injury,* **16**, 182–7

Hope, P.G., Williamson, D.M., Coates, C.J. and Cole, W.G. (1991) Biodegradable pin fixation of elbow fractures in children. A randomised trial. *J. Bone Joint Surg.,* **73B**, 965–8

Horner, S.R., Saasivan, K.K., Lipka, J.M. and Saha, S.S. (1989) Analysis of mechanical factors affecting fixation of olecranon fractures. *Orthopedics,* **12**, 1469–72

Hume, A.C. (1957) Anterior dislocation of the head of the radius associated with undisplaced fracture of the olecranon. *J. Bone Joint Surg.,* **39B**, 508

Landin, L.A. and Danielsson, L.G. (1986) Elbow fractures in children. An epidemiological analysis of 589 cases. *Acta Orthop. Scand.,* **57**, 309–12

Larson, E. and Lyndrup, P. (1987) Netz or Kirschner pins in the treatment of olecranon fractures. *J. Trauma,* **27**, 664–6

Le Vay, D. (1990) *The History of Orthopaedics.* Carnforth: Parthenon

Lou, I. (1949) Olecranon fractures treated in the Orthopaedic Hospital Copenhagen 1936–1947, a follow-up examination. *Acta Orthop. Scand.,* **19**, 166–79

Macko, D. and Szabo, R.M. (1985) Complications of tension-band wiring of olecranon fractures. *J. Bone Joint Surg.,* **67A**, 1396–401

Matthews, J.G. (1981) Fractures of the olecranon in children. *Injury,* **12**, 207–12

McKeever, F.M. and Buck, R.M. (1947) Fractures of the olecranon process of the ulna. *JA.M.A.,* **135**, 1–5

Montgomery, R.J. (1986) A secure method of olecranon fixation; a modification of tension band wiring. *J. R. Coll. Surg. Edinb.,* **31**, 179–82

Müller, M.E., Nazarian, S., Koch, P. *et al.* (1990) *The Comprehensive Classification of Fractures of Long Bones.* pp. 86–95, Berlin: Springer-Verlag

Netz, P. and Stromberg, L. (1982) Non-sliding pins in traction absorption wiring of fractures: a modified technique. *Acta Orthop. Scand.,* **53**, 355–60

Pauwels, F. (1965) Der schenkelhalsbruch, ein mechanisches problem. Gesammelte abhandlungen zur funktionellen anatomie des bewegungsapparates. Berlin: Springer-Verlag

Parker, M.J., Richmond, P.W., Andrew, T.A. *et al.* (1990) A review of displaced olecranon fractures treated conservatively. *J. R. Coll. Surg. Edinb.,* **35**, 392–4

Perkins, G. (1936) Fractures of the olecranon. *B.M.J.,* **2**, 668–9

Rockwood, C.A. and Green, D.P. (1991) *Fractures in Adults,* 3rd edn, vol. 1. Philadelphia: Lippincott

Rowe, C.R. (1965) The management of fractures in elderly patients is different. *J. Bone Joint Surg.,* **47A**, 1043–59

Stoffelen, D.V. and Holdsworth, B.J. (1994) Excision or silastic replacement for comminuted radial head fractures, a long term follow up. *Acta Orthop. Belg.,* **60**, 402–7

Waldram, M.A. and Porter, K.M. (1987) Late treatment of non-union of fracture of the olecranon. *Injury,* **18**, 419–20

Wissing, J.C. and van der Werken, C. (1991) Die Zuggurtungsosteosynthese aus resorbierbarem material (Tension band osteosynthesis of absorbable material). *Unfallchirurgie,* **94**, 45–6

Wolfgang, G., Burke, F., Bush, D. *et al.* (1987) Surgical treatment of displaced olecranon fractures by tension band wiring technique. *Clin. Orthop.,* **224**, 192–204

Zuelzer, W.A. (1951) Fixation of small but important bone fragments with a hook plate. *J. Bone Joint Surg.,* **33A**, 430–6

Index

Abductor pollicis longus muscle, 48
Annular ligament, 44, 45
Anterior interosseous muscle, 48
AO classification system, 12–14
    diaphyseal fractures, 25, 26–7
    distal end fractures, 30–5
    open fractures, 17
    proximal fractures, 19, 21–3, 24–5
    soft tissue injuries, 17
Axis deviation, 141

Bado's classification, 25, 28–9, 110, 139
Barton's fracture, 30
Biceps muscle, 47
Bipolar forearm fracture dislocation, 131–2, 134
Bone mineral substitutes, 76, 77
Boyd's approach, 57–8, 113
Brachial artery, 49
Brachioradialis muscle, 47
Bridging external fixation, distal radial fractures, 69–71
Buckle fractures, children, 145

Carpal alignment, 65, 66
Casting, see Plaster casting
Children, 138–57
    association injuries, 141–5
        arm injuries, 142–3
        forearm injuries, 141–2
        remote injuries, 143–5
        soft tissue injuries, 141, 152
    classification of fractures, 14–15, 139–41
        epiphyseal injuries, 14–15, 139
        fracture displacement, 140–1
        non-physeal injuries, 15–16, 139
    complications, 154–7
        distal growth arrest, 155–7
        proximal growth arrest, 155

epidemiology, 8–10
Galeazzi fractures, 140
injury patterns, 138–9
investigation and assessment, 141
Monteggia fractures, 25, 139–40
postoperative management, 152–4
    3-week look and remanipulation, 153–4
radial head fractures, 162–3
treatment, 145–52, 157, 162
    casting, 146–7, 157
    closed reduction, 146–7, 150–2, 157
    external fixation, 149, 152
    initial management, 145
    internal fixation, 138–9, 149, 152, 157
    intramedullary fixation, 147, 151–2, 157
    non-operative management, 145–6
    olecranon fractures, 180–1
    open reduction, 149, 152, 157
    percutaneous fixation, 147, 150–1, 157
    preparation, 149–50
    soft tissue injuries, 152
    surgical approaches, 150
    surgical techniques, 150–2
    timing of surgery, 149
Chinese finger traps, 146
Classification systems, 12–35
    children's fractures, 14–16, 139–41
        epiphyseal injuries, 14–15, 139
        fracture displacement, 140–1
        non-physeal injuries, 15–16, 139
    diaphyseal fractures, 25, 26–8, 86
    distal end fractures, 25, 25–34
        intra-articular fractures of the distal radius, 34–5
    Essex–Lopresti lesions, 107
    Monteggia fracture, 25, 28, 110
    proximal fractures, 18–25
        AO classification, 21–3, 24–5
        coronoid fractures, 23–4

Classification systems:
   proximal fractures, *contd.*
      olecranon fractures, 19, 170–1
      radial head and neck fractures, 18–19,
         160–1
   soft tissue injuries, 16–17
      closed fractures, 17
      open fractures, 16–17
Closed fractures, soft tissue injury classification,
   17
Colles' fracture, 29, 30, 63
Colton's classification, 19, 20, 171
Compartment syndrome, 47, 92, 120–1
   children, 141, 152, 154
   release, 58, 121
Complications, 119–34
   children, 154–7
      distal growth arrest, 155–7
      proximal growth arrest, 155
   distal radial fractures, 78–80
   Essex–Lopresti lesions, 108
   Galeazzi fractures, 101–2
   infection, 125, 126, 155
   malunion, 119, 128–9, 152, 154–5
   Monteggia fractures, 115
   neurovascular injury, 121–5
      compression neuropathy, 78–80
      reflex sympathetic dystrophy, 80
   non-union, 125–8, 155
   radioulnar dissociation, 130–4
   refracture, 102, 130
   synostosis, 129–30
   tendon rupture, 80
   treatment-related complications, 80, 119–20
   see also Compartment syndrome
Compression neuropathy, 78–80
Computed tomography (CT):
   distal radial fractures, 65–6
   radial head fractures, 161
   radioulnar joint, 46
Coronoid process, 41
   fractures of, 23–4

Darrach–Hughston–Milch fracture, *see* Galeazzi
   fracture
Diaphyseal fractures, 84–95
   children, 139, 149, 157
   classification of, 25, 26–8, 86
   clinical assessment, 86
   epidemiology, 3–5
   fractures of both bones, 88–9
      choice of implant, 89
      reduction, 88–9

   fractures of one bone, 89–95
   open fractures, 95
   surgical treatment:
      implant removal, 95
      indications for, 85–6
      planning of, 86
      surgical approaches, 86–8
   wound closure, 95
Dinner fork deformity, 64
Distal fractures:
   classification of, 25, 29–35
      intra-articular fractures of the distal radius,
         34–5
   epidemiology, 5–8
   see also Distal radial fractures
Distal growth arrest, 155–7
Distal radial fractures, 63–80
   clinical features, 64
   complications, 78–80
      compression neuropathy, 78–80
      malunion, 78
      reflex sympathetic dystrophy, 80
      tendon rupture, 80
      treatment-related complications, 80
   distal third, 102–3
   external fixation, 69–73
      bridging external fixation, 69–71
      non-bridging external fixation, 71–3
   history, 63–4
   initial management, 66–8
   internal fixation, 73–8, 102
      bone substitutes, 76
      displaced intra-articular fractures, 76–8
      dorsal plating, 74, 75
      percutaneous pinning, 76, 150–1
      plating, 73–5
   intra-articular fractures of the distal radius,
      35
   investigation, 64–6
   non-operative management, 68
Dorsal plating, 74, 75
Dorsal radioulnar ligament, 46

Early mobilization, 64
Epidemiology, 1–10
   children, 8–10
   diaphyseal fractures, 3–5
   distal fractures, 5–8
   proximal fractures, 2–3
Epiphyseal injuries, children, 14–15, 139
Essex–Lopresti lesion, 106–8, 132–4, 161
   classification, 107
   clinical presentation, 107

complications, 108
mechanism of injury, 107
radiographic findings, 107
treatment, 107–8
Exposures, *see* Operative exposures
Extensor carpi radialis brevis muscle, 47, 48
Extensor carpi radialis longus muscle, 47, 48
Extensor carpi ulnaris muscle, 48
Extensor digiti quinti tendon, 48
Extensor digitorum communis muscle, 48
Extensor indicis tendons, 48
Extensor pollicis brevis muscle, 48
Extensor pollicis longus muscle, 48
External fixation:
    children's fractures, 149, 152
    distal radial fractures:
        bridging external fixation, 69–71
        non-bridging external fixation, 71–3

Fernandez classification system, 29–30, 31
Fixation, *see* External fixation; Internal fixation;
        Intramedullary nailing; Percutaneous
        fixation
Flexor digitorum profundus muscle, 48
Flexor pollicis longus muscle, 48
Floating elbow, 142, 143, 144
    treatment, 147, 150, 157
Floating radius, 132, 134
Forearm:
    evolution, 37
    innervation, 48–9
    kinesiology, 50–2
        pronation, 50
        rotation, 37, 47, 50
    muscle tendon units, 46–7
    vasculature, 49
Forearm fractures, *see* Diaphyseal fractures
Frykman's classification, 28–9

Galeazzi fracture, 89, 91, 94, 97–102, 131, 132
    children, 140, 157
    classification, 25
    clinical features, 98
    complications, 101–2
    mechanics, 98
    radiographic findings, 98–9
    treatment, 99–101, 157
Galeazzi-equivalent lesion, 98
Greenstick fractures, children, 145
Growth arrest:
    distal, 155–7
    proximal, 155

Gustilo's classification, 17, 141, 152
Head injury, children, 143–5
Henry exposure, 54, 55, 87, 88, 99

Immobilization, distal radial fractures, 68
Implants:
    diaphyseal fractures, 89
    Essex–Lopresti lesions, 108
    removal of, 95
Infection, 125, 126
    children's fractures, 155
Internal fixation, 120
    children's fractures, 138–9, 147–9, 152, 157
    distal radial fractures, 73–8, 102
    Essex–Lopresti lesions, 107–8
    Galeazzi fractures, 99–101
    Monteggia fractures, 112–15
    non-union, 125–8
    olecranon fractures, 170, 174–6, 178–80
    proximal radial fractures, 103–6
    radial head fractures, 161–3
    radioulnar dissociation and, 132–4
    see also Intramedullary nailing; Percutaneous
        fixation
Interosseous arteries, 49–50
Interosseous ligament, 45, 46, 52
Intra-articular fractures of the distal radius, 34–5
    management, 76–7
Intramedullary nailing, 120, 125–7
    children:
        closed reduction and, 147, 151–2, 157
        open reduction and, 149, 152, 157

Kocher's interval, 57

Lateral antebrachial cutaneous nerve, 49
Ligamentotaxis, 69
Lister's tubercle, 44

McConnell-type incision, 58
Magnetic resonance imaging (MRI), 161
Malunion, 119, 128–9, 152
    children's fractures, 154–5
    distal radial fractures, 78
    Galeazzi fractures, 101–2
    Monteggia fractures, 115
Mason's classification, 18–19, 160
Medial antebrachial cutaneous nerve, 49
Median nerve, 48–9
    injury to, 78–80, 121, 122

Metaphyseal injuries, children, 139, 154
Methyl methacrylate cement, 76
Mobile wad of Henry, 48, 49
Monteggia fracture, 89, 92–3, 109–15, 131–3
    children, 25, 139–40, 147, 157
    classification of, 25, 28, 110
    clinical features, 112
    complications, 115
    mechanism of injury, 111–12
    radiological features, 112
    results, 115
    treatment, 112–15, 147, 157
        operative exposure, 57
        transolecranon fracture dislocation, 180
Myositis ossificans, 173

Nerve palsies, 121–3
    posterior interosseous nerve, 115
Neurovascular injury, 121–5
    children, 141–2, 154
    compression neuropathy, 78–80
    reflex sympathetic dystrophy, 80
    see also Compartment syndrome
Nightstick fractures, 109
Non-bridging external fixation, distal radial
        fractures, 71–3
Non-union, 125–8
    children's fractures, 155
    Galeazzi fractures, 101
    Monteggia fractures, 115
    olecranon fractures, 178

Oblique cord, 45
Older's classification, 29, 30
Olecranon, 40–1
Olecranon fractures, 19, 40, 169–81
    associated injuries, 173
    classification, 19, 170–1
    complications, 176–8
        non-union, 178
        reduced range of movement, 177–8
        wire loosening, 177
    investigations, 173
    mechanism of injury, 171
    natural history, 172
    postoperative management, 176
    treatment, 169–70, 178–80
        children, 180–1
        complex fractures, 180
        conservative treatment, 172–3
        displaced simple fractures, 174–6
        initial management, 173

oblique fractures, 180
positioning, 174
surgical approaches, 174
tension band principle, 176, 178–80
timing of surgery, 173–4
transolecranon fracture dislocation, 180
transverse fractures, 178–80
undisplaced fractures, 174
Open fractures:
    children, 141
    diaphyseal fractures, 95
        wound closure, 95
    soft tissue injury classification, 16–17
Operative exposures, 51–8
    children, 150
    diaphyseal fractures, 86–8
    distal radius, 54–7
        alternative volar exposure, 54–5
        dorsal exposures, 55–7
    indirect reduction, 52–3
    olecranon fractures, 174
    radial head fractures, 57, 163–7
    radius, 53–7, 88
        anterior/Henry exposure, 54, 55, 87, 88, 99
        dorsal/Thompson exposure, 53–4
    radius and ulna, 57–8
    skin incision, 52
    ulna, 53, 88

Percutaneous fixation:
    children's fractures, 147, 150–1, 157
    distal radial fractures, 76
Piedmont fracture, *see* Galeazzi fracture
Plaster casting:
    children's fractures, 145–6
        closed reduction and, 146–7
    distal radial fractures, 68
Plastic bowing injuries, children, 145
Plating, 120, 127–8
    children's fractures, 149
    diaphyseal fractures, 95
    distal radial fractures, 73–5
        dorsal plating, 74, 75
    Galeazzi fractures, 99
    olecranon fractures, 176
Posterior antebrachial cutaneous nerve, 49
Posterior interosseous nerve, 47, 49, 53–4, 124
    Monteggia fractures and, 112, 115, 122, 132
Pronator quadratus muscle, 47, 48
Pronator teres muscle, 47
Proximal fractures:
    classification of, 18–25
        AO classification, 21–3, 24–5

coronoid fractures, 23–4
olecranon fractures, 19, 170–1
radial head and neck fractures, 18–19,
160–1
epidemiology, 2–3
radius, 103–6
see also Olecranon fractures; Radial head
fractures
Proximal growth arrest, 155

Quadrate ligament, 46

Radial angle, 65
Radial artery, 49–50, 124
Radial collateral ligament, 44–5
Radial head fractures, 159–67
classification, 18–19, 160
complications, 167
diagnosis, 160
results, 167
treatment, 161–7
operative exposure, 57, 163–6
postoperative treatment, 166–7
Radial nerve, 48–9
Galeazzi fracture complications, 101
injury to, 78–80, 121, 124
Radial shift, 65
Radial shortening, 64–5, 66, 98
Radiographic investigation, *see* X-ray
investigation
Radioulnar joint, 44–6, 103
dissociation, 130–4
radiographic anatomy, 46
see also Essex–Lopresti lesion; Galeazzi
fracture
Radius:
operative exposures, 54–9, 88
anterior/Henry exposure, 54, 55, 87, 88, 99
distal radius, 54–7
dorsal/Thompson exposure, 53–4
radial head, 56
radius and ulna, 57–9
skeletal anatomy, 42–4
see also Diaphyseal fractures; Distal radial
fractures; Galeazzi fracture; Proximal
fractures; Radial head fractures
Reduction, 119–20
children's fractures, 146–9
closed reduction, 146–7, 150–2, 157
open reduction, 149, 152, 157
diaphyseal fractures, 89–95
both bone fractures, 88

distal radial fractures, 66–8, 102
closed reduction, 67–8
displaced articular fractures, 78
Galeazzi fractures, 99–101
indirect reduction, 52
Monteggia fractures, 112–13
proximal radial fractures, 103–6
radial head fractures, 160
radioulnar dissociation and, 132–4
Reflex sympathetic dystrophy, 80
Refracture, 130
Galeazzi fractures, 102
Remodelling process, children, 138–9, 163
Reverse Monteggia fracture, *see* Galeazzi
fracture

Salter–Harris classification system, 14–15, 139
Sigmoid notch, 44
Smith's fracture, 30, 63
Soft tissue injuries:
children, 141, 152
classification of, 16–17
closed fractures, 17
open fractures, 16–17
wound closure, 95
Supinator muscle, 47
Surgical exposures, *see* Operative exposures
Sympathetic maintained pain, 124–5
Synostosis, 129–30

Tendon rupture, 80
Tension band principle, 176, 178–80
Thomson's exposure, 53–4, 113
Torus fractures, children, 145
Triangular fibrocartilage complex, 46
tearing of, 98
Trochlear notch, 38–41
Tscherne's classification system, 17, 18

Ulna:
nightstick fractures, 109
operative exposure, 53, 88
radius and ulna, 58–9
radioulnar joint, 44–5
skeletal anatomy, 37–42
see also Diaphyseal fractures; Distal fractures;
Monteggia fracture; Proximal fractures
Ulnar artery, 49, 53
Ulnar nerve, 48, 53
injury to, 78–80, 121, 123
Ulnohumeral joint, 39–40

Volar radioulnar ligament, 46

X-ray investigation:
  children's fractures, 145

distal radial fractures, 64–5
Essex–Lopresti lesions, 107
Galeazzi fracture, 98–9
olecranon fractures, 173
radial head fractures, 160